Mycoplasma Diseases
of Trees and Shrubs

Mycoplasma Diseases of Trees and Shrubs

edited by

Karl Maramorosch

Waksman Institute of Microbiology
Rutgers University
New Brunswick, New Jersey

S. P. Raychaudhuri

International Union of Forestry Research Organization
W. P. Plant Mycoplasma Diseases
New Delhi, India

ACADEMIC PRESS   1981
A Subsidiary of Harcourt Brace Jovanovich, Publishers
NEW YORK LONDON TORONTO SYDNEY SAN FRANCISCO

ACADEMIC PRESS, INC.
111 Fifth Avenue, New York, New York 10003

United Kingdom Edition published by
ACADEMIC PRESS, INC. (LONDON) LTD.
24/28 Oval Road, London NW1 7DX

Library of Congress Cataloging in Publication Data
Main entry under title:

Mycoplasma diseases of trees and shrubs.

    Papers of the third International Union of Forestry
Research Organizations (IUFRO) Working Party Conference
on Mycoplasma Diseases, held at Rutgers University,
N. J., in Aug., 1979; sponsored by IUFRO's Working
Party on Mycoplasma Diseases of Forest Trees.
    Contents: Isolation, characterization and identification
of spiroplasmas and MLO's / E. A. Freundt -- Yellows
diseases of trees / Carl E. Seliskar and Charles L.
Wilson -- Stubborn disease of citrus / D. J. Gumpf and
E. C. Calavan -- [etc.]
    1. Trees--Diseases and pests--Congresses. 2. Shrubs--
Diseases and pests--Congresses. 3. Mycoplasma diseases in
plants--Congresses. I. Maramorosch, Karl. II. Raychaudhuri,
Satya Prasad, Date. III. International Union of Forestry
Research Organizations (IUFRO) Working Party Conference on
Mycoplasma Diseases (3rd : 1979 : Rutgers University)
IV. International Union of Forestry Research Organizations.
Working Party on Mycoplasma Diseases of Forest Trees.
SB761.M93   634'.04932   81-3534
ISBN 0-12-470220-1   AACR2

PRINTED IN THE UNITED STATES OF AMERICA

81 82 83 84   9 8 7 6 5 4 3 2 1

Contents

Contributors

Numbers in parentheses indicate the pages on which the authors' contributions begin.

Hidefumi Asuyama (135), *Laboratory of Plant Pathology, Faculty of Agriculture, The University of Tokyo, Bunkyo-Ku, Tokyo, Japan*

E. C. Calavan (97), *Department of Plant Pathology, University of California, Riverside, California 92521*

Robert E. Davis (259), *Plant Virology Laboratory, Plant Protection Institute, Agricultural Research, Science and Education Administration, United States Department of Agriculture, Beltsville, Maryland 20705*

Yoji Doi (135), *Laboratory of Plant Pathology, Faculty of Agriculture, The University of Bunkyo-Ku, Tokyo, Japan*

E. A. Freundt (1), *FAO/WHO Collaborating Centre for Animal Mycoplasmas, Institute of Medical Microbiology, Bartholin Building, University of Aarhus, DK-8000 Aarhus C, Denmark*

S. K. Ghosh (231), *Kerala Forest Research Institute, Kerala, India*

H. C. Govindu (299), *College of Agriculture, University of Agricultural Science, Bangalore, India*

D. J. Gumpf (97), *Department of Plant Pathology, University of California, Riverside, California 92521*

Peter Hunt (185), *L.P.T.I Sub-Station, Solok, Sumatera Barat, Indonesia*

Tatsuji Ishiie (147), *Pathology and Entomology Division, The Sericultural Experiment Station, Yatabe, Ibaragi 305, Japan*

Takashi Ishijima (147), *Kyushu Branch, The Sericultural Experiment Station, Ueki, Kumamoto 861-01, Japan*

Robert P. Kahn (281), *Plant Protection and Quarantine Programs, Animal and Plant Health Inspection Service, United States Department of Agriculture, Federal Bldg., Hyattsville, Maryland 20782*

Karl Maramorosch (185), *Waksman Institute of Microbiology, Rutgers University, New Brunswick, New Jersey*

R. Naidu (299), *Citrus Experiment Station, Gonicoppal, Kodagu District, Karnataka State, India*

V. M. G. Nair (325), *Plant and Forest Pathology, College of Science and Environmental Change, University of Wisconsin-Green Bay, Green Bay, Wisconsin 54302*

S. P. Raychaudhuri (315), *International Union of Forestry Research Organization, W. P. Plant Mycoplasma Diseases, C-52 Inderpuri, New Delhi, India; and Department of Plant Pathology, Haryana Agricultural University, Kisser 12500, India*

Narayan Rishi (315), *Department of Plant Pathology, Haryana Agricultural University, Kisser 12500, India*

Carl E. Seliskar (35), *Forest Sciences Laboratory, Northeastern Forest Experiment Station, United States Department of Agriculture, Delaware, Ohio*

Darryl L. Thomas (211), *Agricultural Research and Educational Center, University of Florida, Fort Lauderdale, Florida 33314*

James H. Tsai (211), *Agricultural Research and Educational Center, University of Florida, Fort Lauderdale, Florida 33314*

J. C. Varmah (253), *Forest Research Institute and Colleges, Dehra Dun, India*

E. H. Varney (245), *Department of Plant Pathology, Cook College, Rutgers University, New Brunswick, New Jersey*

Charles L. Wilson (35), *Appalachian Fruit Research Station, SEA, United States Department of Agriculture, Kearneysville, West Virginia 25430*

Foreword

The importance of mycoplasmas as tree pathogens has been increasingly realized in recent years. The Executive Board of the International Union of Forestry Research Organization (IUFRO), therefore, decided at its 1977 meeting in Ibadan/Nigeria to create a special Working Party (S2.06-09). Since then this research unit has been most active under the chairmanship of Dr. S. P. Raychaudhuri.

The first meeting took place in Bangalore, India, in December 1977 since much work had been done in the state of Karnataka by Indian scientists on Sandal Spike caused by mycoplasmas. The proceedings were published in a book entitled "Mycoplasma Diseases of Trees." The second meeting of the Working Party was held in Munich in August 1978 in conjunction with the Third International Congress of Plant Pathology.

This volume contains the edited papers presented at the Third Working Party meeting that was organized by Professor Karl Maramorosch at Rutgers University, New Jersey, in August 1979. Additional chapters by invited contributors have been included in this volume. I am most thankful to Professor Maramorosch and Dr. Raychaudhuri for organizing this workshop and publishing its results.

The IUFRO Working Party on Mycoplasma Diseases has been very actively engaged in developing research in this area. I feel that with its efforts, excellent collaborative research could be initiated at an international level to identify and characterize the MLO diseases of trees and to culture and characterize the pathogens, with the ultimate goal of controlling these disastrous maladies.

Professor Dr. W. Liese
President, IUFRO

Preface

Until 1967 plant mycoplasma diseases were considered, without exception, to be virus diseases. Plant pathologists and forest pathologists did not even suspect mycoplasmas as causal agents of diseases of trees and shrubs. Witches' broom and plant yellows diseases characterized by yellowing of leaves without mosaic, greening of petals, phyllody of floral organs, proliferation of axillary buds, breaking of dormancy, seed sterility, or a combination of these symptoms were grouped together as yellows-type virus diseases. All attempts to transmit the presumptive causative viruses to plants by mechanical inoculation of extracts from diseased to healthy plants failed, however. Extracts from certain plants as well as from insect vectors were successfully injected into specific leafhopper or planthopper vectors, and the injected insects transmitted the disease agents to susceptible plants after an incubation period of several days or weeks. Numerous economically important diseases of woody plants were extensively investigated by plant virologists, but no viruses could be isolated or visualized by electron microscopy techniques.

In 1967 Yoji Doi, Tatsui Ishiie, Michiaki Terenaka, Kiyoshi Yora, and Hidefumi Asuyama published in Japan the first electron micrographs of distinctive pleomorphic bodies discovered in the phloem of mulberry trees with mulberry dwarf disease and Paulownia trees with witches' broom disease. They noted that the mycoplasma-like organisms did not occur in healthy trees nor outside the sieve elements of diseased plants. Professor Asuyama and his associates interpreted the microorganisms to be parasitic mycoplasmas. They reported that mycoplasmas were less numerous during early stages of infection. When they treated mulberry dwarf-diseased plants with tetracycline antibiotics, they noticed temporary remission of disease symptoms and concomitant disappearance of mycoplasmas from the phloem.

The 1967 discovery of mycoplasmas in Japan revolutionized work on yellows-type diseases. Additional evidence accumulated, and in 1972 Robert F. Davis in the United States discovered that some mycoplasmas form spirals. He coined the new term *spiroplasmas* for these microorganisms. From other laboratories came additional evidence for the constant association of mycoplasmas with witches' broom diseases and, within a few years, several spiroplasmas were isolated and grown in pure culture. Spiroplasmas were soon reported as causative agents of diseases affecting invertebrate and vertebrate animals as well as plants. Several spiroplasmas were also isolated from the nectar and surface of flowers, particularly from certain trees. These mycoplasmas did not cause plant diseases.

During the past decade several reviews and books have been published on mycoplasma diseases of plants, and one small volume, edited by one of us (S.P.R.), appeared on mycoplasma diseases of trees, with papers presented at the First IUFRO Working Party Conference on Mycoplasma Diseases in India in 1977. Following the Third IUFRO Mycoplasma Conference at Rutgers University, the organizers decided to publish a treatise on tree and shrub disease caused by mycoplasmas. The foremost world authorities, including the original discoverers of plant mycoplasma diseases from Japan, were invited to contribute chapters to this book.

We hope that this treatise will provide a standard reference work for all interested in plant mycoplasma diseases in forest pathology, in entomology, and in disease control. The exhaustive treatment of recent advances and the critical review of the rapidly expanding field of plant mycoplasma diseases of woody plants update our knowledge of these diseases, as well as the current research on vector transmission and chemotherapy.

We express our sincere gratitude to the contributors for the effort and care with which they have prepared their chapters and to the staff of Academic Press for their part in various aspects of production of this volume.

Karl Maramorosch
S. P. Raychaudhuri

ISOLATION, CHARACTERIZATION, AND IDENTIFICATION
OF SPIROPLASMAS AND MLOs

E. A. Freundt

FAO/WHO Collaborating Centre for Animal Mycoplasmas
Institute of Medical Microbiology
Bartholin Building
University of Aarhus
DK-8000 Aarhus C, Denmark

I.   INTRODUCTION

The spiroplasmas were recognized as a new microbial entity
in 1972-1973 (Davis *et al.,* 1972a, b; Davis and Worley, 1973;
Saglio *et al.,* 1973; Cole *et al.,* 1973). Their very frequent
association with plants and insects, as pathogens or commen-
sals, is now firmly established. They are classified, together with other prokaryotes with which they share the property of lacking a cell wall and cell wall precursors, in the
order Mycoplasmatales, class Mollicutes. The classification
scheme of this group of organisms, and the main criteria that
are used to distinguish between the various taxonomic categories within the class, are presented in Table I.
The term "mycoplasmas" is used as the trivial name for all
members of the Mollicutes and thus includes the spiroplasmas.
"Mycoplasma-like organisms" (MLO's) is the generally accepted
designation for organisms which, although closely resembling
mycoplasmas in their morphology and ultrastructure, have not
yet been otherwise characterized and whose final classification with the Mycoplasmatales has to await future studies.
The failure consistently experienced till now to devise suitable artificial media that will support the growth of these
organisms is the major obstacle to a more detailed characterization by cultural, biochemical, and serological methods.
The number of plant diseases of the yellows-type associated
with nonhelical MLO's, as demonstrated by electron microscopy

1

of the phloem tissue, currently amounts to about 100 (McCoy, 1979). To this may be added a rather considerable number of potential insect vectors that have likewise been shown to harbor MLO's in their tissues (Tsai, 1979).

The vast majority of the plant- and insect-associated mycoplasmas that have been grown thus far belong in the family Spiroplasmataceae. At present, only one genus and species, *Spiroplasma citri* (Saglio *et al.*, 1973) has been recognized in the family. However, a number of other serologically distinct spiroplasmas have been isolated and partly characterized, as, for example, the corn stunt spiroplasma (CSS), honeybee spiroplasma, suckling mouse cataract spiroplasma (SMCS), and spiroplasmas recovered from the surface and the nectar of flowers. Mention should be made also of the *Drosophila* sex-ratio organism which, although as yet noncultivable, has been shown to meet all other criteria for spiroplasmas. From 1 to 6% of the populations of several related species of *Drosophila* have been found to be infected with this transovarially transmitted spiroplasma, which has the effect of eliminating the male sex from the progeny (Whitcomb and Williamson, 1979; Williamson and Poulson, 1979).

Although for obvious reasons the spiroplasmas are the main concern of this chapter, it should be mentioned that members of the genus *Acholeplasma* (Table I) have been isolated with increasing frequency during recent years from plant materials (Gianotti, 1974; Eden-Green, 1978; and others). Acholeplasmas have been shown also to multiply in experimentally infected leafhoppers (Whitcomb *et al.*, 1973). However, presently available evidence does not allow any definite conclusions as to the nature of the relationship between acholeplasmas and plants. For a more detailed description of this genus, the reader should refer to a recent survey by Tully (1979).

II.   ISOLATION AND GROWTH

A.   *GENERAL GROWTH REQUIREMENTS OF MYCOPLASMAS*

With a diameter for the smallest viable cells of about 250 to 300 nm, the mycoplasmas are the smallest organisms that are able to reproduce in artificial, cell-free media. The size of the genome varies for different families within the class from 4.5×10^8 daltons, i.e., one-half the size of that of the smallest bacteria, to 1×10^9 daltons. The value of 4.5×10^8 daltons may thus be assumed to represent the minimal size of a genome that can code for all of the syn-

TABLE I. Taxonomy of the Class Mollicutes

Class: Mollicutes
 Order: Mycoplasmatales
  Family I: Mycoplasmataceae
    1. Sterol required for growth

    2. Sensitive to digitonin (1.5%)

    3. Genome size about 5.0×10^8 daltons

    4. NADH oxidase localized in cytoplasm

   Genus I: *Mycoplasma* (about 60 species currently)

      Do not hydrolyze urea

   Genus II: *Ureaplasma* (single species with serotypes)

      Hydrolyzes urea

  Family II: Acholeplasmataceae

    1. Sterol not required for growth

    2. Resistant to digitonin (1.5%)

    3. Genome size about 1.0×10^9 daltons

    4. NADH oxidase localized in membrane

   Genus I: *Acholeplasma* (eight species currently)

  Family III: Spiroplasmataceae

    1. Helical organisms during some phase of growth

    2. Sterol required for growth

    3. Sensitive to digitonin (1.5%)

    4. Genome size about 1.0×10^9 daltons

    5. NADH Oxidase localized in cytoplasm

   Genera of uncertain taxonomic position

      Anaeroplasma (two species)

      Thermoplasma (single species)

---

thetic capabilities required by an organism that is capable
of reproducing in artificial media and whose metabolism does
not depend on other cell systems.

The most remarkable nutritional requirement that is shared
by a major fraction of all mycoplasmas is that for cholesterol and related sterols, as first demonstrated by Edward and
Fitzgerald (1951) and further defined by Rodwell (Rodwell and
Abbot, 1961; Rodwell, 1963, 1969; Razin and Tully, 1970). In
addition to cholesterol, many mycoplasmas have a requirement
for an exogeneous supply of long-chain fatty acids (Rodwell
and Abbot, 1961; Rodwell, 1967, 1968; Razin and Rottem, 1963;
Razin and Cosenza, 1966; Razin et al., 1966, 1967). On the
other hand, unesterified fatty acids may have a toxic effect
when added even in small quantities to semisynthetic media.
The addition of albumin to such media may be necessary to
counteract this effect. Albumin acts as a buffer that binds
the fatty acids, liberating only minute quantities that are
not sufficient to produce lysis of mycoplasma cells (Razin,
1969). The use of Tween 80 as a water-soluble source of oleic
acid that is slowly hydrolyzed by mycoplasma lipase, and hence
less toxic, is further advantageous (Razin, 1969). Determination of nutritional requirements of mycoplasmas for cholesterol and fatty acids can be easily performed by growth experiments in differential media containing cholesterol and Tween
80 (plus albumin) in various combinations (Razin and Tully,
1970; Edward, 1971). In practice, the demand for lipids is
usually satisfied by incorporating horse serum (or some other
source of animal serum) into the growth medium. Among other
nutritional factors shown to be needed by at least some mycoplasmas may be nucleic acid precursors, certain vitamins (ribo-
flavin, thiamin, and nicotinamide), and glycerol (Rodwell, 1960,
1967; Rodwell and Abbot, 1961). The avian species *Mycoplasma
synoviae* has a special requirement for β-nicotinamide dinucleotide in the reduced form (Frey et al., 1968). The growth
of many mycoplasmas is stimulated by yeast extract which is
used, therefore, as a standard component of most mycoplasma
media. A recent study(Rylance et al., 1979) indicated the
presence in yeast extract of two active components that were
both dialysable and heat-stable, but which have not yet been
chemically identified. Very similar growth-promoting factors
were found by Rylance et al. (1979) in extracts of many animal tissues as well. Glucose is utilized as a source of carbon and energy by all *Acholeplasma* and several *Mycoplasma*
species. Arginine serves as a source of energy for several
other *Mycoplasma* species, and a rather limited number of species are able to catabolize both glucose and arginine (Freundt
et al., 1979; Table II). Formulas of media that will support
the growth of a great variety of different mycoplasmas may

be found in recent surveys (Freundt, 1978, 1979; Freundt *et al.*, 1979).

B. *GROWTH REQUIREMENTS OF SPIROPLASMAS*

  1. *Historical Notes*

      The discovery in the late 1960's of MLO's in the phloem elements of a great variety of diseased plants and in tissues of insects, and the susceptibility of these organisms to tetracycline antibiotics, provided the impetus for intense efforts in many laboratories to obtain growth of mycoplasmas on artificial media. Although most of such studies failed to yield any positive results, claims of the isolation of mycoplasmas soon appeared in the literature. Thus, Hampton *et al.* (1969) reported the isolation from naturally infected pea plants of a mycoplasma found to be antigenically related to certain animal mycoplasmas. However, no convincing evidence was presented, either in this or in several other studies of a similar nature, as to the identity of the isolate(s) with the MLO originally observed in the tissues of diseased plants. Actually, Hampton *et al.* (1976) presented in a recent paper circumstantial evidence that their isolate had been derived from animal rather than from plant sources. In retrospect, an early study by Chen and Granados (1970) indicates that these authors were able to preserve the infectivity of juice extracted from diseased corn plants by maintenance in artificial medium under conditions suggesting the occurrence of some multiplication of the CSS. However, the turning point in the history of plant mycoplasma research was reached in 1971 with the simultaneous isolation by French (Saglio *et al.*, 1971) and Californinan (Fudl-Allah *et al.*, 1972) groups of workers of the causative agent of citrus stubborn disease, and the subsequent classification of this organism in a new taxon, under the name of *Spiroplasma citri* (Saglio *et al.*, 1973). Two observations made in connection with the successful *in vitro* cultivation of *S. citri* had obvious practical implications for the resulting intensified attempts at isolating new spiroplasmas: (1) the demonstration of the importance of a high osmotic pressure of the growth medium, equivalent to that of sieve tube sap, and (2) the finding of a sharp optimum temperature for growth (Saglio *et al.*, 1971, 1973). Yet, a number of years elapsed before conclusive evidence for the growth of CSS could be reported (Chen and Liao, 1975; Williamson and Whitcomb, 1975). Thereafter, following in rapid succession was the isolation and cultivation of the honeybee

spiroplasma (Davis *et al.*, 1976; Clark, 1977), the SMCS (Tully
et al., 1977), and spiroplasmas from the external surface and
nectar of flowers (Davis, 1978; McCoy *et al.*, 1979). Some
strains of this latter group of spiroplasmas are serologically
related to *S. citri* and the honeybee spiroplasma (Davis, 1978)
while others are not (Davis *et al.*, 1978; Williamson *et al.*,
1979). The recovery of isolates that are antigenically and
otherwise indistinguishable from *S. citri* from plant hosts,
other than citrus trees, as reported repeatedly during recent
years, cannot, of course, be accepted as cultivation of new
spiroplasmas, although, on the other hand, such findings may
contribute significantly to our knowledge of the ecology of
S. citri.

2. *Source and Preparation of Material*

The material used for attempted isolation of spiroplasmas
(or MLO's) may consist of sap expressed or extracted from infected plants, surface washings of flowers, or the hemolymph
of insects.

In the case of spiroplasmas contained in the sieve tube
sap, the choice of the plant material to be used for cultivation attempts is very important. Saglio *et al.* (1973) were
guided in their selection of material used for primary isolation of *S. citri* by the observation that very young leaves
(but not mature leaves) of orange seedlings showing early leaf
symptoms of "stubborn" disease were able to transmit the
pathogen with a high frequency of infection. Moreover, they
took advantage of the demonstration of a much higher concentration of mycoplasma-like structures in the sieve tubes of
such leaves when the orange seedlings were grown at 32° rather
than at 24°C.

For the isolation of spiroplasmas from the surface of
flowers, freshly collected whole flowers or bracts and petals
removed from the flowers are either rinsed with liquid growth
medium, or soaked in the medium for 15 to 30 minutes (Davis,
1978; McCoy *et al.*, 1979).

Although the procedures used for the isolation of spiroplasmas from insects are outside the scope of this chapter,
it may be mentioned that the first *in vitro* culture of CSS
was established from hemolymph withdrawn from *Drosophila pseu-
doobscura* females used to pass the organism (Williamson and
Whitcomb, 1975).

The frequent occurrence of spiroplasmas on the surface of
flowers, a source that was not known at the time when *S. citri* and CSS were first isolated, makes a meticulous surface
sterilization particularly important as a preliminary step in

the isolation of such organisms from phloem and phloem tissues. Sterilization may be performed with 70% ethanol, 0.1%
$HgCl_2$ or 5% sodium hypochlorite (singly or combined) followed
by careful rinsing with sterile water. The phloem may then
be obtained either by pressing the sap from cut leaves or
stems or by grinding the plant material in a mortar with a
small volume of liquid medium. Although sap obtained under
sterile conditions by one or the other method may, after
centrifugation and appropriate dilution, be used directly
for inoculation, it is usually safer to pass the fluid through
a 450-nm membrane filter to eliminate any contaminating bacteria or fungi before inoculation into the growth medium. To
minimize the possible inhibitory effect of toxic factors contained in the crude sap prepared from plant tissue (Chen and
Liao, 1975; Liao and Chen, 1975, 1980), it is advisable to
prepare a number of serial dilutions of the sap and to inocu-
late each dilution into several tubes. In the absence of any
evidence of growth during the early stages of attempted isolation, it may be very important to carry out a series of
blind passages at regular, short intervals.

3. Estimation of Growth

    In contrast to the rather heavy turbidity produced by the
growth of most bacteria in liquid media, the turbidity produced by spiroplasmas and most other mycoplasmas is usually
so slight, even when the organisms are multiplying to relatively high titers, that it may not be discernible, or only
barely so, by gross examination of a broth culture. Photo-
metrically determined increase of turbidity may, of course,
be of some help, but does not always correlate with growth.
In consequence, other methods of estimating the growth of
spiroplasmas in liquid medium have to be depended upon.
    A very convenient method is recording of the change of
pH to the acid side that may result from fermentation by
spiroplasmas of glucose and certain other carbohydrates.
Quantitation of the growth on the basis of acid production
may be performed in two ways: (a) by determining the amount
of 0.01 N HCl required to lower the pH of noninoculated medium to the pH value attained in the culture, or (b) by a
tenfold serial dilution technique (Jones et al., 1977). In
the latter more accurate method, the end point of the titra-
tion is defined as the highest dilution at which an acid pH
shift occurs, and the titer of the culture is expressed in
color changing units (CCU), one CCU being defined as the
log_{10} (reciprocal of end point). The time required for the
development of a demonstrable color change of phenol red or

some other indicator is highly variable, depending, in particular, on the actual organism concerned and its adaptation to
the growth medium. For example, whereas *S. citri* produced a
drop of pH from 7.8 to 5.0 in 2 to 3 days during growth at an
early passage level in the sorbitol medium (Saglio *et al.*,
1973), a slight acidic shift in pH on primary isolation of
SMCS did not occur until after 10 to 14 days (Tully *et al.*,
1977). On adaptation, a more significant change of pH may
be produced by SMCS in less than 1 week.

Another metabolic activity of spiroplasmas that may affect
the pH of the growth medium is hydrolysis of arginine, resulting in the release of ammonia (Townsend, 1976). The extent to which the production of ammonia will influence the
pH change depends, of course, on the composition and buffering
capacity of the test medium. Thus, in the experiments of
Townsend (1976) with *S. citri,* the production of acid predo-
minated throughout the growth cycle in a basal medium supplemented with glucose (0.1%) and fructose (0.1%), but without
added arginine. However, in a medium supplemented with arginine (0.42%) in addition to the carbohydrates, a characteristic V-shaped pattern of pH fluctuation was observed, an
initial pH shift to the acid side being followed by a shift
to the alkaline side.

Monitoring of cultures by darkfield or phase-contrast
microscopy is another very useful method for the demonstration of growth (Tully *et al.*, 1977; McCoy *et al.*, 1979) that
may reveal an increased number of motile, helical organisms
even before or in the absence of any discernible color change.
The method used by Liao and Chen (1978) for measuring growth
in their study of the effect of various physical parameters
was based on direct counting of the number of helices under
the darkfield microscope. The suitability of direct microscopy for enumeration of viable cells is limited, however, by
the difficulties involved in recognizing nonhelical and nonmotile viable cells.

Determination of the number of colony-forming units (CFU)
by subcultivation from serial dilutions of liquid cultures
onto solid medium is a sensitive and reliable method in the
case of *S. citri* and other spiroplasmas that readily produce
colonies, but it is not suited, of course, for spiroplasmas
such as CSS and SMCS which do so only inconsistently or not
at all. The tendency of some spiroplasmas to develop small
secondary (satellite) colonies (Davis, 1978) in the vicinity
of primary colonies is another factor that may complicate
quantitative estimates of growth by determination of the number of CFU.

A simple, quick, and very accurate indirect method for
measuring the growth of *S. citri* was used by Saglio *et al.*

(1973, 1979), who counted the radioactivity of trichloroacetic acid-insoluble culture precipitate after incorporation of labeled precursors. Another equally sensitive indirect method based on assays of ATP and energy charge as criteria of growth and metabolism of *S. citri* was also employed by Saglio *et al.* (1979). These indirect methods have the disadvantage that the curves obtained do not parallel the growth curve as determined by the number of CFU during the phase of decline.

4. Composition of Growth Media and Nutritional Requirements

It is interesting to note that the medium used by Saglio *et al.* (1971, 1973) for primary isolation of *S. citri* was a modification of the medium originally devised by Hayflick (Chanock *et al.*, 1962) for isolation of the human pathogen *Mycoplasma pneumoniae*. The major components of Hayflick's medium, that are still widely used for the growth of mycoplasmas of human and animal source, are beef heart infusion broth, horse serum, and fresh yeast extract. With the purpose of increasing the osmotic pressure, sorbitol and sucrose were added by Saglio and his co-workers in concentrations of 7 and 1% (wt/vol), respectively. The medium was further supplemented by glucose and fructose (0.1% each), both of which are fermented by *S. citri*.

As the sorbitol medium proved unable to support the growth of the CSS, a spiroplasma that is serologically fairly closely related to *S. citri,* other approaches had to be made to meet its nutritional requirements. A possible clue for the proper composition of a CSS medium was obtained by the observation of multiplication of this organism in the hemolymph of leafhoppers and other insects (Whitcomb *et al.*, 1973). Since the composition of the hemolymph had served as a basis for the formulation of growth media for insect cell cultures, it was logical to try the growth-promoting effect on CSS of such a medium, as for example Schneider's *Drosophila* medium. This is a complex synthetic medium consisting essentially of (1) inorganic salts, (2) organic acids (including α-ketoglutaric, fumaric, malic, and succinic acids), (3) free amino acids, and (4) carbohydrates. In the M1 medium of Williamson and Whitcomb (1975), that proved an excellent growth medium for CSS as well as for *S. citri,* Schneider's medium makes up about 50% of the total volume, the other half consisting of fetal bovine serum and an autoclavable fraction very similar to Saglio's sorbitol medium. The medium devised by Chen and Liao (1975) for the growth of CSS also contained Schneider's medium, but in a much lower proportion, viz., 0.5 to 100 ml of a horse serum-sucrose broth, plus 1 ml of medium 199. Surprisingly, an even

simpler medium consisting merely of Bacto PPLO broth (Difco),
horse serum, and sucrose in high concentration supports the
growth of CSS (Liao and Chen, 1975, 1977).

As the SMCS could not be grown in any of the earlier media,
advantage was taken of a series of systematic studies carried
out by Jones et al. (1977), as a result of which a few new
formulations (SP-1, SP-4) proved suitable for the growth of
SMCS.

The honeybee spiroplasma, on the other hand, turned out to
be one of the most easily cultivable spiroplasmas, since it
could be grown, even on primary isolation, in an ordinary mycoplasma broth with 20% added calf serum (Clark, 1977).

The stepwise progress thus made since the early 1970's in
cultivation of the spiroplasmas was mostly obtained, as will
have appeared from the above, on a purely empirical basis.
What can be deduced from our knowledge of the composition of
the growth media as to the exact nutritional requirements of
the spiroplasmas is rather sparse and should be supplemented
with analytical data. A consistently recurrent component of
all of the media referred to is animal serum, supplying, among
other things, cholesterol, that has been shown to be required
by at least S. citri (Saglio et al., 1973) and the honeybee
spiroplasma (Davis et al., 1976). The sterol requirements for
the growth of S. citri was confirmed by Freeman et al. (1976)
who found the growth-promoting effect of cholesterol and β-si-
tosterol to be superior to that of stigmasterol and ergosterol.
Palmitic acid, and to some extent oleic, linoleic or linolenic
acids, were shown by the same authors to stimulate the growth
of S. citri, although none of them were essential nutrients.
Mudd et al. (1977) found that S. citri preferably incorporated
palmitic acid and discriminated against linoleic acid during
growth in media that differed in their composition with respect to the sources supplying cholesterol and fatty acids.
Fudl-Allah and Calavan (1972) observed a stimulatory effect of
fresh yeast extract on the growth of S. citri, which was later
confirmed by Jones et al. (1977). The nutritional studies of
Jones et al. (1977) referred to above were primarily designed
with the aim of defining more exactly differences in the growth
requirements of S. citri and CSS. Experiments were carried out
with various modifications of the highly complex Ml medium
(Williamson and Whitcomb, 1975) from which a number of consti-
tuents were systematically deleted, singly or in groups. Several factors were found, thereby, to contribute to the adequacy of the Ml medium for primary isolation and continued
growth in passages of CSS, including the presence of organic
acids and yeast extract, and a favorable salt balance and
osmolarity. Also, fetal bovine serum proved better than horse
serum. However, the single most important growth-promoting

factor for CSS appeared to be α-ketoglutaric acid.   In contrast, the growth of *S. citri* was not diminished by deletion
of all organic acids and was not stimulated by α-ketoglutaric
acid.   Free amino acids, that were not required by CSS, were
important, on the other hand, for the growth of *S. citri* and
could not be satisfactorily replaced by lactalbumin hydroly-
sate.   However, a more recent study by Lee and Davis (1980)
indicating the importance of a proper balance between amino
acids and keto acids for the growth of *S. citri* seems to suggest a less sharp difference between the nutritional requirements of CSS and *S. citri*.

   Very interestingly, spiroplasma medium SP-4 has been found
to improve very considerably, also the rate of recovery of
M. pneumoniae from clinical specimens (Tully *et al.*, 1979).

5. *Physical Growth Conditions*

   The physical parameters that are most important for the
growth of spiroplasmas include osmolarity and pH of the growth
medium, and the growth temperature.

   As mentioned before, the modifications of the Hayflick
medium introduced by Saglio *et al.* (1971, 1973) intended in
particular, by the addition of sorbitol and sucrose, to increase the osmotic pressure to bring it up to about the same
level as found in the sieve tube sap, i.e., about 15 atm or
670 mOsm.   The adequacy of maintaining a high osmolarity of
spiroplasma media was largely confirmed by subsequent studies,
although the dependence on a very high osmotic pressure soon
turned out to be less critical than was thought at first.   In
the case of CSS, an increase of the osmolarity to 700 mOsm, as
achieved by increasing the concentration of sucrose (Chen and
Liao, 1975), resulted in good growth in a medium which, without an excess of sucrose, had earlier supported only a very
limited multiplication of this organism (Chen and Granados,
1970).   However, the M1 medium used by Williamson and Whitcomb
(1975) for their isolation of CSS contained only about one-
third of the amount of sorbitol added to Saglio's medium, and
Liao and Chen (1978), in a more detailed study, demonstrated
the ability of CSS to grow in a wide spectrum of osmolarity,
viz. from 360 to 1120 mOsm, with optimal values in the range
of 610 to 840 mOsm.   The osmotic pressure of the SP-1 medium
which supported good growth of the SMCS was no more than 332
mOsm (Tully *et al.*, 1977).   It should also be mentioned that
the requirements for a high osmotic pressure exhibited by
spiroplasmas on primary isolation and in early subcultures may
be considerably reduced on continued subcultivation.   *Spiroplasma citri*, at least, can be rather easily adapted to grow

in the conventional Hayflick medium. As suggested by Chen and
Davis (1979) the nature of the osmoticum used to increase the
pressure may be important. Thus, the different response of
S. citri and CSS to high concentrations of sorbitol may be
ascribed to a detrimental effect of sorbitol on this latter
organism. The relatively low osmotic pressure of 332 mOsm
required by the SMCS, and the ability of the honeybee spiroplasma to grow in an ordinary mycoplasma broth medium, is
easily explained, on the other hand, by the relatively low
osmotic pressure found in the hemolymph of their natural arth-
ropod hosts.

The pH of spiroplasma media usually varies between 7.4
(Chen and Liao, 1975; Liao and Chen, 1977; Tully *et al.*, 1977)
and 7.8 (Saglio *et al.*, 1973). In a recent study by Liao and
Chen (1978), optimal pH for CSS was 7.2, with no growth occurring at pH lower than 5.4 or higher than 8.0 I.M. Lee
(quoted by Chen and Davis, 1979) found the pH optimum for
S. citri at 7.6, whereas growth was significantly reduced at
pH below 7.0 and above 8.0. The buffering capacity of the
medium is equally important because of the deleterious effect
on the spiroplasmas of the acids produced by fermentation of
glucose and other carbohydrates during growth.

The dependence of the growth on the temperature varies
rather significantly for different spiroplasmas. *Spiroplasma
citri* has a sharp optimum at 32°C, growth at 29° and 35°C representing only 30 and 24% respectively, of that attained at
32°C. Growth at 20° and 37°C is almost nil (Saglio *et al.*,
1973). The *in vitro* dependence of the multiplication of *S.
citri* on the incubation temperature has its close counterpart
in the natural infection: 32°C is the temperature at which the
number of spiroplasma cells reaches the highest concentration
in the citrus leaves (Lafleche and Bove, 1970: Saglio *et al.*,
1973), and the range of 28° to 32°C is about the level that is
most favorable for the expression of citrus stubborn disease
(Chen and Davis, 1979). Although CSS was originally isolated
at 29°C (Chen and Liao, 1975; Williamson and Whitcomb, 1975),
its optimum was determined by Liao and Chen (1978) to be the
same as for *S. citri*, i.e., 32°C, only slight growth being
recorded at 15° and 37°C. The honeybee spiroplasma grows over
the wide range of 18° to 37°C (R. E. Davis, unpublished observation quoted by Chen and Davis, 1979), and the spiroplasmas isolated from the flowers of tulip trees over the range of
22° to 37°C, with an optimum at 34°C (Davis, 1978). The SMCS,
that is able to multiply in ticks as well as in mammals, grows
equally well at 30° and 37°C (Tully *et al.*, 1977).

C. *NONCULTIVABLE SPIROPLASMAS AND MLO'S*

The number of spiroplasmas that have been grown in artificial media represent a very minor proportion only of the total number of MLO's that have been demonstrated by electron microscopy in different plants and insects. Among these, one important organism exhibiting spiral morphology and motility remains to be cultivated — the *Drosophila* sex-ratio organism. Most of the remaining as yet noncultivable MLO's do not appear to possess the morphological characteristics of the spiroplasmas. In spite of the optimism that emerged from the successful cultivation of *S. citri* and other spiroplasmas, the non-helical MLO's have resisted till now any attempt at cultivation in artifical media. Although several claims of cultivation of MLO's have been reported, none of them have been confirmed. Especially in the early days of the era of plant mycoplasmology, the many pitfalls that may be encountered in attempts to cultivate mycoplasmas, such as the risk of contamination from outside sources or the occurrence of pseudocolonies, were not always — as discussed in more detail by Maramorosch (1972) — appreciated by newcomers in the field.

It is pertinent to mention that the problem of apparently noncultivable mycoplasmas is well known also from animal mycoplasmology. The mycoplasmal etiology, for example, of endemic pneumonia of swine was suspected for several years before cultivation of the very fastidious causative organism, *M. hyopneumoniae*. Perhaps more surprising is the phenomenon experienced in many laboratories that mycoplasmas contaminating tissue cultures, although orginally easily cultivable, may become increasingly difficult to grow on artifical media following adaptation to the tissue culture system. The organisms may grow to high titers in the tissue culture system, produce typical cytopathology, and may be demonstrable by immunofluorescence techniques. Yet they cannot always be subcultured by conventional methods on ordinary cell-free mycoplasma media. The hypothesis has been advanced, therefore, that the cell-dependent strains may "turn off" essential metabolic systems which normally permit these organisms to grow on artifical broth and agar media (Hopps *et al.*, 1976). It can only be speculated, if MLO's from plants and insects are noncultivable merely in the sense that the media used for their attempted cultivation are deficient in as yet undefined essential growth factors or whether their multiplication is obligatory cell-dependent. The fact that the use of plant phloem sap as enrichment for MLO media. although supporting the growth of some spiroplasmas (McCoy, 1976), has not solved the problem, may suggest the last-mentioned alternative as the most likely explanation. The frustrations experienced by numerous highly competent investigators in many laboratories all over the

world clearly indicate that there is probably no easy way of
overcoming the obstacles encountered in growing the MLO's.

III.   CHARACTERIZATION OF SPIROPLASMAS

A.   *COLONIAL APPEARANCE*

    The appearance of colonies grown on the surface of solid
media is among the criteria to be determined for classification of an organism with the Mollicutes. Obtaining isolated
colonies on solid medium is also an important step in the
cloning procedure usually used to secure, as far as possible,
pure cultures of mycoplasmas (Subcommittee on the Taxonomy of
Mollicutes, 1979). Solid medium cultures provide the basis,
moreover, for several biochemical and serological tests used
for further characterization of mycoplasmas.
    To facilitate growth on agar plates, incubation should be
made in a moist atmosphere with about 5% CO_2 (candle jar) or
in an anaerobic environment (5% CO_2 in nitrogen). Spiroplasma
colonies are very small, 0.05 to 0.2 mm in diameter. When
examined under a low-power stereo microscope, most fully developed colonies have the fried-egg appearance typical of mycoplasmas: a central hemispherical center in which the organisms grow into the medium and which is surrounded by a peripheral zone of surface growth. (Figs. 1 and 2). It should be
noted that the colonies of some spiroplasmas, as for example
CSS, may not develop the typical fried-egg appearance on primary isolation or in early subcultures. The colonies of SMCS
have entirely failed, under any cultural conditions tested,
to differentiate into a central and peripheral zone.

    The ability of different spiroplasmas to grow at all on
solid media also varies. Whereas *S. citri* is easily cultivated on agar media, difficulties may be encountered with CSS
and SMCS even when multiplying to relatively high titers in
liquid media of identical composition. Also, the motility of
the spiroplasmas may prevent typical colony formation, especially when grown on a medium with a moist surface (Townsend
et al., 1980b).

B.   *CELLULAR MORPHOLOGY, MOTILITY, AND ULTRASTRUCTURE*

    The demonstration by electron microscopy of mycoplasma-
like structures in the sieve tubes of diseased plants provided,

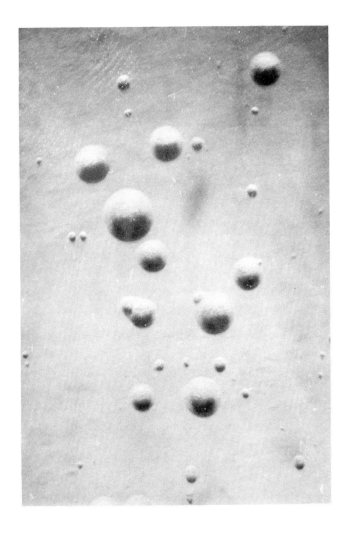

Fig. 1. *Agar culture of spiroplasma colonies. Isolate from aster yellows-diseased lettuce, 13 day-old colonies. x110. Photograph by Dr. F. Kondo.*

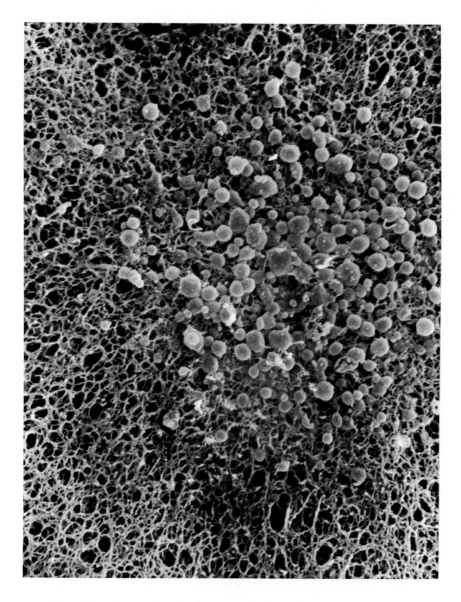

*Fig. 2. Scanning electron micrograph of spiroplasma co-
lony on agar. Isolate from aster yellows-diseased lettuce.
Courtesy Dr. D. M. Phillip, Rockefeller University.*

as mentioned repeatedly in earlier sections of this chapter,
the first clue for the concept that a great many plant diseases hitherto thought to be caused by viruses might rather
be caused by organisms related to the mycoplasmas. The
morphological and ultrastructural properties that are shared
by MLO's, spiroplasmas, and other mycoplasmas include the
pleomorphism of the cells and their lack of a cell wall.
The pleomorphism of nonhelical mycoplasmas and MLO's is characterized by a variation in the shape of the cells ranging from
small spherical to long filamentous structures. The human
pathogen *M. pneumoniae* and a few animal mycoplasmas show
a peculiar gliding motility (Bredt, 1979). The spiroplasmas,
on the other hand, are distinguished by the helical or spiral
morphology after which they have been named, and by different
types of motility (see below).

Examination of the morphology of the spiroplasmas may be
performed by darkfield or phase-contrast microscopy and by
electron microscopy of sap expressed from infected plants,
sections of the phloem tissue, or of cultured cells (Fig. 3).
The very detailed description presented by Cole *et al.* (1973)
of the morphology and ultrastructure of *S. citri* may be regarded as representative for all spiroplasmas. Examination
by darkfield microscopy of broth cultures in the logarithmic
phase of growth revealed a number of short, helical filaments
and some small round cells. The length of the filaments
usually varied from 2 to 4 µm. On prolonged incubation of
postlog cultures, the filaments increased in length and at the
same time showed a progressive loss of helicity together with
a tendency to develop "blebs." At a very late stage, increasing fragmentation and distortion of the filaments were observed
and many round or irregular bodies appeared. Two types of
motility were exhibited by the helical cells: a rapid rotatory
or "corkscrew" motion which could reverse, thereby leading to
a minimal back-and-forth progress, and a slow undulation and
bending of filaments that did not lead to a change of their
position. Loss of helicity on increasing age of the cultures
was associated with loss of motility. Recently, a third type
of spiroplasmal motility, translational locomotion, was demonstrated in viscous liquid medium and on glass surfaces
(Davis, 1978; Daniels *et al.*, 1980). Motility is energy-de-
pendent and is optimal at pH 7, as shown in the study of
Daniels *et al.* (1980), who also found that spiroplasmas dis-
play positive as well as negative chemotaxis. When examined
by electron microscopy, helicity of the filaments is best demonstrated in negatively stained specimens. However, as
shown by Cole *et al.* (1973), preservation of the helical shape
depends to a wide extent on the fixative and the stain used
for the preservation. Whereas helical filaments were pre-

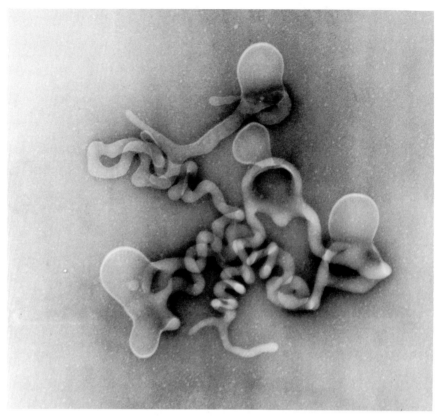

Fig. 3. Negatively stained spiroplasmas. Isolate from Opuntia tuna monstrosa. x39,000. Electron micrograph by Dr. F. Kondo.

served by fixation with formalin or glutaraldehyde made up in medium, helicity was lost upon fixation with glutaraldehyde dissolved in cacodylate buffer or by fixation with osmium tetroxide. Helical morphology was further preserved by stain-ing with ammonium molybdate, but not with sodium phosphotung-state. Helical cells may be seen also by electron microscopy of sectioned material, provided 1 μm thick sections are used. The influence of various external environment factors, such as the composition and pH of the growth medium, on maintenance of the helical morphology of *S. citri* was examined more re-cently by Patel *et al.* (1978). Very interestingly, a nonheli-cal and nonmotile strain of *S. citri,* which failed, moreover, to produce helical cells during more than 40 subcultivations in different media, has been described by Townsend *et al.*

(1977). Helical morphology of the CSS was demonstrated some
years before cultivation of this spiroplasma (Davis et al.,
1972a, b: Davis and Worley, 1973). Micrographs obtained by
stereo electron microscopy of the phloem tissue of corn stunt-
infected plants are particularly impressive (Davis and Worley,
1973). A number of environmental factors, including pH and
lipid composition of the growth medium, have been shown to
influence the helical morphology of spiroplasma cells (Freeman et al., 1976; Patel et al., 1978a).

The most prominent ultrastructural characteristic of the
spiroplasmas is their lack of a cell wall, as demonstrated
by electron microscopy of thin-sectioned material (Fig. 4)
showing the cells to be bounded only by a triple-layered unit
membrane (Cole et al., 1973). A discontinuous layer internal
to and abutting on the cytoplasmic membrane was observed in
some cells of S. citri by these authors, who found some evidence, moreover, in negatively stained specimens for the presence of an outer layer of surface projections. Freeze-etched
preparations of S. citri cells studied by Razin et al. (1973)
showed typical particle-studded fracture faces, and observations made by this method also revealed an envelope-like
structure that was either close to the cell surface or detached from it. The possible function of the additional internal
and external surface structures remains obscure, but they may
be of some importance for the ability of the spiroplasma cell
to maintain its helical morphology. Axial filaments or any
other organelles that might account for the motility of spiroplasma cells have not been observed. However, 3.5 nm fibrils
wound helically along the inner surface of the cytoplasmic
membrane, to which they are apparently fixed by a protein integrated into the membrane, have been demonstrated in several
spiroplasmas (Williamson, 1974; Williamson and Whitcomb, 1974;
Stalheim et al., 1978; Townsend et al., 1980a. Most likely,
these very unusual fibrils play an important role in maintain-
cell shape and for the motility of spiroplasmas. The main
protein of the S. citri membrane is spiralin (Wroblewski et
al., 1976; Wroblewski, 1978) that may be identical with the
protein thought to provide the anchorage points for the
fibrils (Townsend et al., 1980a). The interior of the cells
consists of a fine reticulum interspersed with amorphous
material and distinct ribosomal bodies. The cell lines of the
two strains of S. citri examined by Cole et al. (1973) were
found to be infected with a tailed, type B bacteriophage.

Available descriptions of the morphology and ultrastructure of other spiroplasmas, including CSS and SMCS, are essentially in good agreement with the above description of S. citri (Chen and Granados, 1970; Davis et al., 1972a, b; William-

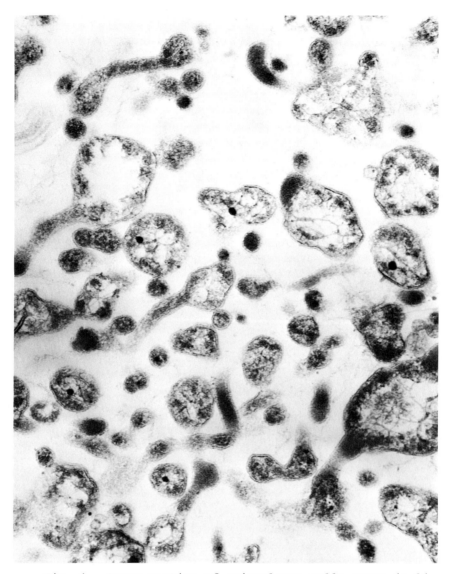

*Fig. 4. Cross section of spiroplasma cells grown in li-
quid medium. Isolate from aster yellows-diseased lettuce.
x 84,000. Courtesy Dr. D. M. Phillip, Rockefeller University.*

son and Whitcomb, 1975; Brinton and Burgdorfer, 1976; Tully
et al., 1977; Bove and Saillard, 1979).
    Before concluding this section, it may be appropriate to
call the attention to the risk of mistaking preparatory arti-

facts for mycoplasmas. The occurrence of pleomorphic structures resembling mycoplasmas in the sap extracted from helathy plants of several species and prepared for electron microscopy by negative staining was pointed out by Wolanski and Maramorosch (1970). These structures occurred in unfixed, but not in fixed, material stained with phosphotungstic acid.

C. FILTERABILITY CHARACTERISTICS

Filtration studies performed by Saglio et al. (1973) with broth cultures of S. citri showed the ability of spiroplasma cells to pass through membrane filters with a pore diameter as small as 220 nm, although in decreasing number with decreasing diameter. The ability of spiroplasmas to pass through filters that will retain the smallest bacteria is a consequence, of course, of the very small dimensions of their minimal reproductive units, but may also be attributed to some extent to the relative plasticity of the cells that is due to the absence of a rigid cell wall. Filterability through filters of 220 to 450 nm pore size is another criteria that has to be fulfilled for classification of an organism with the Mollicutes (Subcommittee on the Taxonomy of Mollicutes, 1979).

D. BIOCHEMICAL CHARACTERISTICS

The biochemical properties of spiroplasmas have been studied most extensively for S. citri (Saglio et al., 1973). It was shown to produce acid from glucose and mannose and to possess phosphatase activity. The tetrazolium reduction test was negative for the type strain (str. Morocco), but positive for another strain tested. Negative reactions were reported for hydrolysis of esculin, arginine and urea, and for digestion of coagulated serum. However, in a more recent study by Townsend (1976), S. citri was shown to hydrolyze arginine with accumulation of ammonia as an end product. The increase, moreover, in concentration of ornithine and citrulline during growth indicated degradation of arginine via the deiminase pathway that is typical of other arginine catabolizing mycoplasmas.

The honeybee spiroplasma appears to be the only other spiroplasma whose biochemical activities have been tested to an extent that allows comparison with S. citri (Davis et al., 1976). It resembles this organism in catabolizing both glucose and arginine, whereas the phosphatase test was found to be negative.

Studies of the biochemical composition of the limiting membrane and adjacent layers of spiroplasma cells have yielded

some interesting results. Most important is the demonstration of the absence in *S. citri* of the major constituent of
the bacterial cell wall, peptidoglycan and its precursors,
diaminopimelic and neuramic acids (Bebear *et al.*, 1974). Isolated membrane preparations of *S. citri* were shown by Razin
et al. (1973) to contain hexosamine in an amount that was a-
bout one-third of that found for *Acholeplasma laidlawii* membranes. As pointed out by Razin *et al.* (1973) the finding of
hexosamine may have some bearing upon the question about the
possible existence of some kind of an envelope or outer layer
enclosing the plasma membrane of *S. citri*. Studies of the
lipid composition of *S. citri* membranes were likewise performed by Razin *et al.* (1973), who found, among other things, free
cholesterol to be the predominant constituent of the neutral
lipid fraction, and by Freeman *et al.* (1976), who showed somewhat similar results. In a study by Patel *et al.* (1976), the
quantitative composition of lipids was found to be identical
for the membranes of *S. citri* and CSS. In another paper
(Mudd *et al.*, 1979), the lipid phosphorus was reported to be
lower, on a protein basis, and cholesterol higher in *S. citri*
than in CSS.

      Investigations of the localization of certain enzymes to
either the membrane or the cytoplasm have revealed interesting
differences between different mycoplasmal genera (Table I).
Thus, nicotinamide adenine dinucleotide oxidase is associated
in *S. citri*, as it is in the sterol-requiring *Mycoplasma* species, with the cytoplasm, whereas in *Acholeplasma* species it
is located in the membrane (Kahane *et al.*, 1977; Mudd *et al.*,
1977). Deoxyribonuclease, on the other hand, is apparently
confined to the membrane of *S. citri*, but has been demonstrated in both fractions of *Acholeplasma* cells (Kahane *et al.*,
1977).

E.   *SENSITIVITY TO ANTIBIOTICS AND OTHER CHEMICALS*

      Historically, the demonstration of the susceptibility of
MLO's to tetracyclines was, as mentioned before, one of the
observations which first suggested a possible relationship between these organisms and the mycoplasmas.
      A systematic testing of the sensitivity of the Morocco and
California strains of *S. citri* to antibiotics and a few other
chemicals was carried out by Saglio *et al.* (1973). Like other
mycoplasmas, both strains were found to be highly resistant to
penicillin "B," not being inhibited by 10,240 IU/ml. In contrast, they were completely inhibited by, for example, erythromycin in concentrations between 0.04 and 0.08 μg/ml to
0.32 μg/ml. Also, in accordance with the nutritional depen-

dence of *S. citri* on sterol, the two strains were shown to be
moderately to highly sensitive to amphotericin B, digitonin,
and polyanethole sulfate. The resistance of *S. citri* to high
concentrations of thallium acetate (up to 1280 µg/ml) is another property which is shared by this organism and other mycoplasmas.

The absolute resistance of spiroplasmas and mycoplasmas
in general to penicillin, and their relative insusceptibility
to thallium acetate in concentrations that will inhibit many
bacteria, are properties that are utilized for the preparation
of selective growth media that will allow the unrestricted
multiplication of mycoplasmas, while inhibiting the growth of
contaminating bacteria. However, the occasional observation
of an inhibitory activity of some penicillins to a few mycoplasmal species, and the variations recorded in the relative
resistance of different mycoplasmas to thallium acetate,
should imply some reservations against an uncritical incorporation of these substances in media used, in particular, for
primary isolation of unknown spiroplasmas and MLO's.

F.   ELECTROPHORETIC ANALYSIS OF CELL PROTEINS

Determination of the patterns of cell proteins by one-
dimensional polyacrylamide gel electrophoresis (PAGE) is extensively used in mycoplasmology in general (Razin and Tully,
1975), and has proved very useful, also, for characterization
and comparison of different strains of spiroplasmas. Thus,
the protein patterns of the Morocco and California strains
of *S. citri* were found to be identical, but clearly distinct
from those of a great many other mycoplasmas tested (Saglio
et al., 1973). The differences demonstrated by Padhi *et al.*
(1977) between the PAGE pattern of *S. citri* and CSS were regarded by these authors as sufficient evidence for their
classification as two different species. Analysis, on the
other hand, by densitometric tracing of the PAGE protein bands
from *S. citri* and spiroplasmas isolated from cactus *(Opuntia
tuna monstrosa)* and yellows-diseased lettuce showed the protein patterns of these three strains to be almost indistinguishable (Padhi *et al.*, 1977). The identity of the cactus and
lettuce spiroplasmas with *S. citri* was subsequently confirmed
by nucleic acid hybridization studies (Christiansen *et al.*,
1980). In another study reported by Davis *et al.* (1978),
PAGE characterization of cell proteins of a number of isolates
from different sources (plants, surface of flowers, and insects) correlated well with a serological grouping (Davis
et al., 1978). The nonhelical strain of *S. citri* isolated by
Townsend *et al.* (1977) was found to lack a single protein band
possessed by other strains of this species.

The number of cell protein bands obtained by one-dimensional PAGE is relatively small, and varies for *S. citri* from about 20 to 50, depending on the technique used (Saglio *et al.*, 1973; Mouches *et al.*, 1979). A much higher resolution can be achieved by two-dimensional analysis by which proteins are first submitted to electrofocusing in cylindrical polyacrylamide gels and further separated by electrophoresis in a slab gel. By this technique about 150 protein spots have been visualized for *S. citri* (Mouches *et al.*, 1979; Bove and Saillard, 1979). Additional useful information can be obtained by two-dimensional co-analysis of mixed protein samples from two different strains. Using this method, the number of comigrating or homologous proteins in *S. citri* and SMCS was shown to be considerably smaller than for *S. citri* and CSS (Mouches *et al.*, 1979; Bove and Saillard, 1979), thereby confirming the very distant relationship of SMCS to *S. citri* and CSS found also by other techniques.

G. ANTIGENIC CHARACTERISTICS

Spiroplasma citri was found, by growth inhibition (GI), metabolism inhibition (MI), and immunofluorescence (IMF) tests, to be antigenically unrelated to any of more than 50 *Mycoplasma* and *Acholeplasma* species against which it was tested (Saglio *et al.*, 1973). The results of serological comparisons between *S. citri* and CSS vary with the technique used. Most investigators have found the two organisms closely related or indistinguishable by the growth inhibition, complement fixation, and ring precipitation tests (Tully *et al.*, 1973; Chen and Liao, 1975; Williamson and Whitcomb, 1975; Tully *et al.*, 1973; Chen and Liao, 1975; Williamson and Whitcomb, 1975; Tully *et al.*, 1976, 1977; Davis *et al.*, 1979). In contrast, *S. citri* and CSS cross-react only to relatively low titers by the metabolism inhibition test (Chen and Liao, 1975; Williamson *et al.*, 1979) and by the spiroplasma deformation test (Chen and Liao, 1975; Williamson and Whitcomb, 1975; Tully *et al.*, 1976; 1977; Williamson *et al.*, 1978, 1979; Davis *et al.*, 1979). The SMCS, however, was found by all of these tests to be only distantly related to either *S. citri* or CSS (Tully *et al.*, 1976, 1977; Williamson *et al.*, 1978, 1979).

The metabolism inhibition and deformation tests appear to be particularly suitable for distinguishing spiroplasmas. The spiroplasma deformation test is a new serological method devised by McIntosh *et al.* (1974) and further developed by Williamson *et al.* (1978). It depends on the ability of homo-

logous antiserum to deform spiroplasma cells to the extent
that they finally lose their helicity and become spheroidal,
as demonstrable by darkfield microscopy. The test is simple
to perform and highly sensitive and specific. A very convenient combined deformation-metabolism inhibition test was described by Williamson *et al.* (1979).

On the basis of antigenicity, Junca *et al.* (1980) in a
very recent study carried out a classification of most of the
presently known spiroplasmas in five serologically distinct
groups: I, *Spiroplasma citri*; II, sex-ratio organism; III,
flower spiroplasma; IV flower spiroplasma; and V, suckling
mouse cataract agent. Group I was further subdivided into
four subgroups (serovars): subgroups I-1, *S. citri*, classical
serovar (strain Morocco = R8A2); I-2, *S. citri*, honeybee serovar; I-3, *S. citri*, corn stunt serovar; and I-4, *S. citri*,
277 F (i.e., a single isolate from the rabbit tick that is
serologically unrelated to SMCS).

H. CHARACTERISTICS OF SPIROPLASMA DNA

The guanine plus cytosine (G+C) content of the DNA of *S.
citri* was found by buoyant density and thermal denaturation
(T_m) to be 26.4 mol% (Saglio *et al.*, 1973; Christiansen *et
al.*, 1979). The values reported for CSS are very similar or
even identical to that of *S. citri*, namely, 25.1 mol% (Christiansen *et al.*, 1979) and 26.3 mol% (T. A. Chen, unpublished
and quoted by Bove and Saillard, 1979), whereas they are definitely higher for SMCS, 29.4 mol% (Bove and Saillard, 1979;
Christiansen *et al.*, 1979).

Spiroplasma DNA has been characterized also by a very recently developed technique that is based on fragmentation of
the nucleic acid molecule by restriction endonuclease and subsequent separation of the fragments by polyacrylamide gel
electrophoresis. However, the use that may be made of this
highly sensitive technique for taxonomic purposes is limited
by the fact that, although the endonuclease restriction frag-
ment patterns of some strains withing a given spiroplasma
species or serogroup are quite similar, others are surprisingly different (Bove and Saillard, 1979).

The genome size of *S. citri* is 10^9 daltons (Saglio *et al.*,
1973), i.e., of the same order or size as for genus *Acholeplasma*, and twice that of other mycoplasma genera (Table I).

DNA-DNA hybridization experiments have been used to determine the degree of genetic relatedness between various
spiroplasmas. Thus, Christiansen *et al.* (1979) and Lee and
Davis (1980) by means of different hybridization techniques
obtained very similar homology values, about 30-33% for *S.

citri and CSS, whereas studies by other workers have yielded
somewhat higher values (C. H. Kiao and T. A. Chen, quoted by
Christiansen *et al.*, 1979; Junca *et al.*, 1980; Rahimian and
Gumpf, 1980). A 45-48% hybridization was demonstrated between
DNA's from *S. citri* and the honeybee spiroplasma (Lee and
Davis, 1980), whereas here again *S. citri* and CSS have been
found to be only very distantly related to SMCS, with homologies of no more than 3-5% (Christiansen *et al.*, 1979; Junca
et al., 1980). In contrast, a 100% homology has been demonstrated between *S. citri* and the spiroplasmas isolated from
cactus and lettuce (Christiansen *et al.*, 1980) which is in
perfect agreement with the results of comparisons made by
other methods between these spiroplasmas.

IV.   IDENTIFICATION

The general principles of identification of mycoplasmas,
and the biochemical and serological methods currently used
for that purpose, were described in great detail in a recent
survey by Freundt *et al.* (1979). With a few notable exceptions, the procedures to be followed for identification of an
unclassified member of the family Spiroplasmataceae do not,
of course, differ essentially from the methods used for mycoplasmas in general. Hence, a very brief consideration of
these methods will suffice here.

Although the characteristics on which the methods of identification are based are those described in the previous sections, the weight carried by each property for the aim of
differentiating between different taxonomic categories of
spiroplasmas is highly variable. The fundamental importance
of attaining growth in artificial media, and preferably on
solid as well as in liquid medium, for proper characterization
and subsequent identification of a mycoplasma was emphasized
earlier. As a further preliminary to a detailed examination,
the organism must be subjected to a cloning procedure which
involves picking an isolated colony from a plate culture and
filtration through a membrane filter with the smallest possible
pore diameter, i.e., 220 to 450 nm (Subcommitte on the Taxonomy
of Mollicutes, 1979).

Identification of an organism, that fulfills the basic requirements for classification with the Mollicutes, as a member
of the family Spiroplasmataceae generally depends on the demonstration of helical morphology and typical motility. Nutritional requirement for sterol may be determined by an indirect method, viz., the digitonin test (Freundt *et al.*, 1979,
p. 395), and further confirmed by the methods of Razin and
Tully (1970) or Edward (1971).

The number of biochemical properties that have proved useful for characterization of mycoplasmas in general at the species level is rather small, and most of them are shared by a great many different species. Consequently the use that can be made of such properties for the purpose of identification is relatively limited. In the case of species of the genus *Mycoplasma* a biochemical screening by a few tests, including tests for catabolism of glucose and arginine, phosphatase activity, and digestion of coagulated horse serum, is nevertheless a great help as a preliminary to final serological identification (Freundt *et al.*, 1979, p. 395). The situation with the spiroplasmas in this respect is very different, or at least appears so from presently available knowledge. Comparative studies of the biochemical properties of the many different groups of spiroplasmas that are known at present have not been carried out systematically. Those spiroplasmas that have been examined for biochemical activities share the properties of catabolizing both glucose and arginine.

It follows, that identification of an unknown isolate as *S. citri*, or its classification with some other group of spiroplasmas, must primarily depend on antigenicity. The serological methods that have been used most extensively to distinguish between different spiroplasmas were briefly summarized in Section III,G, and it was pointed out, also, that the metabolism inhibition and deformation tests are among the techniques that have proved most suitable for *Spiroplasma* identification and systematics. Both tests are highly sensitive and capable of revealing antigenic differences among spiroplasmas. The combination of the two tests into a single test run, as carriedout in microtiter plates (Williamson *et al.*, 1979), adds further to their practicability and simplicity in performance. Also, the growth inhibition test, which is one of the most specific tests in mycoplasmology in general, possess an ideal combination of specificity and sensitivity for separation of spiroplasmas at the species level (Whitcomb, 1980).

Any attempt at identifying new spiroplasma isolates should, of course, take into consideration the classification scheme of Junca *et al.* (1980) referred to in an earlier section. The establishment, in addition to *S. citri*, of four distinct serogroups, that appear to be candidates for new *Spiroplasma* species, and the subdivision of *S. citri* itself into four serovars, is indeed an important achievement in spiroplasma systematics that will also facilitate identification of new isolates.

REFERENCES

Bebear, C., Latrille, J., Fleck, J., Roy, B., and Bove, J. M. (1974). *Spiroplasma citri:* Un mollicute. *Colloq. Inst. Nat. Sante Rech. Med. 33,* 35-42.

Bove, J. M. and Saillard, C. (1979). Cell biology of mycoplasmas. *In* "The Mycoplasmas" (R. F. Whitcomb and J. G. Tully, eds.), Vol. III, Academic Press, New York.

Bredt, W. (1979). Motility. *In* "The Mycoplasmas" (M. F. Barile and S. Razin, eds.) Vol. I, Chapter 5. Academic Press, New York.

Brinton, L. P., and Burgdorfer, W. (1976). Cellular and subcellular organization of the 277F agent, a spiroplasma from the rabbit tick *Haemaphysalis leporispalustris* (Acari: Ixodidae). *Int. J. Syst. Bacteriol. 26,* 554-560.

Chanock, R. M., Hayflick, L., and Barile, M. F. (1962). Growth on artificial medium of an agent associated with atypical pneumonia and its identification as a PPLO. *Proc. Nat. Acad. Sci. U. S. 48,* 41-49.

Chen, T. A. and Davis, R. E. (1979). Cultivation of spiroplasmas. *In* "The Mycoplasmas" (R. F. Whitcomb and J. G. Tully, eds.), Vol. III, Chapter 3, Academic Press, New York.

Chen, T. A. and Granados, R. R. (1970). Plant-pathogenic mycoplasma-like organism: Maintenance *in vitro* and transmission to *Zea mays* L. *Science 167,* 1633-1636.

Chen, T. A. and Liao, C. H. (1975). Corn stunt spiroplasma: isolation, cultivation, and proof of pathogenicity. *Science 188,* 1015-1017.

Christiansen, C., Askaa, G., Freundt, E. A., and Whitcomb, R. F. (1979). Nucleic acid hybridization experiments with *Spiroplasma citri* and the corn stunt and suckling mouse cataract spiroplasmas. *Curr. Microbiol. 2,* 323-326.

Christiansen, G., Freundt, E. A., and Maramorosch, K. (1980). Identity of cactus and lettuce spiroplasmas with *Spiroplasma citri* as determined by DNA-DNA hybridization. *Curr. Microbiol. 4,* 353-356.

Clark, T. B. (1977). *Spiroplasma* sp., a new pathogen in honey bees. *J. Invertebr. Pathol. 29,* 112-113.

Cole, R. M., Tully, J. B., Popkin, T. J., and Bove, J. M. (1973). Morphology, ultrastructure, and bacteriophage infection of the helical mycoplasma-like organism (*Spiroplasma citri* gen. nov., sp. nov.) cultured from "stubborn" disease of citrus. *J. Bacteriol. 115,* 367-386.

Daniels, M. H., Longland, J. M., and Gilbart, J. (1980). Aspects of motility and chemotaxis in spiroplasmas. *J. Gen. Microbiol. 118,* 429-436.

Davis, R. E. (1978). Spiroplasma associated with flowers of the tulip tree (*Liriodendron tulipifera* L.). *Can. J. Microbiol. 24*, 954-959.

Davis, R. E. and Worley, J. F. (1973). Spiroplasma: Motile, helical microorganism associated with corn stunt disease. *Phytopathology 63*, 403-408.

Davis, R. E., Whitcomb, R. F., Chen, T. A., and Granados, R. R. (1972a). Current status of the aetiology of corn stunt disease. *In* "Pathogenic Mycoplasmas, Ciba Foundation Symposium" (K. Elliott and J. Birch, eds.). Elsevier Excerpta-Medica-North Holland, Amsterdam.

Davis, R. E., Worley, J. F., Whitcomb, R. F., Ishijima, T., and Steere, R. L. (1972b). Helical filaments produced by a mycoplasma-like organism associated with corn stunt disease. *Science 176*, 521-523.

Davis, R. E., Worley, J. F., Clark, T. B., and Moseley, M. (1976). New spiroplasma in diseased honeybee (*Apis mellifera* L.): isolation, pure culture, and partial characterization *in vitro*. *Proc. Amer. Phytopathol. Soc. 3*, 304.

Davis, R. E., Lee, I.-M., and Basciano, L. K. (1978). Spiroplasmas: Identification of serological groups and their analysis by polyacrylamide gel electrophoresis of cell proteins. *Phytopathol. News 12(9)*, 415. (Abstr.).

Davis, R. E., Lee, I.-M., and Basciano, L. K. (1979). Spiroplasmas: serological grouping of strains associated with plants and insects. *Can. J. Microbiol. 25*, 861-866.

Eden-Green, S. J. (1978). Isolation of acholeplasmas from coconut palms affected by lethal yellowing disease in Jamaica. *Zentrabl. Bakteriol. Parasitenk. Infektionsk. Hyg., Abt. 1 Orig. A241*, 226.

Edward, D. G. ff. (1971). Determination of sterol requirement for Mycoplasmatales. *J. Gen. Microbiol. 69*, 205-210.

Edward, D. G. ff. and Fitzgerald, W. A. (1951). Cholesterol in the growth of organisms of the pleuropneumonia group. *J. Gen. Microbiol. 5*, 576-586.

Freeman, B. A., Sissenstein, R., McManus, T. T., Woodward, J. E., Lee, I. M., and Mudd, J. B. (1976). Lipid composition and lipid metabolism of *Spiroplasma citri J. Bacteriol. 125*, 946-954.

Freundt, E. A. (1978). Culture media (natural and synthetic): Mycoplasmas. *In* "CRC Handbook Series in Nutrition and Food. Section G: Diets, Culture Media and Food Supplements" (M. Rechcigl, ed.), Vol. III. CRC Press, Cleveland, Ohio.

Freundt, E. A. (1979). Isolation and cultivation of mycoplasmas. - A survey of general principles. *In* "Mycoplasma Diseases of Trees", pp. 59-70. (S. P. Raychaudhuri, ed.). Associated Publishing Company, New Delhi.

Freundt, E. A., Ernø, H., and Lemcke, R. M. (1979). Identifi-
    cation of mycoplasmas. *In* "Methods in Microbiology" (T.
    Bergan and J. R. Norris, eds.), Vol. 13, Chapter 9.
    Academic Press, New York.
Frey, M. L., Hanson, R. P., and Anderson, D. P. (1968). A
    medium for the isolation of avian mycoplasmas. *Amer. J.
    Vet. Res. 29*, 2163-2171.
Fudl-Allah, A. E. A. and Calavan, E. C. (1972). Effect of su-
    gars, tryptone, PPLO broth, yeast extract, and horse se-
    rum on growth of the mycoplasma-like organism associated
    with stubborn disease of citrus. *Phytopathology 62*, 758.
    (Abstr.).
Fudl-Allah, A. E. A., Calavan, E. C., and Igwegbe, E.C. K.
    (1972). Culture of a mycoplasma-like organism associated
    with stubborn disease of citrus. *Phytopathology 62*, 729-
    731.
Gianotti, M. J. (1974). Cultures de Mollicutes a partir de
    plantes atteintes de jaunisse et d'insectes vecteurs.
    Colloq. Inst. Nat. Sante Rech. Med. 33, 99-106.
Hampton, R. O., Stevens, J. O., and Allen, T. C. (1969).
    Mechanically transmissible mycoplasma from naturally in-
    fected peas. *Plant Dis. Rep. 53*, 499-503.
Hampton, R. O., Florance, E. R., Whitcomb, R. F., and Seidler,
    R. J. (1976). Evidence suggesting nonassociation of myco-
    plasma with pea disease. *Phytopathology 66*, 1163-1168.
Hopps, H. E., DelGiudice, R. A., and Barile, M. F. (1976).
    Current status of "non-cultivable" mycoplasmas. *Proc.
    Soc. Gen. Microbiol. III (4)*, p. 143.
Jones, A. L., Whitcomb, R. F., Williamson, D. L., and Coan,
    M. E. (1977). Comparative growth and primary isolation
    of spiroplasmas in media based on insect tissue culture
    formulations. *Phytopathology 67*, 738-746.
Junca, P., Saillard, C., Tully, J., Garcia-Jurado, O., Degorce-
    Dumas, J.-R., Mouches, C., Vignault, J. C., Vogel, R.,
    McCoy, R., Whitcomb, R., Williamson, D., Latrille, J.,
    and Bove, J. M. (1980). Caracterisation de spiroplasmes
    isoles d'insectes et de fleurs de France continentale, de
    Corse et du Maroc. Proposition pour une classification
    des spiroplasmes. *C. R. Acad. Sci. Ser. D290*, 1209-1212.
Kahane, I., Greenstein, S., and Razin, S. (1977). Carbohy-
    drate content and enzymatic activities in the membrane of
    Spiroplasma citri. J. Gen. Microbiol. 101, 173-176.
Lafleche, D. and Bove, J. M. (1970). Mycoplasmes dans les
    agrumes atteints de "Greening", de "Stubborn" ou de mala-
    dies similaires. *Fruits 25*, 455-465.
Lee, J.-M. and Davis, R. E. (1978). Identification of some
    growth-promoting components in an enriched cell-free
    medium for cultivation of *Spiroplasma citri*. Phytopathol.
    News *12(9)*, 414 (Abstr.).

Lee, I.-M. and Davis, R. E. (1980). DNA homologies among spiro-
     plasma strains representing different serological groups.
     Phytoplathology 70, 464.
Liao, C. H. and Chen, T. A. (1975). Inhibitory effect of
     corn stem extract on the growth of corn stunt spiroplasma.
     Proc. Amer. Phytopathol. 2, 53. (Abstr.).
Liao, C. H. and Chen, T. A. (1977). Culture of corn stunt spiro-
     plasma in a simple medium. *Phytopathology 67,* 802-807.
Liao, C. H. and Chen, T. A. (1978). Effect of osmotic poten-
     tial, pH, and temperature on the growth of a helical,
     motile mycoplasma causing corn stunt disease. *Can. J.
     Microbiol. 24,* 325-329.
Liao, C. H. and Chen, T. A. (1980). Presence of spiroplasma
     inhibitory substances in plant tissue extracts. *Can. J.
     Microbiol. 26,* 807-811.
McCoy, R. E. (1976). Plant phloem sap: A potential mycoplas-
     ma growth medium. *Proc. Soc. Gen. Microbiol. III(4),*
     p. 155.
McCoy, R. E. (1979). Mycoplasmas and yellows diseases. *In*
     "The Mycoplasmas" (R. E. Whitcomb, and J. G. Tully, eds.),
     Vol. III, Chapter 8. Academic Press, New York.
McCoy, R. E., Williams, D. S., and Thomas, D. L. (1979). Iso-
     lation of mycoplasmas from flowers. *Proc. R.O.C. U.S.
     Cooperative Sci. Semin. Mycoplasma Dis. Plants, NSC Symp.
     Ser. I,* March 27-31, 1978. pp. 75-81. Nat. Sci. Council,
     Taipei.
McIntosh, A. H., Skowronski, B. S., and Maramorosch, K. (1974).
     Rapid identification of *Spiroplasma citri* and its relation
     to other yellows agents. *Phytopathol. Z. 80,* 153-156.
Maramorosch, K. (1972). The enigma of mycoplasma in plants
     and insects. *Phytopathology 62,* 1230-1231.
Mouches, C., Vignault, J. C., Tully, J. G., Whitcomb, R. F.,
     and Bove, J. M. (1979). Characterization of spiroplasmas
     by one and two dimensional protein analysis on polyacryla-
     mide slab gels. *Curr. Microbiol. 2,* 69-74.
Mudd, J. B., Ittig, M., Roy, B., Latrille, J., and Bove, J.
     M. (1977). Composition and enzyme activities of *Spiroplas-
     ma citri* membranes. *J. Bacteriol. 129,* 1250-1256.
Mudd, J. B., Lee, I.-M., Liu, H.-Y., and Calavan, E. C. (1979).
     Comparison of the membrane composition of *Spiroplasma
     citri* and the corn stunt *Spiroplasma. J. Bacteriol. 137,*
     1056-1058.
Padhi, S. B., McIntosh, A. M., and Maramorosch, K. (1977).
     Characterization and identification of spiroplasmas by
     polyacrylamide gel electrophoresis. *Phytopathol. Z. 90,*
     268-272.
Patel, K. R., Mayberry-Carson, K. J., and Smith, P. F. (1978a).
     Effect of external environmental factors on the morphology
     of *Spiroplasma citri. J. Bacteriol. 133,* 925-931.

Patel, K. R., Smith, P. F., and Mayberry, W. R. (1978b). Comparison of lipids from *Spiroplasma citri* and corn stunt spiroplasma. *J. Bacteriol. 136,* 829-831.

Rahimian, H., and Gumpf, D. J. (1980). Deoxyribonucleic acid relationship between *Spiroplasma citri* and corn stunt spiroplasma. *Int. J. Syst. Bacteriol. 30,* 605-608.

Razin, S. (1969). Structure and function in mycoplasma. *Annu. Rev. Microbiol. 23,* 317-356.

Razin, S. and Cosenza, B. J. (1966). Growth phases of *Mycoplasma* in liquid media observed with phase-contrast microscope. *J. Bacteriol. 91,* 858-869.

Razin, S. and Rottem, S. (1963). Fatty acid requirements of *Mycoplasma laidlawii. J. Gen. Microbiol. 33,* 459-470.

Razin, S. and Tully, J. G. (1970). Cholesterol requirement of mycoplasmas. *J. Bacteriol. 102,* 306-310.

Razin, S. and Tully, J. G. (1975). Identification of mycoplasmas by electrophoretic analysis of cell proteins. WHO Working Document VPH/MIC/75.3.

Razin, S., Cosenza, B. J., and Tourtellotte, M. E. (1966). Variations in mycoplasma morphology induced by long-chain fatty acids. *J. Gen. Microbiol. 42,* 139-145.

Razin, S., Cosenza, B. J., and Tourtellotte, M. E. (1967). Filamentous growth of mycoplasma. *Ann. N. Y. Acad. Sci. 143,* 66-72.

Razin, S., Hasin, M., Ne'eman, Z., and Rottem, S. (1973). Isolation, chemical composition, and ultrastructural features of the cell membrane of the mycoplasma-like organism *Spiroplasma citri. J. Bacteriol. 116,* 1421-1435.

Rodwell, A. W. (1960). Nutrition and metabolism of *Mycoplasma mycoides* var. *mycoides. Ann. N. Y. Acad. Sci. 79,* 499-507.

Rodwell, A. W. (1963). The steroid growth requirement of *Mycoplasma mycoides. J. Gen. Microbiol. 32,* 91-101.

Rodwell, A. W. (1967). The nutrition and metabolism of mycoplasma: Progress and problems. *Ann. N. Y. Acad. Sci. 143,* 88-109.

Rodwell, A. W. (1968). Fatty acid composition of Mycoplasma lipids: Biomembrane with only one fatty acid. *Science 160,* 1350-1351.

Rodwell, A. W. (1969). The supply of cholesterol and fatty acids for the growth of mycoplasmas. *J. Gen. Microbiol. 58,* 29-37.

Rodwell, A. W. and Abbot, A. (1961). The function of glycerol, cholesterol and long-chain fatty acids in the nutrition of *Mycoplasma mycoides. J. Gen. Microbiol. 25,* 201-214.

Rylance, H. J., Marr, W., Malcolm, M. G., Stewart, S. M., and
    Marmion, B. P. (1979). Growth factors for *Mycoplasma
    pneumoniae* in yeast and tissue extracts. *J. Appl.
    Bacteriol. 47,* 341-345.

Saglio, P., Lafleche, D., Bonissol, C., and Bove, J. M. (1971).
    Isolement, culture et observation au microscope electron-
    ique des structures de type mycoplasme associees a la
    maladie au Stubborn des agrumes et leur comparaison avec
    les structures observees dans le cas de la maladie du
    Greening des agrumes. *Physiol. Veg. 9,* 569-582.

Saglio, P., Lhospital, M., Lafleche, D., Dupont, G., Bove,
    J. M., Tully, J. G., and Freundt, E. A. (1973). *Spiro-
    plasma citri* gen. and sp. n.: a mycoplasma-like organism
    associated with "stubborn" disease of citrus. *Int. J.
    Syst. Bacteriol. 23,* 191-204.

Saglio, P. H. M., Daniels, M. J., and Pradet, A. (1979). ATP
    and energy charge as criteria of growth and metabolic ac-
    tivity of Mollicutes: Application to *Spiroplasma citri.
    J. Gen. Microbiol. 110,* 13-20.

Stalheim, O. H. V., Ritchie, A. F., and Whitcomb, R. (1978).
    Cultivation, serology, ultrastructure, and virus-like
    particles of spiroplasma 277F. *Curr. Microbiol. 1,* 365-
    370.

Subcommittee on the Taxonomy of Mollicutes (1979). Proposal
    of minimal standards for descriptions of new species of
    the class *Mollicutes. Int. J. Syst. Bacteriol. 29,* 172-
    180.

Townsend, R. (1976). Arginine metabolism by *Spiroplasma citri.
    J. Gen. Microbiol. 94,* 417-420.

Townsend, R., Markham, P. G., Plaskitt, K. A., and Daniels,
    M. J. (1977). Isolation and characterization of a non-
    helical strain of *Spiroplasma citri. J. Gen. Microbiol.
    100,* 15-21.

Townsend, R., Archer, D. B., and Plaskitt, K. A. (1980a).
    Purification and preliminary characterization of spiro-
    plasma fibrils. *J. Bacteriol. 142,* 694-700.

Townsend, R., Burgess, J., and Plaskitt, K. A. (1980b). Isola-
    tion and characterization of a non-helical strain of
    Spiroplasma citri. J. Bacteriol. 142, 973-981.

Tsai, J. A. (1979). Vector transmission of mycoplasmal agents
    of plant diseases. *In* "The Mycoplasmas" (R. F. Whitcomb
    and J. G. Tully, eds.), Vol. III, Chapter 9. Academic
    Press, New York.

Tully, J. G. (1979). Special features of the acholeplasmas.
    In "The Mycoplasmas" (M. F. Barile and S. Razin, eds.),
    Vol. I , Chapter 16, Academic Press, New York.

Tully, J. G., Whitcomb, R. F., Bove, J. M., and Saglio, P. (1973). Plant mycoplasmas: Serological relation between agents associated with citrus stubborn and corn stunt disease. *Science 182,* 827-829.

Tully, J. G., Whitcomb, R. F., Williamson, D. L., and Clark, H. F. (1976). Suckling mouse cataract agent is a helical wall-free prokaryote (spiroplasma) pathogenic for vertebrates. *Science 259,* 117-120.

Tully, J. G., Rose, D. L., Whitcomb, R. F., and Wenzel, R. P. (1979). Enhanced isolation of *Mycoplasma pneumoniae* from throat washings with a newly modified culture medium. *J. Infect. Dis. 139,* 478-482.

Whitcomb, R. F. (1980). The genus *Spiroplasma. Annu. Rev. Microbiol. 34,* 677-709.

Whitcomb, R. F., and Williamson, D. L. (1979). Pathogenicity of mycoplasmas for arthropods. *Zentrabl. Bakteriol. Parasitenk. Infektionsk. Hyg. Abt. 1 Orig. A245,* 200-201.

Whitcomb, R. F., Tully, J. G., Bove, J. M., and Saglio, P. (1973). Spiroplasmas and acholeplasmas: Multiplication in insects. *Science 182,* 1251-1253.

Williamson, D. L. (1974). Unusual fibrils from the spirochete-like sex ratio organism. *J. Bacteriol. 117,* 904-906.

Williamson, D. L. and Poulson, D. F. (1979). Sex ratio organisms (spiroplasmas) of *Drosophila. In* "The Mycoplasmas" (R. F. Whitcomb, and J. G. Tully, eds.), Vol. III, Chapter 6. Academic Press, New York.

Williamson, D. L. and Whitcomb, R. F. (1974). Helical, wall-free prokaryotes in *Drosophila,* leafhoppers, and plants. *Colloq. Inst. Nat. Sante Rech. Med. 33,* 283-290.

Williamson, D. L. and Whitcomb, R. F. (1975). Plant mycoplasmas: A cultivable spiroplasma causes corn stunt disease. *Science 188,* 1018-1020.

Williamson, D. L., Whitcomb, R. F., and Tully, J. G. (1978). The spiroplasma deformation test, a new serological method. *Curr. Microbiol. 1,* 203-207.

Williamson, D. L., Tully, J. G. and Whitcomb, R. F. (1979). Serological relationships of spiroplasmas as shown by combined deformation and metabolism inhibition tests. *Int. J. Syst. Bacteriol. 29,* 345-351.

Wolanski, B. and Maramorosch, K. (1970). Negatively stained mycoplasmas: fact or artifact? *Virology 42,* 319-327.

Wroblewski, H., Johansson, K.-E., and Hjerten, S. (1977). Purification and characterization of spiralin, the main protein of the *Spiroplasma citri* membrane. *Biochim. Biophys. Acta 465, 275-289.*

Wroblewski, H. (1978). Spiralin: its topomolecular anatomy and its possible function in the *Spiroplasma citri* cell membrane. *Zentrabl. Bakteriol. Parasitenk. Infektionskr. Hyg-Abt. 1 Org. A241,* 179-180.

YELLOWS DISEASES OF TREES

Carl E. Seliskar and Charles L. Wilson

Forest Sciences Laboratory
Northeastern Forest Experiment Station
United States Department of Agriculture
Delaware, Ohio

and
Appalachian Fruit Research Station
SEA, United States Department of Agriculture
Kearneysville, West Virginia

I. INTRODUCTION

The yellows diseases of trees rank among the earliest recognized and most important diseases of plants known to man.
Sandal spike, for example, was first reported in 1898 (McCarthy, 1903), and is the most serious yellows-type disease of
forest trees known in the world today. The disease has spread
progressively over the years, devastating large forest tracts
and threatening the entire sandal industry in South India,
where production of sandalwood oil is of major importance
(Raychaudhuri and Varma, 1980). Even earlier, peach yellows
and peach rosette had already made serious inroads in the
peach industry in the southern United States (Smith, 1888,
1891). The recently described and rapidly spreading lethal
yellowing disease of coconut palm and other species is now
believed caused by a mycoplasma (Beakbane et al., 1972; Heinze
et al., 1972; McCoy, 1972; Plavsic-Banjac et al., 1972; McCoy
and Gwin, 1977). A number of other yellows diseases of trees
are of major importance throughout the world either because
of local impact or widespread distribution. Among these are
Prunus X-disease, pear decline, elm phloem necrosis, apple
proliferation, the witches'-broom diseases of paulownia and
jujube, and the greening and stubborn diseases of citrus.

Yellows diseases of trees are associated with two of the most significant discoveries in recent plant pathological history: the first report of mycoplasma in plants (Doi *et al.*, 1967; Ishiie *et al.*, 1967), and the successful isolation and culturing of a spiroplasma on defined media (Davis *et al.*, 1972a, b; Davis and Worley, 1973). In the first instance, mycoplasma were observed in paulownia trees affected with witches'-broom and mulberry infected with dwarf; in the second instance, *Spiroplasma citri* was successfully cultured from citrus trees affected with stubborn disease.

Generally, yellows diseases are most serious in perennial crops, particularly woody plants, which once infected remain infected for life. Unlike many virus disorders, mycoplasma diseases are often lethal, although death may be slow. With some yellows diseases, infected trees may exhibit chronic symptoms over a more or less normal life span. Most commonly, however, mycoplasmas cause diseases that debilitate infected trees, resulting in a gradual decline and reduction in quantity and quality of fruit. Details of the symptoms caused by yellows diseases are given in the next section.

Our primary purpose in this chapter is to provide resumes of many of the yellows diseases of trees reported in the literature caused by mycoplasmas (MLO) and fastidious bacteria (RLO). Excluded from consideration, however, are sandal spike, lethal yellowing of palm, citrus stubborn and greening diseases, paulownia witches'-broom, and mulberry dwarf. These have been reviewed in considerable detail in other chapters of this volume.

II.   CHARACTERISTICS OF MYCOPLASMA, SPIROPLASMA, AND FASTIDIOUS
      BACTERIA-ASSOCIATED DISEASES OF TREES

A wide variety of symptoms are caused by mycoplasma, spiroplasma, and fastidious bacteria diseases of trees. In general, fastidious bacterial diseases (so called "rickettsialike" bacteria) affect the xylem. However, the citrus greening agent is a phloem inhabitant. Phloem is the primary site for diseases caused by mycoplasmas and spiroplasmas. No clear distinction can be made between symptoms of mycoplasma and spiroplasma diseases. In fact, some presently feel mycoplasmas may be spiroplasmas (Maramorosch, 1979). Generally, however, chlorosis induced by spiroplasmas is milder than that produced by mycoplasma-like organisms. Also, proliferation of growing points is generally less in "witches'-broom" symptoms caused by spiroplasmas than by mycoplasma-like organisms. Citrus stubborn and perhaps western X-disease of *Prunus* (Purcell *et al.*, 1980) are the only tree diseases that are known to be caused by spiroplasmas. Since citrus stubborn will be discussed elsewhere in this volume, only western X will be covered in this chapter.

Many MLO-associated diseases are described collectively as "yellows diseases." This term is somewhat misleading, since often the most salient symptom expressed is not yellowing. Proliferation of lateral shoots ("witches'-brooms") and virescence of floral parts is more characteristic. Other symptoms sometimes associated with these diseases include: phyllody, abnormal fruit and seeds, upright terminal growth, and overgrowth of the stem. Different strains of the MLO or spiroplasm causing X-disease of *Prunus* may induce varying symptoms on the fruit: short pedicels, a conical shape, a dull buckskin effect on the styler end, and insipid taste, undersize fruit, and failure to mature. The girdling brought about by sieve tube necrosis can result in leaf rolling, vein swelling, leaf abscision, arrested fruit development, tree decline, and wilting (Anon., 1976).

Sometimes, as with X-disease and elm phloem necrosis, MLO may affect the root system more than above-ground parts of the tree. In these cases, the trees may have the general symptom of starvation. The foliage may become wrinkled or strap-shaped. Necrosis of the phloem may occur, and there may be dieback of either terminal branches or roots. Flowers can revert to leaves (phyllody), drop, proliferate, become dwarfed, or even enlarge.

Although MLO's have been found associated with the sieve elements in a number of tree diseases, very little is known about the progression of the disease in these tissues. MLO have been found within sieve tubes and sometimes in phloem parenchyma and companion cells (Schneider, 1973). Sieve tube necrosis, in conjunction with hypertrophy and hyperplasia of adjacent parenchyma, is a common symptom at the cellular level. The xylem tissue is generally not affected. An exception is rubbery-wood disease of apple where lignification of xylem elements is reduced. This may be a secondary symptom in response to the diseased phloem.

McCoy (1979) believes that reports of MLO in cells other than sieve elements are due to misidentifications. He feels that sieve elements have been mistaken as parenchyma cells or that host vesicles have been interpreted as mycoplasmas. Since mycoplasmas do not appear to have any mechanism for penetrating cell walls, they may well be confined to sieve tubes.

Plants affected by fastidious bacteria can show extreme symptoms, from enhanced growth and vigor (e.g., phony peach) to wilted and scorched foliage (e.g., almond leaf scorch). A salient characteristic of a number of diseases caused by fastidious bacteria is a scalding or scorching of the foliage. The fruit may wilt or dry up and the plant may die. This is

indicated by such disease names as plum leaf "scald," almond
leaf "scorch," and elm leaf "scorch." At the other end of
the spectrum we have diseases where the plants' health is
apparently enhanced, such as phony peach and citrus greening.

Wilt and scorch symptoms induced by fastidious bacteria
occur primarily in the stem and above-ground portions of the
plant; in greening disease fastidious bacteria occur in the
roots. Also, in phony peach the organism is primarily in the
tree roots. Diseased trees develop shortened terminal internodes which gives them a flattened or umbrellalike canopy.
Foliage is denser and darker green. However, trees produce
progressively smaller fruit. After 3 to 5 years the fruit is
no longer marketable. In trees, fastidious bacteria have been
found primarily in the xylem elements.

Little is known about cellular changes in trees invaded
by fastidious bacteria. Our most detailed knowledge of woody
tissue is of grape invaded by Pierce's disease organism (Schneider, 1973). Here gums and tyloses develop in the lumens of
vessels. Other effects include defunct phelogen and abnormally
small mesophyll cells in leaves. Toxins are known to be produced and may be responsible for the symptoms.

Fastidious-bacterial infections in roots of peach trees
with phony and stems of almonds with scorch can be detected by
a simple chemical test. When unblemished root or stem sections with woody cylinders 1/2 to 3/4 inch in diameter are
immersed in 5 cm^3 of acidified methanol, distinctive purple
spots appear in the wood at the point of infection (Anon.,
1976; Mircetich *et al.*, 1976).

When fastidious bacteria invade the phloem of citrus they
induce a characteristic stunting and clubbing of the foliage.
This "greening disease" is often mild and recovery may be
spontaneous (Bove and Saglio, 1974).

III.   TRANSMISSION OF YELLOWS DISEASES

A. *INSECT TRANSMISSION*

Transmission by insect vectors is doubtless the most important mode of disease spread. Insects with sucking mouth-
parts, which feed in the phloem or xylem, transmit the causal
agents through their normal feeding habits. They obtain the
pathogens from contaminated sap and deposit them directly into
the appropriate vascular tissues of healthy trees. Although
insect vectors have been identified for a number of the described diseases, for many no vector is known. It can be as-

sumed, however, that insects play a major role in the spread
of these diseases.

Leafhoppers (Cicadellidae), both phloem- and xylem-feeding
types, and psyllids (Psylloidea) are the two principal families
of insects involved in transmitting yellows diseases of woody
plants. Pierce's disease of grape is an exception. While it
is known to be transmitted by some 23 species of xylem-feeding
leafhopper vectors, it is also transmitted by three species in
the Cercopidae. A homopterous insect, *Haplaxius crudus,* is
another probable exception; it is suspect as a vector of coconut palm lethal yellowing disease (Tsai, 1979).

Insect vectors that transmit prokaryotes usually do so
persistently for most of their life span. Those organisms that
multiply in the vector are said to be propagative and many prokaryotes that cause yellows disease are believed to be of this
type. Although transovarial transmission has been demonstrated with vectors of several yellows diseases, it apparently
occurs infrequently (Tsai, 1979).

B.   GRAFT TRANSMISSION

While transmission by insect vectors is undoubtedly the
most important, naturally occurring root grafts between trees
provide another means of spread of yellows diseases. Root
grafting is essential for the survival of sandal, which para-
sitizes the roots of other plants. In this way sandal spike
is naturally transmitted to many other hosts and from those
hosts back to sandal. Likewise, natural root grafting among
closely planted elms along streets in cities throughout the
United States provides an excellent opportunity for tree-to-
tree spread of elm phloem necrosis.

Artificial transmission by grafting has long been used
experimentally as important evidence for the virus causation
of diseases in plants. We now know, of course, that many of
the so-called yellows-type virus diseases, which were graft
transmitted, are caused by prokaryotes. Most, if not all,
prokaryotes that are insect-transmitted are also graft-transmissible. Many other diseases for which a vector is not yet
known have also been transmitted by grafting. Since yellows
disease agents occur and multiply in the vascular system of
their hosts, it can be safely assumed that all yellows diseases are graft-transmissible from plant to plant within the
limits of tissue compatibility. A number of forms of grafting are used in experimental work: budding, chip budding, bark
patch grafting, root grafting, and scion grafting are commonly
used with woody plants. Diseases caused by xylem-limited bacteria (RLO) must have xylem included in the grafting tissue
for successful transmission.

C. DODDER TRANSMISSION

Transmission by dodder, *Cuscuta* spp., has been extensively used as an important experimental tool in the transmission of yellows diseases of both woody and herbaceous plants. Dodder is a parasitic plant that obtains its nutrition from other plant hosts through the formation of haustoria. These root-like structures become intimately attached to the conductive tissues between host and parasite and thus are highly effi-cient in transmitting nutrients and other materials in the phloem. The use of dodder in transmission has certain advan-tages over other methods; it is not as limited by compatibility as interspecific grafting and thus may be used as a bridge between widely divergent hosts. *Cuscuta campestris* and *C. sub-inclusa* have very broad host ranges and are particularly valu-able in transmission experiments with woody plants. The use of dodder for transmission of prokaryotes has its limitations, however. These parasites are phloem feeders and cannot be used to transmit xylem-limited prokaryotes. In addition, they have not been successful in transmitting some mycoplasma di-seases.

D. MECHANICAL TRANSMISSION

Although there are a number of reports in technical jour-nals that claim successful mechanical transmission of myco-plasma diseases, none of these claims has been sufficiently substantiated to be accepted by the scientific community as a whole. A review of these reports and a discussion of the mech-anical methods tried are not provided here, but one may wish to consult McCoy (1979) for further detail. Recently, how-ever, the rickettsia-like bacterium (RLO) causing Pierce's disease of grapes was successfully cultured from diseased plants (Davis *et al.*, 1978; Raju *et al.*, in press). Except for *Spiroplasma citri,* the causal agent of citrus stubborn disease (Saglio *et al.*, 1973; Markham *et al.*, 1974), and the corn stunt organism (Chen and Liao, 1975), none have satisfied Koch's postulates.

E. SOIL TRANSMISSION

The only yellows disease ever reported to be transmitted in the soil is larch witches'-broom, caused by a rickettsia-like bacterium (Nienhaus *et al.*, 1976).

IV. DESCRIPTIONS OF YELLOWS DISEASE

A. *ALMOND LEAF SCORCH*

   Almond leaf scorch is also known as "golden death" and
"almond decline." This disease caused by fastidious bacteria
was first reported on a few scattered trees in Los Angeles
County, California. Subsequently, it has been found in 14
counties ranging from San Diego to Gleen (Wilson and Ogaua,
1979). No stone fruits other than almond are known to be affected by the causal agent, which is a xylem inhabitor.
Pierce's disease and phony peach are thought to be caused by
a similar, if not the same, pathogen (Davis *et al.*, 1980).
   Almond scorch has been found on more than 11 commercially
important almond cultivars, including Nonpareil, Mission, Ne
Plus Ultra, IXL, Peerless, Drake, Trembath, and Jordanolo.
   First symptoms generally appear after mid-June, when a
marginal leaf scorch appears on single-terminal shoots or
branches. Symptoms increase each year and after 3 to 8 years
the entire tree shows the characteristic golden-brown appearance. Affected trees produce normal leaves in the spring, but
there is reduced terminal growth and dieback of spur and terminal shoots. Affected leaves do not wilt but develop a marginal burning. The zone between the green and scorched areas
appears as a bright yellow band on some varieties.
   This disease can be diagnosed by placing affected wood in
acidified methanol. Affected tissues develop a characteristic
reddish-purple color (Mircetich *et al.*, 1976). This is the
same procedure used to diagnose phony peach.
   The almond scorch pathogen is transmitted by the same
leafhopper *(Draculacephala minerva)* that transmits Pierce's
disease (Davis *et al.*, 1980). The pathogen causing Pierce's
disease can also be transmitted to healthy almonds and induce
typical leaf scorch symptoms.

B. *APPLE CHAT FRUIT*

   "Chat" is a term originally used by the English for pota-
toes or apples that were small or worthless. "Chat fruit" became a term to define apples that were prevented from growing
to normal size by what was supposed to be a virus. Recent
studies have indicated that the disease may be caused by a
mycoplasma (Beakbane *et al.*, 1971). Both chat fruit and rubbery wood began to appear when the variety Lord Lambourne was
top-worked onto a number of old trees. The disorder was re-

ported by Wallace *et al*. (1944) and by Luckwill and Crowdy
(1950) indicating that viruses were the cause. Budding or
grafting will transmit the chat fruit agent from scion to
stock, or vice versa (Atkinson, 1971).

At present chat fruit or small fruit disease is of minor
importance. It is present in England, New Zealand, and perhaps the United States. Its spread in the field is very slow
and no vectors are known.

Chat-infected fruit appears normal until the end of the
June drop at which time its growth becomes arrested. Affected red cultivars fail to develop bright color; there are no
obvious color differences on green cultivars. Fruit drop is
delayed on chat-affected trees. Symptoms may vary from year
to year on individual trees; trees may have all undersized
apples or a mixture of full and small-sized fruit.

Beakbane *et al*. (1971) reported mycoplasma-like bodies in
phloem tissue of pedicels collected from the Lord Lambourne
variety which showed chat fruit symptoms. Their micrographs
are not convincing and further study is needed to determine
the exact etiology of this disease.

C. APPLE PROLIFERATION

Apple proliferation, a European disease, is also known as
apple witches-broom or *"Hexenbesenkrankeit."* It is found in
Austria, Czechoslovakia, Denmark, Finland, France, Germany,
Greece, Holland, Hungary, Italy, Latvia, Poland, Romania,
Spain, and Switzerland. Although it was originally thought to
have a virus etiology, mycoplasmas and fastidious bacteria have
recently been suspected as the causal agents (Sarie, 1969;
Giannotti *et al*., 1968).

Apple proliferation is an important disease of apples, reducing tree vigor, fruit size, and quality. Ninety-five percent reduction in yield has been recorded. In some cases,
total losses have resulted because fruit was unmarketable.

Apple appears to be the only host of this disease. It has
been recorded on the varieties Golden Pippin, Beauty of Boskoop, and other Dutch varieties, and on Abbondanza and Golden
Delicious in Italy. Belle de Boskoop, Gravenstein, Starking,
Golden Delicious, and Winter Banana appear to be the most susceptible.

The most salient symptom is premature development of axillary buds, which grow rather than remain dormant until the
next year. These abnormal shoots develop near the apex of the
branch giving a "witches'-broom" effect. Leaves on affected
trees are smaller than normal and are finely and irregularly
serrated. Stipules become enlarged. Phyllody has been noticed

in some cases. The most damaging symptom is a reduction in fruit size. Fruit shape is often flattened. Fruit is usually poor in flavor and lacks sugar and acidity.

Apple proliferation has been transmitted experimentally only by grafting between apple varieties and from apple to pear (Pena-Iglesias, 1975). The disease spreads within orchards but a vector has not been found. It has been suggested that spread can occur on pruning tools (Posnette, 1963).

A variety of inclusions have been found in tissues of trees affected with apple proliferation. Besides mycoplasma-like bodies, these include fastidious bacteria (Pena-Iglesias, 1975) and tubular structures (Kralik and Brcok, 1974). Although symptoms suggest that apple proliferation is caused by a mycoplasma-like organism, cytological evidence presented so far is not very convincing.

D. APPLE RUBBERY WOOD

Apple rubbery wood is a unique disease of apples that results in poor lignification of the wood. Affected branches have normal leaves and fruit but lack rigidity. In some cases, young trees are so pliable that they cannot stand erect. Severely affected shoots may be compressed. Characteristically, shoots with apparently normal rigidity commonly arise from the base of symptomatic trees. Normally spreading branches of bearing trees droop when they are affected. Internodes are shortened, growth is reduced, and, in severe cases, the tree is stunted. Fruit usually attain normal size but yield is reduced because the trees are smaller.

Susceptible varieties in England and Europe are Dartmouth Crab, Golden Delicious, James Grieve, Kingston Black, Millers Seedling, and Sturmen. Rubbery wood has been found in the varieties Hasting and Nelson in New Zealand (Atkinson, 1971). The disease moves slowly in the orchard. No vectors are known and little is known about its natural spread. At present, this disease is of little economic importance.

As with apple chat fruit, Beakbane *et al.* (1971) have reported mycoplasmas associated with rubbery wood. Their micrographs are not convincing, however. Since the xylem tissue is primarily affected in this disease, one would suspect that a xylem-inhabiting pathogen may be involved— perhaps a fastidious bacterium.

E. ARECA YELLOW LEAF

A mycoplasma-like organism (MLO) is associated with mature sieve elements in the phloem of arecanut palms *(Areca catechu*

L.) suffering from yellow leaf disease (Nayar and Seliskar,
1978). MLO were most abundant in the phloem of young rachilla,
less abundant in partly expanded leaves, and only empty vesicles were found in old diseased leaves. Although small dense
bodies were encountered, round to ovoid forms (250-400 nm)
were most frequently observed.

Studies on yellow leaf of catechu (YLD) were initiated
about 1959 after the disease had reached serious proportions
in southern India (Menon, 1960, 1963; Nayar, 1971). It remains the most serious disease of betel nut palm in Kerala and
Karnataka states (Nayar and Seliskar, 1978). The fruit of
betel nut palm is an important food crop in India where thousands of acres of these trees are cultivated (Menon, 1963).

Signs of the disease are first manifested in the growing
spindle as translucent spots, followed by necrotic streaks in
the unfolding leaves. Leaves become yellow, starting at the
tips, and then developing throughout the entire leaf. The
disease becomes systemic and gradually encompasses the whole
crown, causing premature shedding of blackened, inedible nuts
(Fig. 1). Eventually, the folded bud leaves in the spindle
become infected and necrotic, followed by collapse and death
of the tree. Death usually occurs in 12 to 18 months after
the initial onset of yellowing symptoms in the crown.

Along with the progression of external symptoms, Nayar
(1968), and Nair and Aravindakshan (1970, 1971) investigated
the abnormal cytology of diseased tissues. In the roots necrotic cells and proliferating tissues are commonly encountered
along with cells in which two to three nuclei are found. The
late stages of infection show extensive disorganization of
tissues, plugging of vascular cells with tyloses, and a separation of sclerenchymatous fibers. Infected leaf cells undergo extensive degeneration, wherein the chloroplasts tend to
clump together as the chlorophyll is destroyed. Brown pig-
ments develop in the dying cells of vascular tissues and necrosis and proliferation of cells occur in the phloem, resulting
in an irregular arrangement of elements in the vascular bundles.

Early investigations reported by Menon (1963) indicated
that the disease could be mechanically transmitted by sap and
by grafting from infected to healthy arecanut and other hosts,
as well as through the soil. In light of current information
on MLO diseases, however, further investigation is needed to
corroborate these findings before conclusions are warranted.
Nayar (1971) also reported the successful isolation and culture of mycoplasma from diseased arecanut trees, as well as
the successful inoculation of healthy arecanut seedlings.
These investigations also need to be repeated and confirmed.

The symptoms of yellow leaf disease in *Areca catechu* L.
closely resemble those observed in palms affected by coconut

Fig. 1. Betel nut palm infected with yellow leaf in advanced stages of decline.

lethal yellowing, and efforts are currently underway to determine the relationship of these two diseases.

F. ASH WITCHES'-BROOM

Hibben and Wolanski (1970, 1971) and Hibben *et al*. (1973) reported mycoplasma-like bodies (MLO) in the mature sieve tubes of the petioles and midribs of ash, *Fraxinus americana* L., infected with witches'-broom. MLO were predominantly spherical, and ranged in diameter from 100 to 500 nm. Ovoid and elongate-to-filamentous forms up to 1000 nm were also noted. The cells in the lower size ranges tended to be densely

granular while the larger bodies contained central fibullar
material and a granular peripheral zone. All cells were bounded by a unit membrane approximately 10 nm thick. Several cells,
apparently in the process of budding, were observed with a
double membrane.

A witches'-broom on Arizona ash, which is identified as
Fraxinus berlandierana A. DC., was first reported in Louisiana
by Plakidas (1949). Later, a similar brooming disease was also reported in Louisiana on green ash, *F. pennsylvanica* Marsh.
(KenKnight, 1963b). Subsequently, witches'-broom was reported
in New York State on white ash, *F. americana* L. by Hibben and
Wolanski (1970, 1971), and in Massachusetts on *F. americana*
L. and *F. pennsylvanica* var. *lanceolata* (Borkk.) Sarg. by
Schall and Agrios (1973). The disease is doubtless much more
widespread than is indicated by these reports, but relatively
little is known about its economic importance. However, the
disease is known to cause decline and death of symptomatic
trees (Plakidas, 1949; Schall and Agrios, 1973). Hibben (1978)
suggests that it may have been a contributing factor in the
serious ash dieback syndrome in the northeastern United States
and Canada.

While the disease occurs naturally on the three species of
ash mentioned, it has also been experimentally transmitted
from naturally infected ash to healthy ash by grafting, including inarch grafting (Plakidas, 1949), and by budding or patch
grafting (Schall and Agrios, 1973). Witches'-broom has also
been transmitted to *Cantharanthus roseus (Vinca rosea)* by
means of dodder, *Cuscuta subinclusa* Dur. & Ailg. Both *C. sub-
inclusa* and *C. campestris* were effective in transmitting the
disease from bunch-diseased periwinkle to healthy carrot, *Dau-
cus carota* L. The symptoms expressed on carrot and periwinkle
were typical of those commonly expressed by yellows-diseased
plants. The disease can also be transmitted from diseased to
healthy periwinkle by cleft grafting, but the disease was not
transmitted from yellowed periwinkle back to healthy white ash
(Hibben and Wolanski, 1971). No natural vector is known.

The brooms formed on infected ash are typical of yellows-
type brooming diseases in trees. Clusters of upright branches
associated with shortened internodes, combined with the precocious development of dormant axillary buds, form witches'-
brooms. These usually appear on small branches attached to
the trunks of trees in advanced stages of decline (Fig. 2).
The leaves on broomed twigs are small, chlorotic, and deformed.
Occasionally, simple leaves are found in place of the normally
pinnately compound leaves (Hibben and Wolanski, 1971).

Fig. 2. White ash, Fraxinus americana L. with witches'-broom on the trunk. Photograph by courtesy of C. R. Hibben, Brooklyn Botanic Garden, New York.

G. BLACK LOCUST WITCHES'-BROOM

Seliskar *et al.* (1973) reported the consistent association of a mycoplasma-like organism with witches'-broom on black locust *Robinia pseudoacacia* L. MLO were observed primarily in the sieve elements of the phloem in leaves, stems, and roots of infected trees. In some cells MLO were abundant while in others they were completely lacking. They appeared to be most numerous in tissues associated with highly symptomatic brooming and were difficult to find in nonbroomed or slightly broomed branches of diseased trees. The MLO are generally spherical or oval. Filamentous forms are sometimes found, usually near the sieve plates; they appear to be associated with cell-to-cell movement through sieve plates (Fig. 3). However, an alternate hypothesis has been advanced to explain the occurrence of narrow filamentous forms of the MLO in the slits of sieve plates (McCoy, 1979). Division by binary fission was apparent in some thin sections. MLO range from 90 to 1,000 nm in the largest dimensions while the commonly occurring oval forms average 200-350 nm in length. The cells are bounded by a plastic unit membrane and contain numerous ribosome-like structures near the periphery and fibrillar material, suggestive of DNA, near the center (Fig. 4).

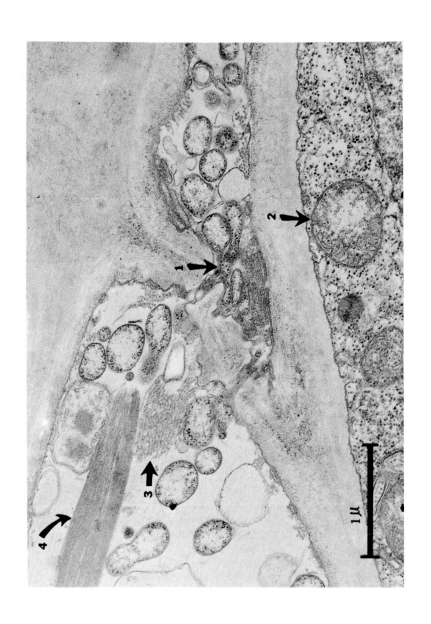

One of the earliest recorded yellows diseases of forest trees, black locust witches'-broom, was described by Waters (1898) on locust stump sprouts in Maryland. This systemic brooming disease is now widespread in the eastern United States and central Europe (Grant and Hartley, 1938; Atanasoff, 1935; Blattny, 1959; Ciferri and Corte, 1960; Ross, 1933). Susceptible hosts include black locust, *Robinia pseudoacacia* L. and possibly also honey locust, *Gleditsia triacanthos* L. (Grant and Hartley, 1938).

The predominant symptom is erect brooms on root and stump sprouts and, less frequently, in the crowns of affected trees (Fig. 5). The leaves in witches'-brooms are much reduced in size, sometimes to the extent that the leaf-bearing twigs have a characteristic shoestring appearance. Not infrequently root sprouts may be completely broomed while the parent tree shows no symptoms of the disease. Large symptomatic trees are seldom found, because on some the brooms die over winter and the trees recover and on others the crowns are completely killed. A tree producing broomed shoots one year may produce normal-appearing shoots the following year. Sometimes pruning healthy limbs on infected trees may stimulate development of brooms while, on the other hand, nonsymptomatic twigs may develop from broomed branches that have been pruned (Grant et al., 1942). Brooms result from the abnormal proliferation of normally dormant axillary buds in the leaf axils of affected shoots. The degree of brooming ranges from mild to severe, depending upon the reduction in size of leaves. Vein clearing is a common symptom on leaves in broomed twigs, particularly in the early stages. If brooming is severe the leaves may become narrowed at the base. Because brooms usually do not form until late summer, they continue growth into late fall or early winter and commonly die back during the winter. Although not usually seen, brooming also occurs on the roots of infected trees, which apparently develop from the suppression of terminal growth and the subsequent development of numerous second-order laterals (Grant, 1939; Grant et al., 1942).

Witches'-broom can be transmitted by scion grafting and, to a lesser extent, by budding (Jackson and Hartley, 1933; Grant et al., 1942). No vector is known.

Fig. 3. Mycoplasma-like bodies in the sieve elements of black locust (1) appear to be moving through sieve plates; mitochondrion (2); P-protein (3); and crystalline protein (4).

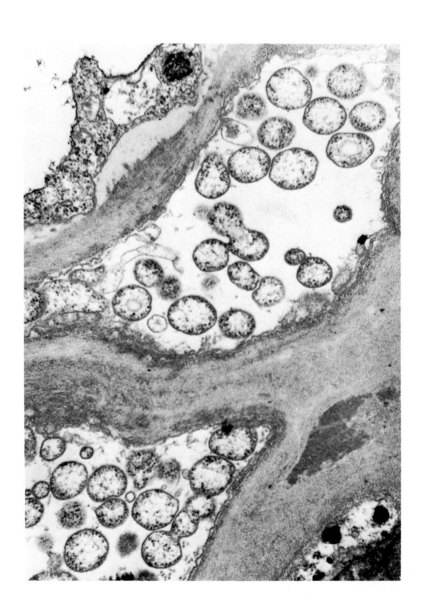

The brooming disease on honey locust, *Gleditsia triacanthos* L., may be the same as witches'-broom on black locust, *Robinia pseudoacacia* L., although this has not been proved. The symptoms and behavior of the disease in root and stump sprouts are identical in all respects.

H. *CHESTNUT YELLOWS*

Okuda *et al.* (1974) reported the consistent association of a mycoplasma-like organism (MLO) with chestnut trees infected with yellows disease. They are phloem-limited and have been observed only in sieve elements of infected tissues. MLO range in size from 70 to 700 nm and contain ribosomes and DNA-like material characteristic of other plant mycoplasmas. The cells are delimited by a plastic unit membrane.

Chestnut yellows was first reported by Shimada and Kouda in 1964 as a transmissible virus disease (Okuda *et al.*, 1974). In 1963 the disease was believed to occur in both the Ibargi and Yamagata Prefectures of Japan (Imazeki and Ito, 1963), but in a more recent report Okuda (1978) states that chestnut yellows occurs only in Ibargi Prefecture. The disease occurs on Japanese chestnut, *Castanea crenata* Sieb. and Zucc., and on Chinese chestnut *C. mollissima* Blume (Imazeki and Ito, 1963) (Fig. 6). Transmission is by grafting. No insect vector is known (Okuda, 1978).

I. *DOGWOOD WITCHES'-BROOM*

A mycoplasma-like organism (MLO) is found in association with broomed silky dogwood, *Cornus amomum* Mill. Pleomorphic MLO, varying in size from 160 to 420 nm, were abundant in leaf phloem tissues of diseased shrubs (Raju and Chen, 1974; Raju *et al.*, 1976).

Witches'-broom symptoms were first reported on silky dogwood, *C. amomum* Mill. and on red osier dogwood, *C. stolonifera* Michx. by Raju *et al.* (1976). The disease is widely distributed in central New Jersey on wild silky dogwood and was also observed on the cultivated red osier dogwood in the shrub collection at Rutgers University.

Fig. 4. Broomed black locust showing oval mycoplasma-like organisms (MLO) in adjacent sieve tubes.

Fig. 6. Chestnut yellows on Japanese chestnut, Castanea crenata Sieb. and Zucc. Photograph courtesy of S. Okuda, Utsunomiya University, Japan.

Fig. 5. Black locust stump sprout with severely broomed branch (right) and nonsymptomatic shoot (left).

The predominant symptom on infected dogwood is the appearance in the spring of characteristic witches'-brooms. These apparently result from the abnormal stimulation of dormant axillary buds and their subsequent proliferation into clusters of shoots bearing chlorotic and stunted leaves. Phyllody also occurs on flowering portions of broomed branches (Raju et al., 1976). Histological examination of young leaves taken from brooms showed no apparent differences between palisade and spongy parenchyma (Raju and Chen, 1974).

Temporary remission of symptoms on symptomatic cuttings taken from broomed shrubs was achieved with applications of tetracycline compounds. Aqueous sprays of tetracycline, oxy-tetracycline, and chloromphenicol applied at the rate of 1000 ppm on alternate days for 3 weeks suppressed symptoms in diseased plants. Symptoms reappeared, however, 2-3 weeks after treatments were terminated (Raju et al., 1976).

J. ELM LEAF SCORCH (ELS)

Elm leaf scorch (ELS) of American elm, *Ulmus americana* L. was formerly considered to have a virus etiology (Wester and Jylkka, 1959), but recent evidence reported by Sherald and Hearon (1978), and Hearon et al. (in press) suggests that the disease is caused by a xylem-limited fastidious bacterium (rickettsia-like bacterium) (Fig. 7) similar to that reported as the cause of Pierce's disease of grape (PD) and almond leaf scorch (ALS) (Goheen et al., 1973; Hopkins and Mollenhauer, 1973; Mircetich et al., 1976; Lowe et al., 1976). The elm leaf scorch organism is serologically related to the PD-ALS bacterium. As yet, the elm bacterium has not been grown on artificial media, but it is transmitted by scion grafting and chip budding (not by bark patches). ELS bacteria are found in the main veins of scorched elm leaves and sometimes in veins of nonsymptomatic leaves of infected trees. The bacteria are rod-shaped, ca. 0.3-0.4 nm x 0.9-2.4 nm, and rounded at the ends. Constrictions in the walls of the longer forms suggest division by binary fission. The cells contain ribosomes, DNA-like strands, and occassionally round osmiophilic bodies and mesosome-like bodies. Cell walls are often rippled, about 35 nm thick, sometimes consisting of five distinct layers. Small, dense bodies with rippled walls also occur in diseased tissues and may represent degenerating forms of the bacterium.

Leaf scorch is found in the southeastern United States where it was first described on American elm, *U. americana* (Wester and Jylkka, 1959). New evidence, however, suggests the same disease also occurs on sycamore *(Platanus spp.)* and two species of red oak *(Quercus spp.)* in the Washington, D.C.

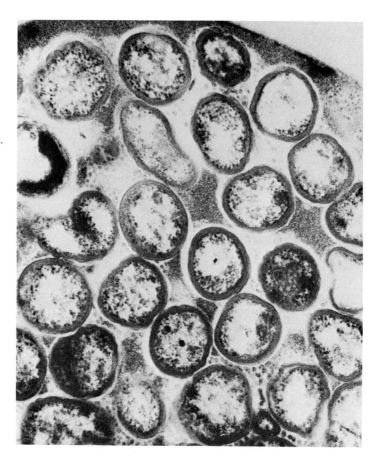

Fig. 7. Cross-sectional view of rickettsia-like bacteria (RLO) in xylem cell of a leaf from American elm, Ulmus americana L., infected with leaf scorch. A few cells show rippled walls and associated tubular structures. Photograph courtesy of J. L. Sherald, USDA.

area. Bacteria, similar to those in diseased elm, were consistently associated with the xylem elements of both sycamore and oak. Fimbriae-like structures, projecting from the bacterial cell walls were much more numerous on the sycamore and oak bacteria than on the elm bacteria, however. Sometimes these bacteria were pointed at one end.

The seasonal development of symptoms of leaf scocrch is essentially similar on elm, sycamore, and oak (Hearon *et al.*, in press). In mid- to late summer necrotic scorching begins

Fig. 8. Shoot of scorch-diseased American elm, Ulmus
americana L. Symptoms first appear at the bases of infected
shoots and then progress toward the tips. Outer margins of
leaves are necrotic. Photograph courtesy of J. L. Sherald,
USDA.

along the leaf margins and in the interveinal tissues and progresses throughout the leaf. In elm and sycamore, symptoms
appear first on leaves at the bases of branches and then progress toward the tips. As a result, branches are commonly
tufted with green leaves at the tips by late summer (Fig. 8).
Scorched leaves may roll inward and absciss prematurely. Symptoms may appear on a single limb or on many limbs throughout
the crown, causing gradual deterioration and decline.

It is noteworthy that Davis et al. (1978) demonstrated a
serological relationship between PD-ALS, phony peach, and
alfalfa dwarf, while Hearon et al. (in press) have shown a
serological relationship between PD-ALS and bacteria isolated
from scorched elm and oak. More recently, Raju et al. (in
press) demonstrated a close serological relationship between
PD bacteria and bacteria associated with plum leaf scald (PLS),
phony peach (PP), and young tree decline of citrus (YTD).

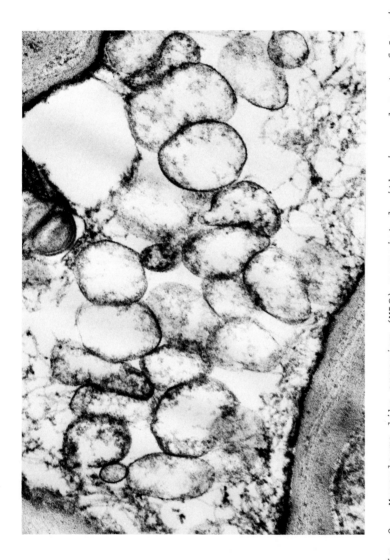

Fig. 9. Mycoplasma-like organism (MLO) associated with sieve element of American elm, Ulmus americana L. infected with phloem necrosis.

K. ELM PHLOEM NECROSIS

The agent causing elm phloem necrosis (PN) appears to be a phloem-inhabiting mycoplasma (MLO) ranging in size from 200 to 1000 nm and enclosed in a tripartite membrane approximately 8 nm thick. Mycoplasmas as seen in the sieve elements with the transmission electron microscope are pleomorphic with spherical and oval forms predominating (Fig. 9). Filamentous forms also occur as well as spherical bodies with extensions or blebs. A form of division suggesting binary fission is sometimes seen. MLO contain ribosomes and DNA-like strands characteristic of animal mycoplasmas. Occassionally, clear areas in the center of some MLO contain reticula suggestive of nucloids or vesicles (Wilson et al., 1972). Braun (1975) reported seeing evenly distributed particles on both fracture faces of the MLO membrane when it was examined in freeze etchings of PN-infected periwinkle phloem. He also reported intracellular vesicles in some mycoplasmas, particularly those believed to be senescent.

Known only in the United States, elm phloem necrosis was first reported by Swingle (1938), following an investigation of an epiphytotic that killed many American elms in the Ohio Valley (Swingle, 1938). Most yellows-type diseases studied before 1967 were attributed to viruses, and he concluded that PN was caused by a virus. In early reports (Garmen, 1899; Forbes, 1912) there is evidence that the disease was present in the Ohio River Valley long before this time, perhaps as early as 1882. From the mid-1930's to the mid-1940's numerous epiphytotics were reported throughout the Midwest (Leach and Valleau, 1939; Bretz, 1944a,b; Caldwell, 1945; Tehon, 1945; Carter, 1945; Larsh, 1945; Bretz and Swingle, 1946)(Fig. 10). Although the disease situation became somewhat clouded in the decade that followed because of the rapid spread of Dutch elm disease in the central states, PN seemed to become enphytotic throughout most of its range. However, a new outbreak in Alabama was reported in 1959 (Curl et al., 1959) and another in Mississippi in 1966 (Filer, 1966). These were followed by reports of epiphytotics in New York state (Sinclair, 1971) and Pennsylvania (Sinclair et al., 1971; Merrill and Nichols, 1972). It was observed in New Jersey in 1973 (Weber et al., 1974), in Michigan in 1975 (Hart, 1978), and in Massachusetts in 1976 (Holmes and Chater, 1977). Since herbarium records indicate that the range of Scaphoideus luteolus Van Duzee, the leaf-hopper vector of PN, extends beyond that of the disease, it can be expected that its range will continue to expand.

Elm phloem necrosis is second only to Dutch elm disease (DED) in importance (Fig. 10). In addition to the epiphytotics previously noted, Carter and Carter (1974) best typified

Fig. 10. Leafhopper vector, Scaphoideus luteolus Van Duzee shown on elm leaf in both nymph and adult stages.

the relative importance of these two diseases in an extensive study of disease spread in the twin cities of Champaign and Urbana, Illinois, over a 29-year period. Of the original population of 14,103 elms, 21.23% were killed by elm phloem necrosis in 29 years. Dutch elm disease killed 78.44% in 22 years. It is interesting to note that where both diseases occur together, the presence of phloem necrosis results in an increase in the incidence of the Dutch elm disease. Apparently, the bark beetle vectors of DED colonize and breed in dead or dying elms, including those attacked by phloem necrosis (Campana and Carter, 1955). On the other hand, the leafhopper vector of phloem necrosis does not colonize dead elms and thus no increase in its incidence can result from DED (Campana, 1958). It can be expected, instead, that the incidence of phloem necrosis will actually be reduced by the removal of elms from the population by DED.

Elm phloem necrosis occurs naturally on most American elm species, including American elm, *Ulmus americana,* L.; winged elm, *U. alata* Michx.; cedar elm, *U. crassifolia* Nutt; September elm, *U. serotina* Sarg; and *U. rubra* Muhl., red or slippery elm (Swingle, 1942; Sinclair, 1972; Sinclair and Filer, 1974; Sinclair *et al.,* 1976). Naturally infected hybrids of *U. rubra* and *U. pumila* were also observed (Braun and Sinclair, 1979). *Ulmus thomasii* Sarg., cork or rock elm, was susceptible to infection by graft transmission in preliminary inoculation

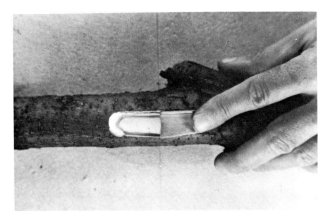

Fig. 12. Phloem necrosis-infected American elm, Ulmus americana L., showing discoloration in the inner bark of small tree. Such bark commonly gives off an odor of wintergreen if confined in the hand for a few moments.

Fig. 11. Urban epiphytotic of elm phloem necrosis on American elms in midwestern United States. Natural root graft transmission of PN may occur with closely spaced trees.

tests and natural infection probably also occurs (R. U. Swingle, personal communication, 1963).

Although field susceptibility of certain exotic elms has
not been observed, a recent report of graft inoculation studies indicates that some species are susceptible to PN (Braun
and Sinclair, 1979). These include the European species *Ulmus carpinifolia* Gleditsch (and *U.* x *hollandica*), *U. laevis*
Pall., and *U. parvifolia* Jacq., an Asiatic species. Braun
and Sinclair (1979) also reported the successful transmission
of PN from *Ulmus americana* and *U. parvifolia* seedlings to
periwinkle via dodder, *Cuscula epithymum* Murray. This was the
only dodder species of a number tested that successfully transmitted PN from elm to periwinkle, but another species, *C.
ceanothi* Behr., was also able to transmit PN from periwinkle
to periwinkle. PN could not be taken back from periwinkle to
elm, however (Braun and Sinclair, 1979). Repeated attempts
to transmit PN to "urban elm," a Netherlands selection *U.
hollandica* Mill "Vegeta" x *U. carpinifolia* Gleditsch, and
U. pumila L., a promising new DED-resistant elm, by bark
grafting have failed, as well as attempts to transmit PN to
Siberian elm, *U. pumila* L. (C. E. Seliskar, unpublished data).

Under natural conditions elm-to-elm transmission of PN
mycoplasma occurs by means of a leafhopper, *Scaphoideus luteolus* Van Duzee, which feeds on phloem in leaf tissue (Baker,
1948, 1949, 1959) (Fig. 11). Since localized epiphytotics of
the disease have recently occurred in areas where this leafhopper is either nonexistent or occurs in very low numbers,
other vectors are suspected (Gibson, 1973). Field observations of the spread of the disease in urban situations strongly suggest that it can also be transmitted short distances
through naturally occurring root grafts. Bark and scion grafting are commonly used to transmit PN experimentally.

The symptoms of PN generally follow the same pattern in
most American elm species. External symptoms are first apparent in the crown, where the foliage becomes sparse, light
green, then yellow, and usually involve the entire crown.
Leaves are often rolled and drooped. Root decline precedes
or develops concurrently, resulting in the death of the small
fibrous roots, followed by the progressive deterioration of
larger roots. A characteristic discoloration, not found in
other elm diseases, commonly appears in the inner phloem of
large roots, stems, and some branches of affected trees. Even
twigs in the extremities of the crown of small trees may show
pronounced discoloration. Phloem tissues are first yellow,
then butterscotch, and finally brown or black (Fig. 12). Infected tissues have a faint odor of wintergreen which is especially noticeable if such tissues are confined in the hand
for a few moments. Phloem discoloration often precedes the
appearance of crown symptoms (Swingle, 1938). In American

elms, trees usually die within 12 to 18 months following the
onset of symptoms. However, this may vary from 3 weeks in
small trees to as long as 3 years in very large elms. The
latent period of infection in graft-inoculated nursery trees
may range from 6 to 24 months (Swingle, 1942).

The histological changes associated with PN disease in
American elm were first investigated by McLean (1943, 1944).
He observed that hypertrophied parenchyma cells and nuclei
appear near mature sieve tubes in the primary phloem, followed
by hyperplasia and collapse of the sieve elements and companion cells. McLean's observations were corroborated by Braun
and Sinclair (1976) who further elaborated on the anatomical
changes taking place in diseased tissues by using both light
and electron microscopy. They described the likely sequences
taking place in phloem deterioration, the abundant deposition
of callose in sieve elements, and the inconsistent associat-
tion of MLO with progressive pathological changes. The histological syndrome in PN is not unlike that reported for other
yellows-type diseases (Schneider, 1973).

While the symptoms in most American elm species are similar to those just described, Sinclair (1972; Sinclair et al.,
1974; Sinclair and Filer, 1974; and Sinclair et al., 1976)
reported that PN symptoms in red (slippery) elm, *Ulmus rubra*
Muhl., are characterized by the production of witches'-brooms,
a maple syrup odor in necrotic phloem and in leaves of brooms,
and a slower rate of decline in diseased trees. Chlorosis
and brooming were also reported in naturally infected *U. pumila* x *U. rubra* hybrid seedlings (Braun and Sinclair, 1979). In
graft-inoculated nursery-grown exotic elm seedlings, stunting
and brooming occurred on *U. carpinifolia* Gleditsch (including
U. x *hollandica*) and *U. parvifolia* Jacq. Chlorosis, epinasty
and stunting were observed on *U. laevis* Pall coated cuttings.
Symptoms of chlorosis, brooming, stunting, and bloom suppression were noted in periwinkle (*Cantharanthus roseus* Don) when
parasitized with dodder *(Cuscuta epithymum)* bridged to infected *U. americana* or *U. parvifolia*. They found no apparent
symptoms in graft-inoculated *U. glabra* Huds. or *U. pumila* L.

At present no satisfactory control is known for elm PN in
American elm species. When DDT was released for civilian use
in 1944 a great deal of research was initiated on its use for
the control of elm PN. Effective spray formulations were
developed to control feeding by *Scaphoideus luteolus* Van Duzee, the leafhopper vector of PN, and were widely recommended
and used throughout the range of the disease. These were
usually applied in combination with sprays for control of the
European elm bark beetle, *Scolytus multistriatus,* the primary
vector of the Dutch elm disease (Whitten and Swingle, 1948;
Whitten, 1949; Swingle et al., 1949; Carter, 1950; Anon.,

1952; Baker and May, 1951; Whitten, 1954; Whitten and Baker,
1973). Methoxychlor was later introduced as a substitute for
DDT in the control of both PN and DED (Schreiber and Peacock,
1974).

Following the discovery of MLO in association with PN
(Wilson *et al.*, 1972), and the sensitivity of MLO to tetracycline antibiotics (Doi *et al.*, 1967), Sinclair *et al.* (19-
74, 1976) obtained temporary remission of PN symptoms in a
small percentage of diseased elms with tetracycline under
field conditions. Results with antibiotics for the control of
PN in Mississippi were more favorable (Filer, 1973; Sinclair
et al., 1976), but more field trials are needed before antibiotics can be recommended for PN control.

While a number of American elm clones were found to be
resistant to PN (Whitten and Swingle, 1948) none have proved
resistant to the Dutch elm disease. No resistance to PN has
been found in American elm selections showing resistance to
DED. However, Siberian elm, *Ulmus pumila* L. and "urban" elm,
a hybrid of *U. pumila* and a Netherlands selection (*U. hollandica* Mill x *U. carpinifolia* Gleditsch), appear to be highly resistant to both PN and DED (C. E. Seliskar, unpublished data).

L.   JUJUBE WITCHES'-BROOM IN INDIA

Jujube witches'-broom is believed to be caused by a MLO.
The case for a MLO relationship in the etiology of jujube
witches'-broom is based on symptomatology and sensitivity to
tetracycline antibiotics, which are characteristic of most
yellows type diseases. Diseased tissues have not been examined with the electron microscope. The relationship between
this disease and the Korean jujube witches'-broom described
by Kim (1965) is unknown.

It was first described on jujube (*Ziziphus mauritiana* Lam.)
in India by Pandey *et al.* (1976). The information that follows is based solely on this citation, the only report of the
disease thus far. The disease occurs in parts of India where
jujube, a forest tree, is commercially important for its fruit.
In 1973, a disease survey revealed that one-third of the jujube plants grown in the locality of Poona (Maharashtra State)
were infected.

Ziziphus mauritiana Lam. is the only known host. The
causal MLO can be transmitted by budding and cleft- and ap-
proach-grafting. Attempts to graft-transmit the disease to
other species of *Ziziphus,* including *Z. numularia, Z. venoplia,
Z. rugosa, Z. sativa,* and *Z. xylocarpa* failed. These species
proved not to be symptomless carriers when attempts to graft-
transmit from these hosts back to healthy *Z. mauritiana* also

failed. It is interesting to note that while jujube witches'-broom could not be transmitted to *Z. oenoplia,* this species is known to be a host for sandal spike mycoplasma, in which leaf roll symptoms, but not witches'-brooms, are produced (Hull and Plaskitt, 1970). Attempts to transmit it mechanically with sap from diseased jujube leaves in 0.05 *M* phosphate buffer at pH 7 to *Z. mauritiana,* other *Ziziphus* species, and a number of herbaceous hosts were unsuccessful.

Under natural conditions the disease is characterized by dieback, brooming, and phyllody (if infection occurs a year before normal flowering). Few fruits are produced and infected trees do not flower following the first year of infection. Eventually, the developing stems and branches become so spindly and weak that they require support to grow upright. Witches'-brooms develop from the excessive production of numerous proliferating axillary twigs on which chlorotic and highly stunted leaves are produced. Abnormally small leaves are produced within 70 to 90 days after graft inoculation, and symptoms typical of those observed in the field appear after the first year.

Diseased seedlings of *Z. mauritiana* recover after the application of tetracycline. Sprays applied twice a week for 10 weeks at concentrations of 500 and 1000 ppm caused a remission of symptoms. Symptoms reappeared in 35 to 40 days after treatment was discontinued, but were again suppressed when the antibiotic spray was reapplied. Penicillin and other antibiotics were ineffective in suppressing symptoms.

M. *JUJUBE WITCHES'-BROOM IN KOREA*

Witches'-broom of jujube (*Ziziphus jujuba* Mill.) in Korea is caused by a mycoplasma-like organism. Observed with the electron microscope, cells are pleomorphic, delimited by a single unit membrane, and contain ribosome-like granules and DNA-like strands. Rounded forms predominate but they range in shape from small round bodies to elongated oval forms. They range from 125 to 970 nm in length. Micrographs suggest that they may divide by binary fission similar to bacteria, or by budding. They occur abundantly in the sieve elements of inner phloem of affected trees (La and Chang, 1979).

Like many yellows diseases, witches'-broom of jujube was long thought to be caused by a virus. It was first observed about 1950, (Hong and Kim, 1960), and is now widely distributed throughout Korea. Early investigators (Hong and Kim, 1960; Hong, 1960; Hong and Hah, 1961) referred to the disorder as the shoot cluster disease of the Chinese date tree, a var-

iety of jujube. Since 1965 it has been referred to as the
witches'-broom disease of jujube (Kim, 1965).

The disease has spread progressively throughout Korea,
causing extensive mortality in both isolated areas and large
tracts. Up to 80% of the jujube trees in Korea have been killed or are infected with witches'-broom. La and Chang (1979)
state that this disease is the major limiting factor in the
cultivation of jujube in Korea.

Two varieties of jujube, *Ziziphus jujuba* var. *inermis* Reh-
der and *Z. jujuba* var. *hooneusis* Lee are the primary hosts.
Ziziphus mauritania, the only known host for jujube witches'-
broom in India, is not cultivated in Korea.

Witches'-brooms develop on diseased trees as a result of
the precocious development of proliferating secondary shoots
which have an overabundance of abnormally small, chlorotic
leaves (Fig. 13). Phyllody is characteristic of diseased trees
and symptomatic limbs do not bear fruit. Symptoms are at first
limited to one or more branches but then the disease spreads
progressively throughout the entire crown. Trees of all ages
are susceptible and die within a few years after symptoms first
appear (La and Chang, 1979).

The means of spread of jujube witches'-broom under natural
conditions is unknown. Experimentally, the disease agent is
readily transmitted by scion grafting and budding. The percentage of successful transmission is markedly reduced when
diseased scions are cut and grafted in the spring as opposed
to using similar scion wood collected the previous fall and
kept in cold storage until grafting the following spring. Apparently the mycoplasma in above-ground portions of diseased
trees are inactivated or killed by low winter temperatures
(Kim, 1965). This hypothesis has also been advanced for other
mycoplasma diseases, such as pear decline (Schneider, 1970)
and elm phloem necrosis (Braun and Sinclair, 1976).

While this disease is similar to jujube witches'-broom in
India they cannot be presumed to be the same. Of the various
species of *Ziziphus* graft-inoculated with scions from diseased
Z. mauritiana in frost range studies conducted in India, only
Z. mauritiana proved to be susceptible (Pandey *et al.*, 1976).
The Korean witches'-broom has been reported only on *Z. jujube*
Mill., and this species was not included in these transmission
tests.

Promising results were obtained in the treatment of symptomatic trees with tetracycline antibiotics gravity-injected
at the rate of 500 ml of 1000 ppm solution per tree. One injection in April provided complete remission of symptoms for
one growing season, while an additional treatment in October
prevented reoccurrence of symptoms the following growing season. From these results, La *et al.* (1976) conclude that oxy-

Fig. 13. Jujube witches'-broom (Korea) is characterized by tight clusters of twigs forming brooms (right). A healthy branch is shown (left).

tetracycline HCl injections are practical for the control of jujube witches'-broom in Korea.

N. LARCH WITCHES'-BROOM

A rickettsia-like organism (RLO) was consistently asso-ciated with degenerating cells of roots and leaves of broomed larch trees, *Larix decidua* Miller (Nienhaus *et al.*, 1976). RLO were common in parenchyma cells adjacent to vascular bund-les and in phloem tissues, but rarely in xylem vessels or in-tercellular spaces. Root specimens collected in winter con-tained cells which were frequently packed with RLO. They ranged from 0.6 to 2.0 μm in length and 0.3 to 0.5 μm in dia-meter. They contained ribosome-like structures in the elec-tron-dense portions of the cytoplasm and DNA-like strands in other parts, as well as dark and light globular inclusions. RLO had wavy cell walls, characteristic of this group of organ-isms.

Since a MLO was sometimes detected along with RLO in di-seased larch, Nienhaus *et al.* (1976) speculate that MLO may also play a role in the disease syndrome. The disease has been reported only from forests near Bonn, Germany (Nienhaus *et al.*, 1976).

Diseased trees have witches'-brooms in the crowns and are severely stunted. The diameter of diseased trees is only a-bout half that of healthy trees of comparable age. Needles are also reduced in size. Heavily broomed trees sometimes die prematurely (Nienhaus *et al.*, 1976).

The disease is apparently transmitted through the soil un-der natural conditions. It was artificially transmitted to a number of plant hosts in pots by introducing minced roots from diseased trees to the soil. After 4 weeks, numerous RLO were observed in the roots of *Larix decidua*, *Chenopodium amaranti-color*, *Phaseolus vulgaris*, *Quercus robur*, and *Cantharanthus roseus* D. Don *(Vinca rosea)*. RLO were less abundant in *Cheno-podium quinoa*, *Datura stromonium*, *Nicotiana glutinosa*, *Vitis vinifera*, and *Gomphrena globosum*. Cells were sparse in *Polypodium vulgare* and *Pseydotsuga taxifolia*. Potted larch seedlings began to show symptoms after 6 to 7 months. Normal-ly dormant side buds were stimulated to grow, as in the initial phases of broom development, and root development was inhibited.

O. BUNCHY TOP OF PAPAYA

Bunchy top of papaya was first reported in Puerto Rico by Martorell in 1931 (Cook, 1975). It was long thought to have a viral etiology, but recent studies by Story and Halliwell (1969) indicate that a mycoplasma-like agent may be the cause.

Adsuar (1946) demonstrated that the causal agent could be transmitted by grafting, and by a leafhopper in the genus *Empoasca*.

Story and Halliwell (1969) found complete remission of symptoms of the disease 40 days after drenching plants with chlortetracycline and/or tetracycline.

P. PHONY PEACH

Phony or phony peach is a disease of peaches in the south-ern United States caused by a fastidious xylem-inhabiting bac-terium. Phony peach was first observed in Georgia about 1885 (Neal, 1920). In 1929, Hutchins reported that this disease was caused by a virus. However, recently, Hopkins *et al.* (1973) demonstrated that a xylem-inhabiting fastidious bac-terium was the incitant.

By 1932 about 35 million peach trees were infected with phony and the disease was found in all the southern Atlantic States below Virginia, in all the Gulf States, Tennessee, Arkansas, Oklahoma, Missouri, and Illinois. Subsequently, it

has been detected in Kentucky, Maryland, and Pennsylvania.
The incidence of phony reached its peak in the 1940's. It is
currently epiphytotic again in Georgia, presumably because of
relaxation in the eradication program. Over 2 million phony-
diseased trees have been removed by control agencies since
1929 (Anon., 1976).

Although first observed on peach, the phony peach organism
can attack other members of the genus *Prunus,* such as apricots
(P. hortulana), Mexican plums, *(P. domestica)* and in the Chick-
asaw plum *(P. angustifolia),* which is a natural reservoir of
the pathogen. Roguing of wild plum is one disease-suppressing
procedure. Recently it has also been reported that a bacterium
closely related to the phony pathogen is present in Johnson
grass *(Sorgum halepense),* a weed commonly found in peach orchards throughout the Southeast (Wells *et al.,* in press).

Interestingly, phony does not cause early death of affected trees. It induces marked dwarfing of new growth and fruits.
The result is a smaller crop of undersized fruit. Phony trees
are typically more compact and flattened than normal trees and
foliage tends to be greener. Phony trees leaf out and flower
earlier than normal.

The phony bacterium is limited primarily to the xylem of
roots of naturally infected trees. If diseased trees develop
from infected scions, the bacterium is carried to all portions
of the tree. If trees are infected by grafting diseased scions
on branches, the bacterium moves rapidly downward and invades
all portions of the roots.

The phony peach organism can be transmitted by grafting.
Also, six species of leafhoppers (Cicadellidae) have been reported as vectors (Turner and Pollard, 1959). All are large,
long-lived species which overwinter as adults. Once they acquire the bacterium they can transmit it the remainder of their
lives.

The organism causing phony peach appears to be the same as
or very similar serologically to that associated with Pierce's
disease of grapevines and almond scorch (Raju *et al.,* in press).

Although serological procedures show a great similarity
among these xylem-inhabiting organisms in a diverse group of
hosts, their relationship needs further study.

Q. PEACH ROSETTE

Peach rosette, like peach yellows, was known in the literature even before the turn of the century (Smith, 1891). It is
widespread and has caused serious losses to peach, *Prunus per-
sica* Batsch, in the southeastern United States. Most recently,
the disease was reported to be causing serious damage in peach

orchards in northeastern Arkansas (Slack and Kim, 1977). The
disease spreads rapidly in peach orchards which are near stands
of rosette-infected wild plum, *P. angustifolia* Marsh. An insect vector, probably a leafhopper, is presumed to effect transmission, but as yet none has been identified.

Recent studies have demonstrated that a mycoplasma was associated with the phloem sieve elements of peach and almond
(Prunus dulcis (Mill.) D. A. Webb), which had been graft-inoculated with buds from rosette-infected peach (Kirkpatrick
et al., 1975). They were also found in periwinkle, *Cantharanthus roseus* Don. *(Vinca rosea)* which had been parasitized with
dodder *(Cuscuta campestris* Yunck) and bridged from diseased
peach. MLO were ovoid to filamentous and ranged in size from
80 to 1000 nm. Peripheral ribosomes and a centrally located
DNA-like network were enclosed within a unit membrane.

In addition to the species mentioned, peach rosette has
been experimentally transmitted to Manchu cherry (Kirkpatrick
et al., 1975), apricot *(P. armeniaca* L.), and sour cherry
(P. cerasus L.)(Smith, 1972).

The predominant symptom of peach rosette is the production
of numerous compact tufts of leaves, each having an abnormally
large number of yellowish-green leaves. Infected trees usually
die within a year following the onset of symptoms. Rosetted
peach trees usually produce no fruit.

A partial remission of symptoms was achieved when tetracycline antibiotics were injected to rosette-infected peach
and almond seedlings. Basal trunk injections with tetracycline hydrochloride were more effective than injections with
other tetracycline compounds. Foliar sprays were ineffective
(Kirkpatrick *et al.,* 1975). More recently, injections with
oxytetracycline compounds at the rate of 400 mg per tree for
2 years resulted in remission of symptoms and restoration of
yield to near-normal (Perason, 1977).

R. PEACH YELLOWS

Peach yellows and little peach are thought to be caused by
closely related mycoplasma strains (Anon., 1976). Peach yellows was one of the first "virus-like" diseases studied by
plant pathologists. The disease has been found in the Atlantic Coastal and Appalachian states and as far north as Ontario,
Canada. It is known as far west as Michigan and Illinois, but
does not appear to have become established west of the Mississippi River. Several outbreaks occurred in the 19th and 20th
centuries in the United States, but it has been of minor importance in recent years.

Peach yellows appears to have a wide host range since all
species of *Prunus* tested appear to be susceptible including
plum, almond, apricot, and cherry.

*Fig. 14. Yellows-infected peach trees develop clusters
of upright shoots with chlorotic leaves, forming brooms.*

Yellows-affected trees take on a bushy appearance from
slender, branched willowy shoots that grow upright from the
main limbs (Fig. 14). At budbreak, leaf buds of affected trees
open prematurely and normally dormant lateral buds unfold.
Affected leaves are narrow, chlorotic, develop red spots, and
roll inward. Terminals die back and trees may die within 2 or
3 years.

Little peach, which is thought to be caused by a strain
of the peach yellows agent, has been more prevalent than yellows in recent years. Affected trees have greener and more
compact foliage during the early part of the growing season.
Leaves are leathery and tend to droop and roll downward, becoming chlorotic as the season progresses.

Jones *et al*. (1974b) found small (124 nm diameter), filamentous, dark-staining bodies predominately in the phloem parenchyma of yellows-affected peach. C. E. Seliskar (unpublished data) confirmed these observations but found the mycoplasma to range from 100 to 1000 nm in length (Fig. 15). In

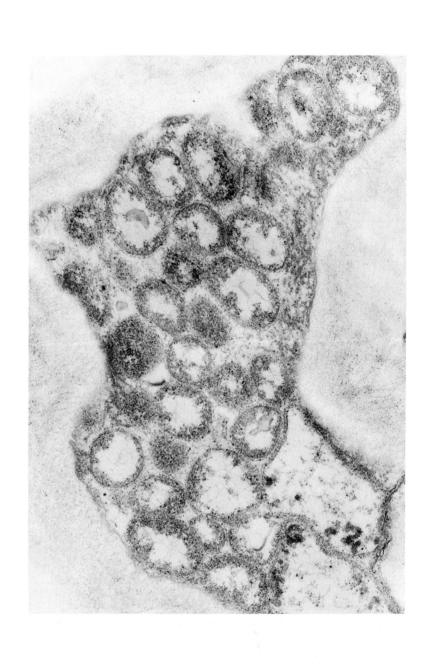

some instances, he observed a ring of material that surrounded the mycoplasma membrane.

The peach yellows agent has been transmitted by grafting and the plum leafhopper (Macropsis trimaculata). Some plum cultivars are symptomless carriers. It is suspected that the disease is often spread from plum to peach by the insect vector.

There is good evidence that little peach is a strain of peach yellows, since both diseases have the same host range and distribution. Also, the agents causing both diseases cross-protect against one another, are vectored by the same insect, and are inactivated at the same temperature.

Because of the unexplained cyclic occurrence of peach yellows, it remains a threat to the peach industry. Suggestions that X-disease may be caused by a culturable spiroplasma rather than a mycoplasma (Purcell et al., 1980) raise the possibility of spiroplasma etiology for peach yellows.

S. PEAR DECLINE

Pear decline is a major disease of pear that appears to be caused by a mycoplasma. This is based on the occurrence of mycoplasma-like bodies in phloem tissue of affected trees and the response of diseased trees to treatment with tetracycline antibiotics (Nyland and Moller, 1973). Recent association of pear decline with X-disease raises the possibility of a spiroplasma etiology for both diseases (Nyland and Raju, 1980).

Pear decline was first recognized in 1959 in California as a bud-union disorder affecting French pears (Pyrus communis) grafted on certain rootstocks (Batjer and Schneider, 1960). Later, the disease was found to have different expressions on other hosts such as P. communis "Comice," where curling and purpling of the leaves was observed.

After its first discovery in California, pear decline was found in Connecticut (McIntyre et al., 1979) and a similar disease has been described in Czechoslovakia (Blattny and Vana, 1974) and Germany (Seemuller, 1976), where it is now a major disease. McIntyre et al. (1979) observed the disorder in 15

Fig. 15. Mycoplasma associated with peach infected with yellows are found in phloem sieve elements. MLO are delimited by a trilaminar unit membrane and contain ribosomes and a central DNA-like fibrillar network.

orchards in Connecticut in 1977; 29% of 4850 trees examined
had symptoms. They observed the disease in 42% of 1696 Bart-
lett trees and 20% of 2173 Bosc trees.

Symptoms of pear decline exhibit three basic forms: tree
collapse, tree decline, and leaf curl. Tree collapse and decline result from damage to phloem tissue at the graft union.
This occurs most commonly on *Pyrus serotina* and *P. ussuriensis*
rootstocks. Tolerant rootstocks (seedlings of *P. communis*,
P. betulaefolia, or *P. calleryana*) generally are associated
with leaf curl symptoms. Leaves curl downward, turn purple
or dark red, and drop prematurely. The main symptoms of tree
decline are poor shoot and spur growth, dieback of lateral and
terminal shoots, upward rolling of leaves, reduced leaf size,
bright red leaf color, and premature leaf drop.

The pear decline agent is graft-transmissible and vectored
in nature by a psyllid, *Psylla pyricola* (Jensen *et al.*, 1964).
Mycoplasma-like bodies have been found in the vector (Hibino
et al., 1971) as well as in diseased phloem in the tree. Because MLO were absent in diseased secondary sieve tubes, Schneider (1977) postulates that toxins produced by the MLO are
responsible for necrosis in these areas.

Since X-disease appears to be more prevalent in areas near
pear orchards with pear decline, a relationship between the
two diseases is suspected in California (Nyland and Raju, 1980).
Pear decline responds well to therapy with tetracycline antibiotics (Nyland and Moller, 1973).

T. PECAN BUNCH DISEASE

Pecan bunch is a yellows-type witches'-broom disease first
reported by Cole (1937) near Shreveport, Louisiana. Since
then the disease has progressively spread throughout eight
southern states and is now serious on native pecan, *Carya illinoensis* (Wangenh.) K. Koch., and a number of its commercial
horticultural varieties.

Pecan bunch was formerly assumed to have a virus etiology,
but recent cytological investigations have demonstrated the
association of a mycoplasma-like organism with diseased phloem
tissues of bunch-diseased pecan trees in Louisiana (Seliskar
et al., 1974)(Fig. 16). Dale (1977) corroborated these results when he observed MLO in bunched pecan in Arkansas. MLO
are found abundantly in sieve elements but sparsely in other
elements in the phloem. As in some other yellows diseases of
trees (Seliskar *et al.*, 1973), they were most abundant in
phloem tissues of severely broomed twigs. MLO are difficult
to find in nonsymptomatic twigs, even those in close proximity
to witches'-brooms. They are mostly oval to spherical with

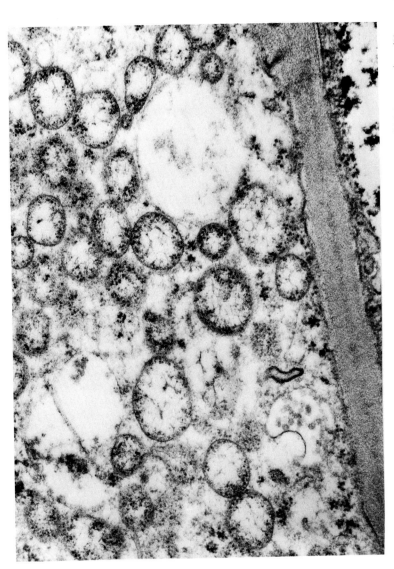

Fig. 16. Mycoplasma in sieve elements of native pecan, Carya illinoensis (Wangenh.) K. Koch.

filamentous forms sometimes observed near sieve plates. The
elongated form may be necessary for cell-to-cell movement.
Dumbbell-shaped forms are also found, and suggest reproduction
by binary fission. Cells range in length from 80 to 800 nm but
the predominantly round forms ranged from 150 to 420 nm. The
viability of cells less than 125 nm is open to question (Clyde,
1968). Morphologically, they are similar to many other plant
mycoplasmas reported in the literature in that they are enclosed
in a plastic unit membrane about 10 nm thick and contain ribosomes and fibrillar material resembling DNA.

Pecan rosette, a disorder caused by zinc deficiency, causes
brooms on the terminals of affected trees similar to those
caused by bunch disease (Alben et al., 1932). Rosette, however, is readily corrected with zinc sprays (Demaree et al.,
1933), while bunch disease is infectious and is not affected
by applications of zinc. The leafing out of brooms about 2
weeks before normal foliage is a positive indication of brooming caused by bunch disease (KenKnight, 1962).

At the time bunch disease was reported on native pecan it
was also observed on nearby water hickory, Carya aquatica
(Michx. f.) Nutt. There is good reason to believe that the
disease originally spread from this species to nearby pecan
of orchards (Cole, 1937; KenKnight, 1970a). In Louisiana the
disease is also found on other native hickories, including
C. ovata (Mill.) K. Koch., C. cordiformis Wangenh.) K. Koch.,
and C. tomentosa (Poir.) Nutt. (Cole, 1937).

The severity of symptoms varies widely in native pecan
seedlings, as well as in horticultural cultivars. One or more
witches'-brooms, each consisting of several or more slender,
willowy shoots, characteristically appear on the branches or
trunks of bunch-diseased trees (Fig. 17). Leaflets are often
chlorotic, sometimes broader than normal, and distorted (Ken-
Knight, 1963a). The shoots are more upright and come into
leaf about 2 weeks before normal foliage in the spring, simpli-
fying detection at this time of year (Fig. 18). Broomed shoots
also continue growth late in the fall and are commonly injured
or killed by low winter temperatures (Cole, 1937; KenKnight,
1962). Severely broomed branches on infected trees produce
no nuts but nonsymptomatic branches on the same trees may produce a normal crop (KenKnight, 1963a). Very susceptible varieties with brooms throughout the crown become chlorotic,
weaken, and finally die.

Among the many commercial pecan varieties planted in the
southern United States, Schley, Desirable, Mahan, Dependable,
Barton, and Elliott are highly susceptible (KenKnight, 1970b),
while Burkett, Van Deman, Pabst, and Moneymaker are intermediate, and Success and Stuart are highly resistant (Ken-
Knight 1963a, 1964). Information on the susceptibility of

Fig. 18. Pecan trees infected with
bunch disease come into leaf about 2 weeks
before normal foliage in the spring.

Fig. 17. A severely affected pecan, Carya
illinoensis (Wangenh.) K. Koch showing numerous
brooms in the crown.

additional cultivars is given in a report by KenKnight (1968).

Systemic spread of the disease within infected trees is usually slow and brooms may be confined to one or a few limbs for years. Once the disease has spread throughout the crowns of a few trees, it spreads rapidly from tree to tree and whole orchards may be devastated in a few years (KenKnight, 1963a). Rate of spread may be correlated, at least in part, with seasonal rainfall (KenKnight, 1964, 1970a; Rose and Rose, 1957).

Bunch disease can be transmitted experimentally by scion grafting (Cole, 1937), but no natural vector is known.

Sanitation procedures, including planting new trees well isolated from sources of infection, pruning to remove all brooms on infected trees, planting resistant varieties, and top-working symptomatic susceptible varieties appear to be the most satisfactory methods for control (KenKnight, 1970a).

U. PLUM LEAF SCALD

Plum leaf scald is a disease found in North and South America, apparently the result of infection by a fastidious bacterium. The disease was first reported in 1935 in the delta region of the Parana River, Argentina, and has subsequently been found in Brazil, Paraguay, and the United States (French and Kitajima, 1978). Fastidious xylem-limited bacteria have consistently been found associated with the disease in both North and South America (Kitajima *et al.,* 1975).

Plum scald causes severe losses in Japanese plum *(Prunus salicina),* the most intensively cultured fruit crop in Argentina. The disease also affects other *Prunus* species such as *P. domestica, P. cerasifera, P. insititia,* and *P. americana. Prunus amygdalus,* although susceptible to the plum scald agent, remains symptomless.

In Argentina, symptoms usually appear in late December or early January. A slight and irregular chlorosis can be seen at the leaf margins, which become brown and dry. A diffuse, chlorotic band separates the green and necrotic areas. Upper branches become brittle; plants look as if they had been scorched and may die within a few years.

The plum leaf scald agent is graft-transmissible and an insect, *Ornenis cestri,* in the family Flatidae, may be a vector (Kitajima *et al.,* 1975).

V. PRUNUS X-DISEASE

Prunus X-disease is apparently caused by a highly variable mycoplasma or spiroplasma with a wide host range and distribution. Because of the varying expressions of this disease in

different hosts, a number of synonyms have been used, such as
western X disease, eastern X-disease, wilt and decline, cherry
buckskin, peach leaf casting yellows, peach yellow leaf roll,
and small bitter cherry. Prunus X was long thought to be
caused by a virus, but recent indications are that mycoplasmas
(Jones et al., 1974a; Garanett and Gilmer, 1971; McBeath et
al., 1972) or spiroplasmas (Purcell et al., 1980) may be the
causal agent.
       Prunus X was originally described in California in 1931 as
cherry buckskin (Rawlins and Horne, 1931). In 1933 it was
discovered in Connecticut (Stoddard, 1934). Within a short
period the disease was found throughout the United States. It
has not been detected in the peach-growing areas of South
Carolina, Georgia, Arkansas, or Texas. Initially, a distinction was made between western and eastern X-disease However,
it is now generally accepted that they are caused by identical
or closely related strains of the same organism.
       X-disease has caused considerable economic losses on peach,
nectarine, Japanese plum, sour cherry, and sweet cherry. It
has been transmitted experimentally to almond, apricot, mahaleb
cherry, Korean cherry, western sand cherry, bitter cherry,
holly leaf cherry, Manchu cherry, wildgoose plum, and several
plum hybrids (Anon., 1976). Common and western chokecherry are
the most common wild reservoirs of the disease. Outbreaks of
X-disease in the northern United States appear to be cyclic,
at intervals of about 8 to 10 years. It is speculated that
these cycles may correspond to buildups in the vector population.
       Peach trees with X-disease develop leaves with yellow to
red blotches, defoliate prematurely, and rarely produce fruit.
X-diseased sour cherry trees propagated on mahaleb rootstocks
wilt and die suddenly in late summer. Trees on mazzard rootstock survive for several years and produce small green cher-
ries.
       Chokecherries infected the first year show a transient de-
lay in growth but appear almost normal until 6 to 8 weeks after growth commences. Then the foliage gradually becomes
orange or red. Affected leaves droop and a second flush of
growth may occur in July or August. During succeeding years,
the growth of affected tress becomes rosetted with dull leaves
that are often yellow instead of red. Infected trees decline
rapidly and often die in 3 to 4 years.
       Although the X-disease agent cannot be transmitted mecha-
nically, it is readily transmitted by budding and grafting.
It has been moved from a variety of herbaceous hosts with
dodder (Cuscuta). A number of leafhoppers are known vectors,
including Callodonus geminatus, Scaphytopius acutus, and Fie-
beriella florii (Wolfe and Anthony, 1953). The X-disease

agent is reported to be lethal to *Calodonus montanus* (Whitcomb *et al.*, 1968).

Studies on the relationships of the various strains of the
X-disease agent have been hampered by inability to culture the
causal agent. Recent possible success in this area may clarify the host range and relationship of this agent to other pathogens (Purcell *et al.*, 1980). A. L. Jones (personal communication) has indicated that culturing studies are encouraging,
but feels that recent claims about the successful culturing
of an X-disease-associated spiroplasma should be viewed cau-
tiously.

W. WALNUT BUNCH DISEASE

Walnut bunch disease was presumed to be virus-caused when
Hutchins and Wester (1947) reported its transmissibility by
grafting. However, considerable doubt as to its virus etiology was raised when Seliskar (1973, 1976) reported that detailed cytological investigations revealed a mycoplasma-like
organism in the phloem of bunch-diseased walnut trees (Fig.
19). Subsequently, these results were corroborated by Mac-
Daniels *et al.* (1975). MLO are commonly found in the sieve
elements of leaf petioles and twigs of brooms, in which they
may become densely packed. More frequently, however, they are
arranged in groups, often near the sieve plates. They are
more abundant in severely broomed branches and twigs than in
branches with mild brooming. MLO were seldom seen in unbroomed
branches of diseased trees. The abundance of MLO in diseased
tissues appears to be related to seasonal development and the
degree of symptom expression. They become increasingly abundant in the sieve tubes as the season progresses, and may play
a role in the physiological functions and cell-to-cell movement of nutrients (MacDaniels *et al.*, 1975). MLO are pleomorphic and vary in shape from oval or nearly spherical to
tubular. While they range from 150 to 1000 nm in length, the
predominant oval or spherical forms average 350-400 nm thick
and contain ribosomes and DNA-like strands characteristic of
animal mycoplasmas. Dumbbell-shaped cells are frequently observed and suggest multiplication by binary fission.

The first report of walnut bunch appears to have been in
1932 when Waite reported it in Delaware (Waite, 1933). McKay
and Crane (1951) suggest that the disease was known long before this -- at least as early as 1914, to judge from photo-
graphs in USDA files dating back to that time. Good historical accounts of the disease are presented in reviews by Becker (1940, 1961), McKay and Crane (1951), Berry and Gravatt
(1955), and MacDaniels and Welch (1964).

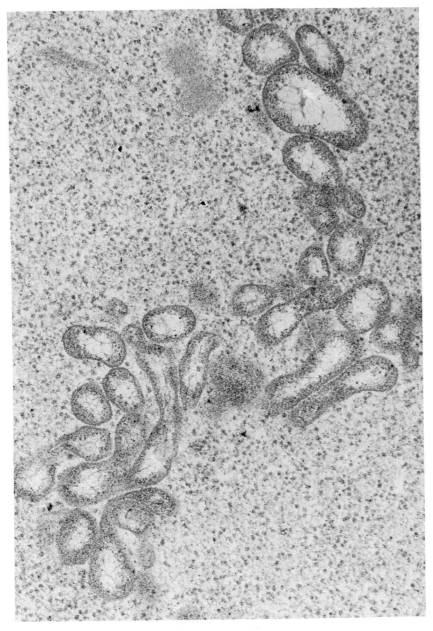

Fig. 19. Mycoplasma-like organism (MLO) in sieve elements
of phloem in bunch-diseased black walnut, Juglans nigra L.

Walnut bunch disease is widely distributed throughout the eastern and central United States, but has not been reported elsewhere. The disease attacks eastern black walnut (*Juglans nigra* L.), Japanese walnut (*J. sieboldiana* Maxim and its variety, *J. sieboldiana* var. *cordiformis,* heartnut), Persian walnut (*J. regia* L.), Manchurian walnut (*J. mandshurica* Les. Hoz.), and butternut (*J. cinerea* L.). A single case of brooming has also been reported on bitternut hickory (*Carya cordiformis*) near a broomed walnut in upstate New York (MacDaniels and Welch, 1964) but it showed no symptoms 10 years later (Mac-Daniels *et al.,* 1975). A number of attempts to detect mycoplasma in broomed shellbark and shagbark hickories in the eastern United States have thus far failed (C. E. Seliskar, unpublished data).

The disease is very serious on Asiatic walnuts, in which it causes severe witches'-brooms and extensive mortality. Butternut and Persian walnut are somewhat less susceptible, while black walnut is least affected. On cultivated black walnut the disease apparently causes the nuts to be poorly filled or shriveled and discolored, and thus is important to the nut growing industry (MacDaniels and Welch, 1964; Mac-Daniels *et al.,* 1975). Should the disease become established on commercially grown Persian walnut in California, it could threaten a major nut industry.

Witches'-broom is a systemic disease characterized by the appearance of more or less distinctive bunches of wiry twigs that form brooms on trunks and large branches (Fig. 20). Witches'-brooms vary widely in number and size with the species affected. They are very abundant on diseased Asiatic walnut (Fig. 21) but often sparse or nonexistent on black walnut. In the latter species, symptoms may be completely masked when brooms die out and do not reappear. Twigs in brooms commonly exhibit epinasty and leaves are dwarfed and chlorotic. New brooms, which are usually produced late in the growing season, develop from the proliferation of normally dormant axillary buds. Since they continue growth late into the fall, after normal growth stops, they are often killed back by early freezes. Nut production and quality of nuts may be severely affected and wood becomes brittle, making limbs highly susceptible to storm damage. Dieback is a common symptom on most species and mortality occurs in the most susceptible walnuts, particularly the Asiatic varieties and butternut.

The disease can be transmitted by bark patch-grafting (Hutchins and Wester, 1947), by budding (McKay and Crane, 1951), by chip-budding (Carling and Millikan, 1974), and by scion grafting (C. E. Seliskar, unpublished data). No vector is known.

Fig. 20. A witches'-broom on Japanese walnut, Juglans sieboldiana Maxim, infected with bunch disease.

Fig. 21. Severely broomed Japanese walnut, Juglans sieboldiana Maxim.

Although no detailed investigations have been conducted,
at least some control of the disease can be achieved by sanitation, particularly the prompt removal of diseased trees. In
black walnut the disease, while systemic, spreads slowly; thus,
if the disease appears in one or two limbs it may be arrested
by removal of the infected limbs. This is not true for other
walnut species, however (McKay and Crane, 1951; Berry and
Gravatt, 1955). In addition, care should be taken in grafting
walnut planting stock so that only disease-free trees are used
as a source of budding or grafting material (Anon., 1947).

X. WILLOW WITCHES'-BROOM (SALIX YELLOWS)

A mycoplasma-like organism has been observed in the midribs of leaves in broomed willow trees. MLO were bounded by
triple-layered unit membranes, ca. 10 nm thick, and contained
DNA-like strands varying in electron density, and ribosomes
that were smaller than those in host cells. Some bodies contained vacuoles. An opaque layer of undetermined character
was commonly detected around the plasma membrane of MLO. Filamentous projections from spherical bodies were frequently
detected, along with hourglass-shaped forms, apparently un-
dergoing binary fission (Holmes et al., 1972). While dimensions of the MLO were not given, the micrographs shown in the
report of this disease indicate that they fall within the
usual range of plant mycoplasmas, ca. 100 to 800 nm, with oval
forms about 250 to 400 nm.

The only account of this disease is given by Holmes et al.
(1972). Brooming has been detected in wand willow (Salix rigida Muhl.) in widely scattered areas in southern New Hampshire, New York, and Massachusetts. Witches'-broom has also
been observed in Virginia on S. nigra Marsh. (C. E. Seliskar,
unpublished data)(Fig. 22).

The witches'-broom on willow is a typical yellows disease
characterized by the appearance of brooms in the crowns of
affected trees. Brooms consist of numerous, spindly, upright
branches bearing small leaves. The proliferation of shoots
results from the abnormal development of dormant axillary
buds. These often die back in winter.

The disease can be transmitted experimentally by grafting
diseased scions to healthy plants. It has been transmitted
from S. rigida to S. caprea L. by this method (Holmes et al.,
1972). Its natural means of spread is unknown.

Y. YOUNG TREE DECLINE

Young tree decline or sand hill decline is a serious disease on rough lemon (Citrus jambhiri) rootstock in Florida.
It is apparently caused by a xylem-inhabiting fastidious

Fig. 22. Witches'-broom on black willow, Salix nigra Marsh.

bacterium. The etiology of this disease is poorly understood. Although it is apparently caused by a transmissible entity, transmission has been erratic (Feldman *et al.*, 1977).

Symptoms of this disease are not expressed until after trees come into bearing. The most characteristic symptom is leaf mottling similar to that caused by zinc deficiency. Leaves roll and wilt during periods of stress. Twigs die back and vessel plugging is apparent. Feldman *et al.* (1977) have found bacteria in the vascular fluid of citrus trees affected by young tree decline.

V.   CONCLUSIONS AND DISCUSSION

Following Doi *et al.'s* (1967) discovery that mycoplasmas were associated with yellows-type diseases, research accelerated on other fastidious prokaryotes as agents of plant disease. A number of tree diseases that had been thought to be caused by viruses were reinvestigated (Wilson and Seliskar, 1976). From these investigations, new mycoplasma- or spiroplasma-associated diseases have been defined. Also, fastidious bacteria that were strictly confined to either the xylem or phloem were discovered associated with some of these diseases.

Heavy dependence has been placed on electron microscopy in establishing the possible role of mycoplasmas, spiroplasmas, and fastidious bacteria in tree diseases. Some of these

agents have been cultured, and we tend to assume that this
verifies the pathogenicity of those that have not. Because
mycoplasmas and spiroplasmas are nondescript in ultrathin sections, their association with disease needs to be viewed cri-
tically. Without evidence other than their appearance in
association with disease in electron micrographs, the door
needs to be kept open to other possible causes.

   The serological similarity of those spiroplasmas and fastidious bacteria that have been cultured is surprising, since
they are associated with diverse hosts. The specialized nutritional requirements of these organisms would seem to indicate a limited rather than a broad host range. This is the
case of other obligate or near obligate parasites of plants.
Our present serological tests may not be sufficiently sensitive to detect such specificity if it exists.

REFERENCES

Adsuar, J. (1946). Transmission of papaya bunchy top by a
     leafhopper of the genus *Empoasca*. *Science 103*, 316.
Alben, A. O., Cole, J. R., and Lewis, R. D. (1932). New devel-
     opments in treating pecan rosette with chemicals. *Phy-
     topathology 22*, 979-981.
Anonymous. (1947). A new walnut and butternut virus disease.
     U. S. Bur. Plant Ind. Rep. 146/47, 62.
Anonymous. (1952). Control of Dutch Elm Disease and elm phloem
     necrosis. *U. S. Dept. Agr. Leaflet 329*, 11 pp.
Anonymous. (1976). Virus diseases and noninfectious disorders
     of stone fruits in North America. *Agr. Handb. No. 437*,
     433 pp.
Atanasoff, D. (1935). Old and new virus diseases of trees and
     shrubs. *Phytopathol. Z. 8*, 197-223.
Atkinson, J. D. (1971). "Diseases of Tree Fruits in New Zea-
     land," 406 pp. A. R. Shearer, Gov. Printer, Wellington,
     New Zealand.
Baker, W. L. (1948). Transmission by leafhoppers of the vi-
     rus causing phloem necrosis of American elm. *Science 108*,
     307-308.
Baker, W. L. (1949). Studies on the transmission of a virus
     causing phloem necrosis of American elm with notes on the
     biology of its insect vector. *J. Econ. Entomol. 42*(5),
     729-32.
Baker, W. L. (1950). Recent developments of transmission of
     elm phloem necrosis disease. *Entomol. Soc. Wash. Proc.
     52*, 52.
Baker, W. L. and May, C. (1951). Phloem necrosis in elm.
     Plants Gard. 7(2), 129-130.

Batjer, L. P. and Schneider, H. (1960). Relation of pear de-
    cline to rootstocks and sieve-tube necrosis. *Proc. Amer.
    Soc. Hort. Sci. 76*, 85-97.
Beakbane, A. B., Mishra, M. D., Posnette, A. F., and Slater,
    C. H. W. (1971). Mycoplasma-like organisms associated with
    chat fruit and rubbery wood diseases of apple *Malus domes-
    tica* (Borhh) compared with those in strawberry with green
    petal disease. *J. Gen. Microbiol. 66*, 55-62.
Beakbane, A. B., Slater, C. H. W., and Posnette, A. F. (1972).
    Mycoplasmas in the phloem of coconut, *Cocos nucifera* L.,
    with lethal yellowing disease. *J. Hort. Sci. 47*(2), 265.
Becker, G. (1940). My observations on witches-broom disease.
    Northern Nut Growers Rep. 31, 106-109.
Becker, G. (1961). Living with walnut broom disease. *Mich.
    Nut Growers Assoc. Newsl. March, 1961*, pp. 6-12.
Berry, F. H. and Gravatt, E. Flippo (1955). Walnut bunch di-
    sease. *Bull. Calif. Dep. Agr. 44*, 63-67.
Blattny, C. (1959). Viroza a z virozy podezrela nemoc Trnovni-
    ku Akatu (*Robinia pseudoacacia* L.) (A virosis and a sus-
    pected virus disease of *R. pseudoacacia*). *Ann. Acad.
    Techcosl. Agr. 5*, 291-294.
Blattny, C. and Vana, V. (1974). Pear decline accompanied
    with mycoplasma-like organisms in Czechoslovakia. *Biol.
    Plant (Prague) 16*, 474-475.
Bove, J. M. and Saglio, P. (1974). Stubborn, greening, and
    related diseases. *Proc. 6th Conf. Int. Organ. Citrus
    Virol.* pp. 1-11.
Braun, E. J. (1975). Freeze-etch and thin-section electron
    microscopy of mycoplasmas in phloem of *Vinca rosea*.
    (Abstr. #NE-11). *Proc. Amer. Phytopathol. Soc. 3*, 301.
Braun, E. J., and Sinclair, W. A. (1976). Histopathology of
    phloem necrosis in *Ulmus americana. Phytopathology 66*,
    598-607.
Braun, E. J. and Sinclair, W. A. (1979). Phloem necrosis of
    elms: Symptoms and histopathological observations in
    tolerant elms. *Phytopathology 69*, 354-358.
Bretz, T. W. (1944a). Observations on phloem necrosis of elm
    in Kentucky, Ohio, Indiana, and Illinois. *Plant Dis.
    Rep. 28*(35), 1056-1057.
Bretz, T. W. (1944b). Phloem necrosis of elms in Missouri.
    Plant Dis. Rep. 28(30), 929-931.
Bretz, T. W., and Swingle, R. U. (1946). Known distribution of
    phloem necrosis of the American elm. *Plant Dis. Rep. 30*,
    156-159.
Caldwell, R. M. (1945). Indiana phloem necrosis. *Hoosier
    Hort. 27*(8), 127-128.
Campana, R. (1958). Elms susceptible to Dutch elm disease and
    elm phloem necrosis. *Trees Mag. 19*(1), 6, 19.

Campana, R. J. and Carter, J. C. (1955). Spread of Dutch elm
    disease in Illinois in 1954. *Plant Dis. Rep. 39,* 245-248.
Carling, D. E. and Millikan, D. F. (1974). Graft transmission
    of an infectious entity associated with bunch disease in
    Juglans. (Abstr.). *Proc. Amer. Phytopathol. Soc.* (1974,
    publ. 1975) *1,* 124(En).
Carter, J. C. (1945). Dying of elms in Illinois. *Plant Dis.
    Rep. 29*(1), 23-26.
Carter, J. C. (1950). Status of oak wilt and elm phloem nec-
    rosis in the midwest. *Arborist News 15,* 45-51.
Carter, J. C. and Carter, L. R. (1974). An urban epiphytotic
    of phloem necrosis and Dutch elm diseases. 1944-1972.
    Bull. Ill. Nat. Hist. Surv. 31, 113-143.
Chen, T. A., and Liao, C. H. (1975). Corn stunt spiroplasma:
    isolation, cultivation, and proof of pathogenicity.
    Science 188, 1015-1017.
Ciferri, R., and Corte, A. (1960). Gli "scopazzi virosici"
    della *Robinia pseudoacacia,* malattia nuova per l'Italia.
    Virus witches'-brooms of *R. pseudoacacia,* a new disease
    for Italy. *Atti. Ist. Bot. Univ. Pavia Ser. 5, 17,*
    122-128, 3 pl.
Clyde, W. A., Jr. (1968). An experimental model for human my-
    coplasma disease. *J. Biol. Med. 40,* 436-443.
Cole, J. R. (1937). Bunch disease of pecans. *Phytopathology
    27,* 604-612.
Cook, A. (1975). Papaya bunch top. "Diseases of Tropical and
    Subtropical Fruits and Nuts," pp. 317. Hafner, New York.
Curl, E. A., Hyche, L. L., and Marshall, N. L. (1959). An
    outbreak of phloem necrosis in Alabama. *Plant Dis. Rep.
    43,* 1245-1246.
Dale, J. L. (1977). Mycoplasma-like organism observed in pe-
    can with bunch disease in Arkansas. *Plant Dis. Rep. 61,*
    319-321.
Davis, M. J., Stassi, D. L., French, W. J., and Thomson, S. V.
    (1978). Antigenic relationship of several ricketssia-like
    (sic.) bacteria involved in plant diseases. *Proc. 4th
    Int. Conf. Plant Pathol. Bacteria, Angers, France,* pp. 311-
    315.
Davis, M. J., Thomson, S. V., and Purcell, A. H. (1980). Etio-
    logical role of the xylem limited bacterium causing
    Pierce's disease in almond leaf research. *Phytopathology
    70,* 472-475.
Davis, R. E., and Worley, J. F. (1973). Spiroplasma: Motile,
    helical microorganism associated with corn stunt disease.
    Phytopathology 63, 403-408.
Davis, R. E., Whitcomb, R. F., Chen, T. A., and Granados, R.
    R. (1972a). Current status of the aetiology of corn stunt
    disease. *In* "Pathogenic Mycoplasmas," (K. Elliott and

J. Birch, eds.). Ciba Found. Symp. pp. 205-225. Associated Science Publishers, New York.
Davis, R. E., Worley, J. F., Whitcomb, R. F., Ishijima, T.,
and Steere, R. L. (1972b). Helical filaments produced by
a mycoplasma-like organism associated with corn stunt
disease. *Science 176*, 521-523.
Demaree, J. B., Fowler, E. D., and Crane, H. L. (1933). Report
of progress on experiments to control pecan rosette.
Proc. 32nd Nat. Pecan Growers Assoc., pp. 90-99.
Doi, Y., Teranaka, M., Yora, K., and Asuyaina, H. (1967). Mycoplasma - or PLT group-like microorganisms found in the
phloem elements of plants infected with mulberry dwarf,
potato witches'-broom, aster yellows, or paulownia
witches'-broom. *Ann. Phytopathol. Soc. Jap. 33,* 259-266.
Feldman, A. W., Hanks, R. W., Gord, G. E., and Braun, G. E.
(1977). Occurrence of a bacterium in YTD-affected as well
as in some apparently healthy citrus trees. *Plant Dis.
Rep. 61,* 546-550.
Filer, T. H., Jr. (1966). Phloem necrosis of American elm in
the Mississippi delta. *Plant Dis. Rep. 50,* 751.
Filer, T. H., Jr. (1973). Pressure apparatus for injecting
chemicals into trees. *Plant Dis. Rep. 57,* 338-341.
Forbes, Stephen A. (1912). What is the matter with the elms
in Illinois? *Univ. Ill. Agr. Exp. Sta. Bull. No. 154,*
22 pp.
French, W. J. and Kitajima, E. W. (1978). Occurrence of plum
leaf scald in Brazil and Paraguay. *Plant Dis. Rep. 62,*
1035-1038.
Garanett, A. C. and Gilmer, R. M. (1971). Mycoplasma associated with X-disease in various *Prunus* species. *Phytopath-
ology 61,* 1036-1037.
Garmen, H. (1899). The elms and their disease. *Ky. Agr. Exp.
Sta. Bull. 84,* 51-75.
Giannotti, J., Marchou, G. and Vago, C. (1968). Micro-organismes
de type mycoplasme dans les cellules liberiennes de *Malus
sylvestris* L. atteinte de la maladie des proliferations.
C. R. Acad. Sci. D267, 76-77.
Gibson, L. P. (1973). An annotated list of the Cicadellidae
and Fulgoridae of elm. *U. S. Dep. Agr. Forest Serv. Res.
Pap. NE-278,* 5 pp.
Goheen, A. C., Nyland, G., and Lowe, S. K. (1973). Association of rickettsialike organism with Pierce's disease of
grapevines and alfalfa dwarf and heat therapy of the disease in grapevines. *Phytopathology 63,* 341-345.
Grant, T. J. (1939). Systemic brooming of *Robinia pseudoacacia*
and other virus-like diseases of trees. *Phytopathology
29,* 8.

Grant, T. J. and Hartley, C. (1938). A witches'-broom on black
    locust and a similar disease on honey locust. *Plant Dis.
    Rep. 22,* 28-31.
Grant, T. J., Stout, D. C. and Readey, J. C. (1942). Systemic
    brooming, a virus disease of black locust. *J. Forest 40,*
    253-260.
Hart, H. H. (1978). Occurrence of elm phloem necrosis in
    Michigan. *Plant Dis. Rep. 62,* 872-873.
Hearon, S. S., Sherald, J. L., and Kostka, S. J. Association
    of a xylem-limited bacteria with elm, sycamore, and oak
    leaf scorch. *Can. J. Bot., in press.*
Heinze, K., Petzold, H., and Marwitz, R. (1972). Contribu-
    tion to the aetiology of lethal yellowing disease of co-
    conut palm. Beitrag zur Atiologie der todlichen Vergil-
    bung der Kokospalme. *Phytopathol. Z. 74*(3), 230-237.
Hibben, C. R. (1978). Ash dieback in the northeast: Report on
    severity and causes. *Metro. Tree Impr. Alliance (METRIA)
    Proc. 1,* 87-96.
Hibben, C. R., and Wolanski, B. (1970). Dodder transmission of
    a mycoplasma from ash trees with yellows-type symptoms.
    Phytopathology 60(9), 1295.
Hibben, C. R., and Wolanski, B. (1971). Dodder transmission of
    a mycoplasma from ash witches'-broom. *Phytopathology 61,*
    151-156.
Hibben, C. R., Hagar, S. S., and Karpel, Mary-Ann (1973). In-
    fection of declining ash trees by virus and mycoplasma-
    like bodies: Identification and implications. (Abstr.
    #0930). *Proc. 2nd Int. Congr. Plant Pathol. Abstr. Papers,
    Minneapolis, Minn., Sept. 5-12.*
Hibino, H., Kaloostian, G. H., and Schneider, H. (1971). My-
    coplasma-like bodies in the pear psylla vector of pear
    decline. *Virology 93,* 34-40.
Holmes, Francis O., Hirumi, Hiroyuki, and Maramorosch, Karl.
    (1972). Witches'-broom of willow: Salix yellows. *Phyto-
    pathology 62,* 826-828.
Holmes, Francis W. and Chater, Clifford S. (1977). Elm phloem
    necrosis in eastern Massachusetts. *Plant Dis. Rep. 61,*
    626-628.
Hong, S. W. (1960). A study of virus disease on Chinese date
    tree. II. On the anatomical effects of the shoot cluster
    disease on the vascular structure of the infected plants.
    Korean J. Bot. 3(2), 29-34.
Hong, S. W., and Hah, Y. C. (1961). A comparative investiga-
    tion of free amino acids in healthy and virus diseased
    Chinese date tree. *Korean J. Bot. 4*(1), 9-12.
Hong, S. W., and Kim, C. J. (1960). A study of virus disease
    on Chinese date tree. I. On the external and the in-
    ternal morphological characteristics of disease infected
    plants. *Korean J. Bot. 3*(1), 32-38.

Hopkins, D. L., and Mollenhauer, H. H. (1973). Rickettsia-
    like bacterium associated with Pierce's disease of grapes.
    Science 179, 298-300.
Hopkins, D. L., Mollenhauer, H. H., French, W. J. (1973).
    Occurrence of a rickettsia-like bacterium in the xylem of
    peach trees with phony peach. *Phytopathology 63*, 1422-1423.
Hull, R. and Plaskitt, A. (1970). Electron microscopy of al-
    ternate hosts of Sandal spike pathogen and of tetracycline-
    treated spike-infected Sandal trees. *J. Indian Acad. Wood
    Sci. 1*(1), 62-64.
Hutchins, L. M. (1929). Phony disease of the peach. (Abstr.)
    Phytopathology 19, 107.
Hutchins, Lee M., and Wester, Horace V. (1947). Graft-trans-
    missible brooming disease of walnut. (Abstr.) *Phytopathol-
    ogy 37*(1), 11.
Imazeki, Rokuya, and Ito, Kazuo. (1963). Dangerous Forest
    Diseases in Japan. *U. S. Dep. Agr. Forest Serv. Misc.
    Publ. No. 939*, pp. 46-54.
Ishiie, T., Doi, Y., Yora, K., and Asuyama, H. (1967). Sup-
    pressive effects of antibiotics of tetracycline group on
    symptom development in mulberry dwarf disease. *Ann. Phy-
    topathol. Soc. Jap. 33*, 267-275.
Jackson, L. W. R. and Hartley, C. (1933). Transmissibility of
    the brooming disease of black locust. *Phytopathology 23*,
    83-90.
Jensen, D. D., Griggs, W. H., Gonzales, C. Q., and Schneider,
    H. (1964). Pear decline virus transmission by pear psy-
    lla. *Phytopathology 54*, 1346-1351.
Jones, A. L., Hooper, G. R., Rosenberger, D. A., and Chevalier,
    J. (1974a). Mycoplasmalike bodies associated with peach
    and periwinkle exhibiting symptoms of peach yellows.
    Phytopathology 64, 1154-1156.
Jones, A. L., Hooper, G. R. and Rosenberger, D. A. (1974b).
    Association of mycoplasmalike bodies with little peach
    and X-disease. *Phytopathology 64*, 755-756.
KenKnight, G. T. (1962). Bunch disease of pecans. *Proc. Texas
    Pecan Grow. Assoc. 41*, 67-74.
KenKnight, G. T. (1963a). Bunch disease of pecans. *Amer.
    Fruit Grow. 83*(3), 26-28.
KenKnight, G. T. (1963b). Progress report on bunch disease of
    pecan and witches'-broom diseases of trees in woods near
    pecan orchards. *Proc. Texas Pecan Grow. Assoc. 42*, 94-
    98.
KenKnight, G. T. (1964). Research on bunch disease of pecan.
    Proc. Texas Pecan Grow. Assoc. 43, 77-79.
KenKnight, G. T. (1968). Resistance of pecan to scab and
    bunch disease in Louisiana. *Plant Dis. Rep. 52*(3), 307-
    309.

KenKnight, G. T. (1970a). Bunch disease and its control. *U.S. Dep. Agr. Pecan Lab. Circ. Lett. Shreveport, La.* 4 pp. (Mimeo.).

KenKnight, G. T. (1970b). Dooryard pecan trees in Louisiana. *U. S. Dep. Agr. Pecan Lab. Circ. Lett. Shreveport, La.* 4 pp. (Mimeo.).

Kim, Chong Jin. (1965). Witches'-broom of jujube tree, *Zizyphus jujube* Mill. Var. *inermis* Rehd. (Part 3). *Korean J. Microbiol. 3*(1), 1-6.

Kirkpatrick, Hugh C., Lowe, S. K., and Nyland, G. (1975). Peach rosette: The morphology of an associated mycoplasmalike organism and the chemotherapy of the disease. *Phytopathology 65,* 864-870.

Kitajima, E. W., Bakarcic, M., and Fernandez-Valiela, M. V. (1975). Association of rickettsialike bacteria with plum leaf scald disease. *Phytopathology 65,* 476-479.

Kralik, O. and Brcok, J. (1974). Tubular structures associated with mycoplasma in proliferation infested apple trees. *Biol. Plant. 16,* 78-79.

La, Y. J. and Chang, Moo Ung. (1979). Association of mycoplasma-like organisms with witches'-broom disease of jujube. *Proc. R. O. C. U. S. Cooperative Sci. Seminar Mycoplasma Dis. Plants 1,* 21-24.

La, Y. J., Brown, W. M., Jr. and Moon, D-S., (1976). Control of witches'-broom disease of jujube with oxytetracycline injection. *Korean J. Plant Prot. 15*(3), 107-110.

Larsh, H. W. (1945). Elm phloem necrosis in Arkansas and Oklahoma. *Plant Dis. Rep. 29*(27), 699-700.

Leach, J. G. and Valleau, W. D. (1939). Two reports on phloem necrosis of elm. *U. S. Bur. Plant Ind. Plant Dis. Rep. 23,* 300-301.

Lowe, S. K., Nyland, G., and Mircetich, S. M. (1976). The ultrastructure of the almond leaf scorch bacterium with special reference to topography of the cell wall. *Phytopathology 66,* 147-151.

Luckwill, L. C. and Crowdy, S. H. (1950). Virus diseases of fruit trees. Observations on rubbery wood, chat fruit and mosaic in apples. *Progr. Rep. Rep. Long Ashton Res. Sta. 1949,* pp. 68-79.

McBeath, J. H., Nyland, G., and Spurr, A. R. (1972). Morphology or mycoplasmalike bodies associated with peach X-disease in Prunus species. *Phytopathology 62,* 935-937.

McCarthy, C. (1903). Report on spike disease in sandalwood in Coorg. *Indian Forest. 29,* 21.

McCoy, R. E. (1972). Remission of lethal yellowing in coconut palm treated with tetracycline antibiotics. *Plant Dis. Rep. 56,* 1019-1921.

McCoy, R. E. (1979). Mycoplasmas and yellows disease. *In* "The Mycoplasmas", Vol. 3, pp. 227-262. (M. F. Barile, S. Ragino, J. G. Tully, and R. F. Whitcomb, eds.), Academic Press, New York.

McCoy, R. E. and Gwin, G. H. (1977). Response of mycoplasma-like organism-infected *Pritchardia, Trachycarpus,* and *Veitchia* palms to oxytetracycline. *Plant Dis. Rep. 61,* 154-158.

MacDaniels, L. H. and Welch, D. S. (1964). The walnut bunch disease problem. *Northern Nut Grow. Assoc. 55th Rep. (1964),* 41-48 pp.

MacDaniels, L. H., Johnson, W. T., and Braun, E. J. (1975). The black walnut bunch disease syndrome. *66th Annu. Rep. Northern Nut Grow. Assoc., Aug. 10-13, 1975,* pp. 71-85.

McIntyre, J. C., Schneider, H., Lacy, G. H., Dodds, J. A., and Walton, G. S. (1979). Pear decline in Connecticut and response of diseased trees to oxytetracycline infusion. *Phytopathology 69,* 955-958.

McKay, John W. and Crane, Harley L. (1951). Bunch disease of black walnut. *North. Nut Grow. Assoc. Ann. Rep. 41*(1950), 56-62.

McLean, D. M. (1943). An experimental and histological study of phloem necrosis, a virus disease of American elm. *Doc. Diss. Ohio State Univ.,* 110 pp.

McLean, D. M. (1944). Histo-pathologic changes in the phloem of American elm affected with the virus causing phloem necrosis. *Phytopathology 34,* 818-826.

Maramorosch, K. (1979). Aster yellows spiroplasma ATCC 29747. (Abstr. 20). *Annu. Meet. Amer. Soc. Microbiol.*

Markham, P. G., Townsend, R., Bar-Joseph, M., Daniels, M. J., Plaskitt, A., and Meddins, B. M. (1974). Spiroplasmas are the causal agents of citrus little-leaf disease. *Ann. Appl. Biol. 78,* 49-57.

Menon, R. (1960). Biochemical studies on the yellow leaf disease of arecanut palms. *Arecanut J. 121,* 16-21.

Menon, R. (1963). Transmission of yellow leaf disease. *Phytopathol. Z. 48*(1), 83-88.

Merrill, W. and Nichols, L. P. (1972). Distribution of elm phloem necrosis in Pennsylvania. *Plant Dis. Rep. 56*(6), 525.

Mircetich, S. M., Lowe, S. K., Moller, W. J., and Nyland, G. (1976). Etiology of almond leaf scorch disease and transmission of the causal agent. *Phytopathology 66,* 17-24.

Nair, R. B. and Aravindakshan, M. (1970). On the occurrence of tyloses in the yellow leaf disease of affected Areca Palms. Effect of the yellow leaf disease on the plant characters of *Areca catechu* L. *Agr. Res. J. Kerala 8*(1), 58; 61-62.

Nair, R. B. and Aravindakshan, M. (1971). On the leaf anatomy
    of yellow leaf diseased areca palm *Areca catechu* L. *Agr.*
    Res. J. Kerala 9(1), 31-32.

Nayar, R. (1968). Histopathogenic studies in *Areca catechu*
    L. infected with yellow leaf disease. *Phytopathol. Z. 61*
    (1), 34-37.

Nayar, R. (1971). Etiological agent of yellow leaf disease of
    Areca catechu. Plant Dis. Rep. 55(2), 170-171.

Nayar, R. and Seliskar, C. E. (1978). Mycoplasma like organ-
    isms associated with yellow leaf disease of *Areca catechu*
    L. (in Kerala and Karnataka). *Eur. J. Forest Pathol. 8*(2),
    125-128.

Neal, D. C. (1920). Phony peach: a disease occurring in middle
    Georgia. *Phytopathology 10*, 106-110.

Nienhaus, F., Brussel, H. and Schinzer, U. (1976). Soil-borne
    transmission of rickettsia-like organsims found in stunted
    and witches'-broom diseased larch trees *(Larix decidua).*
    Z. Pflanzenkrank. Pflanzensch. 83(6), 309-316.

Nyland, G. and Moller, W. J. (1973). Control of pear decline
    with a tetracycline. *Plant Dis. Rep. 57,* 634-637.

Nyland, G. and Raju, B. C. (1980). An epidemic of peach yellow
    leaf roll (X-disease) in northern California: Association
    with pear orchards (Abstr.) *Proc. 61st Annu. Meet. Pac.*
    Div. AAAS, p. 13.

Okuda, S. (1978). "Plant diseases due to mycoplasma-like
    organisms in Japan." Compiled by Food and Fertilizer Tech-
    nology Center for the Asian and Pacific Region. *FFTC*
    Book Ser. No. 13, pp. 24-28.

Okuda, S., Doi, Y., and Yora, K. (1974). Mycoplasma-like
    bodies associated with chestnut yellows. *Ann. Phytopath-*
    ol. Soc. Jap. 40(5), 464-468.

Pandey, P. K., Singh, A. B., Nimbalkar, M. R., and Marathe,
    T. S. (1976). A witches'-broom disease of jujube from
    India. *Plant Dis. Rep. 60,* 301-303.

Pearson, R. C. (1977). Control of X-disease of peach. *Proc.*
    Ann. Meet. N. Y. State Hort. Soc. (1977) *122,* 176-183.
    (New York State Agri. Exp. Sta., Highland, N.Y.) (From
    Hort. Abstr. 48, 2121.).

Pena-Iglesias, A. (1975). Apple proliferation disease in Spain:
    graft transmission and detection of mycoplasma and ricket-
    tsia-like organisms in infected tissue. *Proc. 9th Int.*
    Symp. Fruit Tree Virus Dis. Acta Hort. 44, 193-197.

Plakidas, A. G. (1949). Witches'-broom, a graft-transmissible
    disease of Arizona ash *(Fraxinus berlandieriana).* (Abstr.)
    Phytopathology 39, 498-499.

Plavsic-Banjac, B., Hunt, P. and Maramorosch, K. (1972). Mycoplasmalike bodies associated with lethal yellowing disease of coconut palms. *Phytopathology 62*, 298-299.

Posnette, A. F. (1963). Virus diseases of apples and pears. *Commonw. Agr. Bur. Farnham Royal Bucks, Engl. Tech. Commun. No. 30*, 141 pp.

Purcell, A. H., Raju, B. C. and Nyland, G. (1980). Transmission by injected leafhoppers of spiroplasma isolated from plants with X-disease. *Proc. 61st Annu. Meet. Pac. Div. AAAS*, p. 13.

Raju, B. C., and Chen, T. A. (1974). Mycoplasmalike organisms associated with *Cornus stolonifera* showing witches'-broom symptoms. (Abstr.). *Proc. Amer. Phytopathol. Soc. 1*, 142.

Raju, B. C., Chen, T. A., and Varney, E. H. (1976). Mycoplasma-like organisms associated with a witches'-broom disease of Cornus amomum. *Plant Dis. Rep. 60*, 462-464.

Raju, B. C., Goheen, A. C., Nyland, G., Nome, S. F., Docampo, D., Wells, J. M., Weaver, D., and Lee, R. F. (1980). Serological relationship of rickettsialike bacterium isolated from or associated with plant diseases. *Can. J. Microbiol. 26*, in press.

Rawlins, T. E. and Horne, W. T. (1931). "Buckskin" a destructive graft infectious disease of the cherry. *Phytopathology 21*, 331-335.

Raychaudhuri, S. P. and Varma, A. (1980). Sandal spike. *Rev. Plant Pathol. 59(3)*, 99-104.

Rose, D. K. and Rose, . (1957). Bunch disease of black walnuts. *48th Annu. Rep. Northern Nut Grow. Assoc. 48*, 23-27.

Ross, H. (1933). Uber nicht parasitare Hexenbesen an Robinia pseudacacia. *Ber. Deut. Bot. Gesellsch. 7*, 292-300.

Saglio, P., Lhospital, M., Lafleche, D., Dupont, G., Bove, J. M., Tully, J. G., and Freundt, E. A. (1973). *Spiroplasma citri* gen. and sp. n.: a mycoplasma-like organism associated with "stubborn" disease of citrus. *Int. J. Syst. Bacteriol. 23*, 191-204.

Sarie, A. (1969). Mikoplazme kao uzroenici biljnih bolesti. *Zast Bilja 20*, 245-248.

Schall, R. A. and Agrios, G. N. (1973). Graft transmission of ash witches'-broom to ash. *Phytopathology 63(2)*, 206-207.

Schneider, H. (1970). Graft transmission and host range of the pear decline causal agent. *Phytopathology 60(2)*, 204-207.

Schneider, H. (1973). Cytological and histological aberrations in woody plants following infection with viruses, mycoplasmas, rickettsias, and flagellates. *Annu. Rev. Phytopathol. 11*, 119-146.

Schneider, H. (1977). Indicator hosts for pear decline: Symptomatology, histopathology, and distribution of mycoplasma-like organisms in leaf veins. *Phytopathology 67*, 592-601.

Schreiber, L. R. and Peacock, J. W. (1974). Dutch elm disease and its control. *U. S. Dep. Agr. Forest Serv., Agr. Inf. Bull. 193*, 15 pp.

Seemuller, E. (1976). Demonstration of mycoplasmalike organism in the phloem of trees with pear decline or proliferation symptoms by fluorescence microscopy. *Phytopathol. Z. 85*, 368-372.

Seliskar, C. E. (1973). Association of a mycoplasmalike organism with walnut bunch disease. (Abstr. #0933). *Proc. 2nd Int. Congr. Plant Pathol. Abstr. Papers, Minneapolis, Minn., Sept. 5-12.*

Seliskar, C. E. (1976). Mycoplasmalike organism found in the phloem of bunch-diseased walnuts. *Forest Sci. 22*(2), 144-148.

Seliskar, C. E., Wilson, C. L., and Bourne, C. E. (1973). Mycoplasmalike bodies found in phloem of black locust affected with witches'-broom. *Phytopathology 63*, 30-34.

Seliskar, C. E., KenKnight, G. T., and Bourne, C. E. (1974). Mycoplasmalike organism associated with pecan bunch disease. *Phytopathology 64*, 1269-1272.

Sherald, J. L. and Hearon, S. S. (1978). Bacterialike organisms associated with elm leaf scorch. (Abstr.) *Phytopathol. News 12*, 73.

Sinclair, W. A. (1971). Elm phloem necrosis in New York state. Memorandum Sept. 8, 1971. 3 p.

Sinclair, W. A. (1972). Phloem necrosis of American and slippery elms in New York. *Plant Dis. Rep. 56*(2), 159-161.

Sinclair, W. A. and Filer, T. H., Jr. (1974). Diagnostic features of elm phloem necrosis. *Arborists News 39*, 145-149.

Sinclair, W. A., Campana, R. J., Manion, P. D., Fry, W. E., Merrill, W., and Nichols, L. P. (1971). Elm phloem necrosis in Pennsylvania. *Plant Dis. Rep. 55*(12), 1085.

Sinclair, W. A., Braun, E. J., and Larsen, A. O. (1974). Phloem necrosis of elms: Newly recognized symptoms and retardation by tetracycline antibiotics. (Abstr.). *Proc. Amer. Phytopathol. Soc. 1*, 144.

Sinclair, W. A., Braun, E. J., and Larsen, A. O. (1976). Update on phloem necrosis of elms. *J. Arbor. 2*(6), 106-113.

Slack, D. A. and Kim, K. S. (1977). Peach rosette disease. *Ark. Farm Res. 26*(6), 15. (Univ. Ark. Agr. Exp. Sta., Fayetteville, Ark.).

Smith, E. F. (1888). Peach yellows: A preliminary report.
  U. S. Dep. Agr. Bot. Div. Bull., 254 pp.
Smith, E. F. (1891). Additional evidence on the communicability
  of peach yellows and peach rosette. *U. S. Dep. Agr. Div.
  Veg. Pathol. Bull. 1*, 65 p.
Smith, K. M. (1972). "A Textbook of Plant Virus Diseases,"
  3rd ed., 684 pp. Academic Press, New York.
Stoddard, E. M. (1934). Progress report of investigations on
  a new peach trouble. *Conn. Pomol. Soc. Proc. 43*, 115-117.
Story, G. E. and Halliwell, R. S. (1969). Association of my-
  coplasma-like organism with the bunchy top disease of
  papaya. *Phytopathology 59*, 1336-1337.
Swingle, R. U. (1938). A phloem necrosis of elm. *Phytopathol-
  ogy 28*, 757-759.
Swingle, R. U. (1942). Phloem necrosis, a virus disease of the
  American elm. *U. S. Dep. Agr. Circ. 640*. (Bur. Plant
  Ind. U. S. Dep. Agr.).
Swingle, R. U. (1963). *In* Internationally Dangerous Forest tree
  Diseases. *U. S. Dep. Agr. Forest Serv. Misc. Publ. No.
  939*. pp. 83-84.
Swingle, R. U., Whitten, R. R., and Young, H. C. (1949). The
  identification and control of elm phloem necrosis and
  dutch elm disease. *Ohio Agr. Exp. Sta. Spec. Circ. 80.*,
  11 pp.
Tehon, L. R. (1945). American elms die by tens of thousands
  from phloem necrosis disease epidemic. *Greenk. Rep. 13*(2),
  17-19.
Tsai, J. A. (1979). Vector transmission of mycoplasmal agents
  of plant diseases. *In* "The Mycoplasmas" (R. R. Whitcomb
  and J. G. Tully, eds.), Vol. III, Chapter 9. Academic
  Press, New York.
Turner, W. F. and Pollard, H. N. (1959). Insect transmission
  of phony peach disease. *U. S. Dep. Agr. Tech. Bull. 1193*,
  27 pp.
Waite, M. B. (1933). Notes on some nut diseases with special
  reference to the black walnut. *Northern Nut Grow. Assoc.
  Annu. Rep. 23*, 60-67.
Wallace, T., Swarbrick, T., and Ogilvie, L. (1944). Some new
  troubles in apples, with special reference to variety
  Lord Lambourne. *Fruitgrower 98*, 427.
Weber, P. V. V., Sinclair, W. A., Peterson, J. L., and Davis,
  S. H., Jr. (1974). New in New Jersey: Elm phloem necrosis.
  Plant Dis. Rep. 58, 387-388.
Wells, J. M., Weaver, D. J., and Raju, B. C. (1980). Distri-
  bution of rickettsia-like bacteria in peach, and their
  occurrences in plum, cherry, and some perennial weeds.
  Phytopathology 70, in press.

Wester, H. V. and Jylkka, E. W. (1959). Elm scorch, graft transmissible virus of American elm. *Plant Dis. Rep. 43,* 519.

Whitcomb, R. F., Jensen, D. D., and Richardson, J. (1968). The infection of leafhoppers by western X-disease virus. IV. Pathology in the alimentary tract. *Virology 34,* 69-78.

Whitten, R. R. (1949). Control of insect carriers of the Dutch elm disease and elm phloem necrosis. *Trees Mag. 9*(3), 6-25.

Whitten, R. R. (1954). Dutch elm disease and elm phloem necrosis. *Ent. Soc. Amer. North Central Br. Proc. 9,* 36.

Whitten, R. R., and Baker, W. L. (1973). Recent experimental results on control of vectors of two elm diseases. *Arborist's News 13*(6), 41-5.

Whitten, R. R. and Swingle, R. U. (1948). The status of research on two epidemic elm diseases. *Nat. Shade Tree Conf. Proc. 24,* 113-120.

Wilson, C. L. and Seliskar, C. E. (1976). Mycoplasma-associated diseases of trees. *J. Arbor. 2*(1), 6-12.

Wilson, E. E. and Ogaua, J. M. (1979). Fungal, bacterial, and certain nonparasitic diseases of fruit and nut crops in California. *Agr. Sci. Publ. Univ. Calif.,* 190 p.

Wilson, C. L., Seliskar, C. E. and Krause, C. R. (1972). Mycoplasmalike bodies associated with elm phloem necrosis. *Phytopathology 62,* 140-143.

Wolfe, H. R. and Anthony, E. W. (1953). Transmission of western X-disease virus with sweet and sour cherry by two species of leafhoppers. *J. Econ. Entomol. 46,* 1090-1092.

STUBBORN DISEASE OF CITRUS

D. J. Gumpf and E. C. Calavan

Department of Plant Pathology
University of California
Riverside, California 92521

I. HISTORICAL BACKGROUND

The disease now called stubborn was first noticed about
1915 in Washington navel orange trees in orchards of the East
Highland Orange Company, near Redlands, California (Fawcett
et al., 1944). From 1915-1917 several nonproductive trees
were observed and some were topworked by E. R. Waite in 1921
using carefully selected buds. In spite of the carefully
chosen budwood, the growth of these scion buds was slow and
exhibited growth characteristics shown by the original tree.
This type of growth was described by Waite as "stubborn."
Stubborn was first used as the disease name in 1929 by Perry
(Fawcett and Klotz, 1948a). Perry also concluded after further study that something was transmitted from the stubborn
trees into the carefully selected good buds used for topwork-
ing.
    In 1931, Reichert and Perlberger (1931) described a new
severe disease that affected the Palestine citrus industry in
the summer of 1928. The disease, "little leaf," was thought
to have been a result of drought. Subsequent work has shown
many similarities between little leaf and stubborn and at
present both names apply to the same disorder.
    Still another disorder of navel orange was observed in the
Redlands area about the same time as stubborn. This disorder,
formerly "pink nose" but now called acorn, was present more
than 50 years ago (Calavan, 1969). Acorn-shaped fruits could
be found in numerous orchards in San Bernardino and eastern
Los Angeles counties in the 1920's. Prior to 1940, acorn-
fruit condition was also associated with "crazy top" growth
of some grapefruit trees (Fawcett *et al.,* 1944). Stubborn

and the acorn-fruit disorder were considered to be different
conditions until 1940 when J. C. Johnston of the University
of California Agricultural Service detailed the similarity of
foliage and branch characteristics of trees with crazy top,
stubborn, and acorn fruit (Fawcett et al., 1944; Fawcett and
Klotz, 1948a).

The first transmission experiment with stubborn material
was initiated in 1938 by Fawcett. Fourteen trees at the Citrus Experiment Station, Riverside, California, were grown from
buds collected from three stubborn trees at Redlands. In
1942, seven of the 14 trees were topworked with buds taken
from healthy trees. The scions that grew from the seven
healthy buds as well as those from the seven infected buds
showed leaf and branch symptoms of stubborn; later, they
produced acorn fruit (Fawcett et al., 1944). Because of this
infectious nature, Fawcett stated that stubborn was probably
caused by a virus. Haas et al. (1944) describing the acorn
disease also suggested that stubborn and acorn disease are
identical. Two years later, Fawcett (1946) reiterated the
virus nature of stubborn. During the mid-1940's stubborn,
which was spreading very slowly if at all, was overshadowed
by tristeza, which became epidemic in parts of southern California. During this period, however, stubborn became an increasingly important problem in navel oranges and grapefruit
(Fawcett and Klotz, 1948b). By the mid-1950's the disease
was so common in Southern California and Arizona that research
was greatly accelerated. Most of the studies that followed
dealt with the transmission (Calavan and Christiansen, 1962),
indexing (Calavan and Christiansen, 1965), and perpetuation
(Carpenter and Allen, 1966) of the disease. These studies
demonstrated that very young leaves from affected seedlings
were good sources of the pathogen, still thought to be a virus. The additional finding that the stubborn pathogen was
present in functioning phloem of young, growing leaves, as
reported by Calavan, Olson, and Christiansen at the 1969 meeting of the International Organization of Citrus Virologists
(Calavan et al., 1972), together with the discovery of mycoplasmalike bodies in phloem tissues of plants previously
thought to have been virus diseases (Doi et al., 1967), accelerated studies on phloem of stubborn-affected citrus as a
favorable site for a mycoplasmalike pathogen. Within a year
mycoplasma-like bodies were reported in sieve tubes of citrus
affected with stubborn disease (Igwegbe and Calavan, 1970;
Igwegbe et al., 1971; Lafleche and Bove, 1970). Mycoplasma-
like organisms were also shown to be associated with little-
leaf disease of citrus in Israel (Zelcer et al., 1971), an
additional argument for its common identity with stubborn.
These reports provided the first real evidence that the causal

agent of stubborn and little leaf is not a virus but a phloem-
confined microorganism.

Attempts to isolate and cultivate the microorganism associated with stubborn followed closely its discovery in affected tissue. Using young leaves from stubborn-affected Madam
Vinous sweet orange seedlings, Saglio *et al.* (1971) and Fudl-
Allah *et al.* (1972) were able, in 1970, to isolate and culture
a mycoplasmalike organism in liquid broth as well as on solid
medium. Once cultured, the organism was characterized by Saglio *et al.* (1973), who also proposed the name *Spiroplasma cit-*
ri. Acceptance of *S. citri* as the causal agent of stubborn-
affected trees, absence from healthy trees, and the failure
to isolate any other type of pathogen disease had to await the
successful completion of Koch's postulates, i.e., experimental
inoculation of this cultured agent into healthy citrus, production of disease, and recovery of *S. citri* from the experimentally infected plants. This necessary proof was soon to
follow. Daniels *et al.* (1973) and Markham *et al.* (1974) both
working with the leafhopper *Euscelis plebejus* (Fallen) micro-
injected with cultured *S. citri,* demonstrated the transmission
and reisolation of *S. citri* to white clover and citrus, respectively. In studies with the leafhopper *Scaphytopius nitri-*
dus (De Long), which breeds on citrus, Kaloostian and Pierce
(1972) and Kaloostian *et al.* (1975) demonstrated that this
leafhopper could acquire *S. citri* by direct feeding on stubborn-infected sweet orange seedlings and transmit the organism
to disease-free periwinkle plants. *Spiroplasma citri* was subsequently cultured from plants showing disease symptoms. Using a membrane feeding technique, Rana *et al.* (1975) found
that *Circulifer tenellus* (Baker) as well as *S. nitridus* could
acquire *S. citri* from 5% sucrose suspensions, and then transmit it to sweet orange seedlings. *Spirosplasma citri* was subsequently isolated from plants with disease symptoms. Since
these studies, investigations with *S. citri* have extended far
beyond citrus because of the organism's wide host range and
relationships with at least three leafhopper species, *C.*
tenellus, S. nitridus, and *S. acutus delongi* (Kaloostian *et*
al., 1979b).

II. GEOGRAPHICAL DISTRIBUTION AND ECONOMIC IMPORTANCE

The occurrence of stubborn disease of citrus in California
was once spotty and scattered except in a few orchards where
a very high percentage of uniformly affected trees would indicate the use of diseased budwood. In recent years the disease has become a major problem in all the important warm
inland valley orange- and grapefruit-producing areas of Cali-

fornia and Arizona. An increasing number of diseased trees
have been detected in nursery and young orchard plantings. The
total damage caused by stubborn disease in California has never been fully assessed. Calavan (1969) did, however, estimate that about one million trees were affected at that time.
Removal of many stubborn affected trees and orchards, particularly in southern California may have reduced the total number of stubborn diseased citrus trees in this state. Symptoms
of experimentally (Calavan, 1969) as well as naturally (Fawcett and Klotz, 1948b) stubborn-affected trees vary from slight
to strong, the production of fruit decreasing with increased
symptom severity. These variations in symptom severity have
been interpreted by Calavan (1969) to result from different
strains of the causal agent.

The first report of stubborn (little leaf disease of citrus) outside California and Arizona was from Palestine (Reichert, 1930) where it caused heavy damage when it appeared in
epidemic form in young citrus groves (Reichert and Perlberger,
1931). Numerous studies have established that stubborn disease is a serious problem widely distributed in many other
Mediterranean and Mideast countries including Algeria, Corsica, Cyprus, Egypt, Iran, Iraq, Jordan, Lebanon, Morocco, Sardinia, Sicily, Syria, and Tunisia (Anonymous, 1970; Bruno,
1964; Cassin, 1965; Chapot, 1959; Chapot and Cassin, 1961;
Childs and Carpenter, 1960; Nour-Eldin, 1967; Scaramuzzi and
Catara, 1968; Vogel and Bove, 1974). In Peru and northwestern
Mexico, trees with stubborn-like symptoms have been seen but
not confirmed by culturing (Calavan, unpublished observations).
In all countries affected, the percentage of diseased trees
varies greatly among individual orchards with many diseased
trees difficult to detect (Pappo and Bauman, 1969). Visual
surveys indicate from less than 1% to over 50% of trees in
affected orchards appear stubborn diseased in California and
Morocco (Calavan and Carpenter, 1965; Anonymous, 1970).

III. SYMPTOMATOLOGY

Symptoms of citrus stubborn disease have been described
by several authors, including Fawcett *et al.* (1944), Chapot
and Delucchi (1964), Calavan and Carpenter (1965), Calavan
(1968, 1979), and Calavan and Oldfield (1979). Fawcett *et al.*
(1944) noted that stubborn causes some degree of stunting in
navel orange trees and an abnormal kind of growth characterized
by shortened internodes between leaves, compact, brushlike appearance, and multiple. buds. These and other symptoms are detailed below.

Fig. 1. Normal (left) and stubborn-diseased (right) trees of Valencia orange. The compact growth and bushiness are typical symptoms but the stunting is unusually severe for this variety.

A.   TREES

Citrus trees affected by stubborn disease usually are slightly to severely stunted and have excessive numbers of shoots, shortened internodes, and multiple axillary buds. Shoots tend to grow conspicuously upright and the strongest flush of growth commonly occurs in fall rather than in spring. An increase in positive heliotropism has been noted in many stubborn-diseased trees (Chapot and Delucchi, 1964; Schneider, 1966; Calavan, 1968). Diseased trees often become denser than normal, so that their overall appearance, especially in Valencia orange, is one of compact bushiness (Fig. 1).

Twig dieback occurs in some severely affected trees. Such trees are especially sensitive to heat and frost damage.

Most diseased trees in California and Israel apparently were infected within 5 years after propagating (Pappo and Bauman, 1969; Calavan et al., 1979). Primary infections in older trees have rarely been noted except in very hot areas. Partially affected trees may retain normal branches for 10 or

more years or the disease may involve the entire tree within
one to several years. The gross aspect of stubborn-diseased
trees is variable and subject to cyclic and seasonal changes.

B.   LEAVES

The average size of leaves on most stubborn-diseased trees
is smaller than normal (Calavan and Christiansen, 1966), which
condition led Reichert 1930) to call the disease little leaf.
The shortened intervals between leaves cause much of the foliage to appear bunchy, some almost rosetted. Some leaves are
abnormally cupped and many show a variety of chlorotic or mottled leaf patterns resembling various minor element deficiencies, particularly those of zinc, iron, and manganese, soon
after becoming fully expanded (Calavan and Carpenter, 1965;
Olson, 1969; Calavan and Oldfield, 1979). Such patterns occur intermittently and somewhat irregularly in most affected
trees. Leaves fall prematurely, giving some severely affected trees an open appearance which contrasts with the compact
bushiness of most stubborn trees. Thick leaves with prominent
veins are present in many affected trees (Hilgeman, 1961).
Schneider (1966) found necrotic sieve tubes and excessive
phloem formation inconsistently present in leaf veins of stubborn-diseased trees in Morocco.

Leaf symptoms associated with stubborn disease are rarely
diagnostic except, at about 30°C, in the greenhouse the distal
portions of nearly expanded leaf blades of stubborn sweet
orange show a characteristic mottling markedly different from
that usually present in orchard trees (Fig. 2; Calavan, 1969;
Calavan and Oldfield, 1979). Under hot, dry conditions many
leaves on stubborn-diseased trees have blunted or heart-shaped
tips (Calavan, 1976), but excessive heat causes similar leaf
distortions in healthy trees (Cochran and Samadi, 1976; Reuther *et al.*, 1979).

C.   FRUIT

Fruits on stubborn-diseased citrus trees may be severely
deformed to normal and are mostly smaller than fruits of comparable healthy trees.

Stubborn disease reduces yields in most citrus trees. The
average yield of diseased Valencia orange trees at Riverside,
California was 44-74% less, depending on rootstock, than from
healthy trees (Calavan, 1969). Yields of navel orange were
reduced up to 100%, depending on climate and disease severity
(Calavan and Oldfield, 1979).

1.   External Symptoms

Abnormalities attributable to stubborn disease include
lopsidedness, a curved columella, "acorn" shape (Fig. 3) with

Fig. 2. Leaf symptoms induced by stubborn disease in Madam Vinous sweet orange. Distal and marginal mottling and clearing are characteristic of stubborn but rarely occur in the field. Leaves at center and right simulate deficiency symptoms and may occur seasonally on affected field trees.

abnormally small oil glands and thin peel on the stylar end but near normal peel on the stem end, a bluish cast of a portion of the rind especially in very hot areas, irregular or inverse yellow or orange coloration with retention of green at the stylar end (Fig. 4), premature fruit drop or, sometimes, mummification of fruit on the tree. Affected grapefruit fruits may have a two-toned color pattern, with brighter than normal yellow on the stylar end. Frequently the bright yellow area extends beyond the fruit's equator and sometimes reaches the button on one side. Diseased sweet orange trees usually bear paler fruits than normal trees.

2. *Internal Symptoms*

Most affected fruits have an insipid, sour, or abnormally bitter flavor; affected grapefruit is of especially poor qual-

Fig. 3. Acorn-shaped and normal fruits of Marrs orange
(courtesy of Dr. J. B. Carpenter).

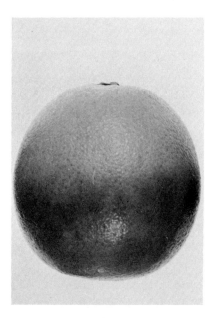

Fig. 4. Inverse coloration or stylar-end greening of
stubborn-diseased navel orange fruit (photograph reproduced
from Wallace, 1979 with permission of University of California Division of Agricultural Sciences).

Fig. 5. Seed abortion in Valencia orange. Normal seeds from healthy fruit (lower row); aborted seeds from stubborn-affected fruit showing dark chalazal area through abnormally thin seed coats (upper row).

ity. Seedy sweet orange varieties, such as Valencia, show excessive seed abortion in stubborn-affected fruits (Carpenter *et al.*, 1965), (Fig. 5); affected seeds are smaller, darker, and the seed coats thinner. The albedo of affected fruits may develop a gray or blue tint, especially in very hot areas, and the vascular system in the albedo is often conspicuously darkened. The tissues beneath the button turn prematurely yellow or orange. Affected oranges and grapefruits frequently develop a cheesy peel.

D. *VARIETAL SUSCEPTIBILITY*

Stubborn disease occurs most frequently in grapefruit, sweet orange, tangelo, mandarin orange, pummelo, and shaddock. Carpenter (1959) found symptoms on 26 grapefruit, 10 sweet orange, 4 tangelo, and several varieties belong to other species. Calavan and Christiansen (1961) transmitted the stubborn pathogen by tissue grafts to all important commercial varieties of grapefruit, sweet orange, tangelo, mandarin, lemon, and lime grown in California. They found that Troyer citrange and trifoliate orange were not severely damaged by stubborn but none of the citrus varieties showed much tolerance to infection. In areas where natural spread has occurred,

stubborn disease is fairly common in grapefruit, sweet orange
and tangelo. Natural infection of mandarin trees occasionally
occurs. Naturally infected lemon or lime trees have not been
found by us. The reason for the apparent absence of natural
infection in these species has not been investigated.

Citrus varieties affected in Morocco (Chapot, 1959; Chapot
et al., 1962) belong mostly to the same species as those
severely damaged in California. Navel oranges are apparently
the most sensitive of the commercially grown sweet orange
varieties. Satsuma orange, a mandarin type, is extremely sensitive, but rarely has stubborn disease in California.

E. VARIABILITY

Symptoms of stubborn disease vary with host variety, nutritional health, age, season, wind, humidity, temperature,
and isolate of S. citri involved.

1. Temperature

Spiroplasma citri is a warm temperature organism which,
in culture, grows well at $28^{\circ}-33^{\circ}C$ (Saglio et al., 1971;
Fudl-Allah et al., 1972). Citrus also grows well and reacts
strongly to infection by S. citri at these temperatures. Bove
et al. (1974) found that controlled temperatures of $27^{\circ}-32^{\circ}C$
induced early severe symptoms in inoculated plants, whereas
$22^{\circ}-24^{\circ}C$ delayed disease response and permitted only the development of very mild nonspecific symptoms during 43 weeks
incubation. Severely affected plants moved from warm to cool
temperatures soon improved but slightly affected plants changed
from cool to warm temperatures developed severe leaf symptoms.
Olson and Rogers (1969) found that high temperatures, $35^{\circ}C$
day/$27^{\circ}C$ night permitted the mottled-leaf symptom to develop
within 2 months after inoculation; at $27^{\circ}C$ day/$23^{\circ}C$ night,
inoculated plants were symptomless for 5 months. Our observations in California indicate that symptoms of stubborn disease
are most severe in the hot southern desert, slightly less severe in the San Joaquin Valley and least severe in the cool
coastal plain. The cool climate of Corsica apparently sup-
presses symptoms in many stubborn-diseased trees (Vogel and
Bove, 1974).

Although stubborn disease causes several abnormalities of
citrus leaves and fruits, heat alone can induce some of these
and may favor their development in stubborn-diseased trees.
Cochran and Samadi (1976) noted atypical growth and ovoid,
blunt, rounded and heart-shaped leaves on healthy young citrus
trees in southern Iran where maximum temperatures of $44^{\circ}-48^{\circ}C$
are common in July and August. Reuther et al. (1979) subjected

healthy young citrus to controlled temperatures of 38°C 14
hr/day for 10 weeks. The heat-treated trees produced small
upright, cupped leaves, many of them blunted and a few cordate
in shape. Blunt, notched or cordate leaves and rosetting can
also result from boron deficiency (Smith and Reuther, 1949).
We conclude that blunt or cordate leaves have little or no
value for diagnosing stubborn disease.

2. Pathogen isoaltes

Variations in symptom severity among stubborn-diseased
citrus trees have been noted by Calavan and Christiansen (1961),
Calavan (1969), and Carpenter and Calavan (1969). They were
considered due, at least in part, to differences among various
strains of the pathogen (Calavan, 1969). Graft transmissions
to young plants resulted in symptoms like those on the source
plants but the severity of symptoms among naturally infected
trees in the orchard varied greatly, even when conditions were
standardized (Calavan, unpublished results). An isolate of
S. citri transmitted to a small seedling sweet orange by mem-
brane-fed Circulifer tenellus by Rana et al. (1975) produced
only mild symptoms and slight stunting whereas several other
isolates, from different citrus sources, recently transmitted
by leafhopper vectors have caused moderate symptoms with conspicuous stunting and small mottled leaves in similar plants
(G. N. Oldfield, personal communication). Isolate C-189, from
a severely diseased tree, has not been transmitted experimentally by vectors. The evidence for variability due to differences in isolates appears strong but more comprehensive and
definitive experiments are needed to improve our understanding
of such phenomena.

We have found field diagnosis of stubborn disease often
difficult and sometimes inaccurate, even when done by those
most familiar with the disease. Individual symptoms can be
confused with effects from other agents, so that only when
most symptoms of the syndrome occur can these symptoms be assigned to stubborn disease with a high degree of reliability.
Culturing the pathogen or serological verification should be
used for confirmation.

IV. THE PATHOGEN, SPIROPLASMA CITRI

The first indication that stubborn disease might be caused
by a mycoplasma-like organism rather than a virus came in 1968

when Igwegbe (1970) discovered that tetracycline suppressed
the disease symptoms in sweet orange seedlings. Subsequent
electron microscopy of ultrathin sections of tissue from stubborn-affected and healthy citrus by Igwegbe and Calavan (1970)
and Lafleche and Bove (1970) revealed mycoplasma-like structures only in the sieve tubes of stubborn-infected tissue.
In 1970, groups in France (Saglio et al., 1971) and in California (Fudl-Allah et al., 1972, cultured in liquid and on
agar media a very pleomorphic prokaryote from young leafy
shoots of stubborn infected sweet orange. Saglio et al. (1973)
fully characterized the organism and proposed for it the name
Spiroplasma citri. It is characteristically a helically filamentous (Fig. 6), motile, cell wall-less prokaryote which un-
dergoes morphological changes during culture in vitro, a result of culture age (Cole et al., 1973).

A.   TAXONOMIC POSITION

    The causal agent of stubborn disease, Spiroplasma citri,
is the only recognized representative of the family Spiroplasmataceae, placing it in the order Mycoplasmatales and class
Mollicutes (Freundt and Edward, 1979). The isolation and
characterization of numerous other spiroplasmas has indicated
the need to determine and develop criteria useful for species
and strain differentiation. Serological techniques, polyacrylamide gel electrophoresis, and DNA homology are approaches
which show great promise in the area of differentiation and
are under further study.
    The serological procedures employed most often to study
relatedness among the spiroplasmas are growth inhibition, deformation, metabolic inhibition, and the enzyme-linked immunosorbent assay (ELISA). Serological grouping of spiroplasma
strains by several laboratories using a combination of the
above procedures have been in very close agreement (Table I).
All of these laboratories recognize a serological group I
which includes four subgroups: (1) S. citri strains; (2) honeybee spiroplasmas; (3) corn stunt strains; and (4) the tick
spiroplasma (Davis and Lee, 1980; Williamson and Tully, 1980).
There is no cross reaction between these group I spiroplasmas
and any of the other major spiroplasma serological groups.
    Additional data from DNA G+C content, polyacrylamide gel
electrophoresis (PAGE) of cellular proteins, and especially
DNA homology all confirm the grouping of S. citri strains
based on serological relationships (Lee and Davis, 1980; Bove
et al., 1980; Rahimian and Gumpf, 1980; Davis and Lee, 1980).

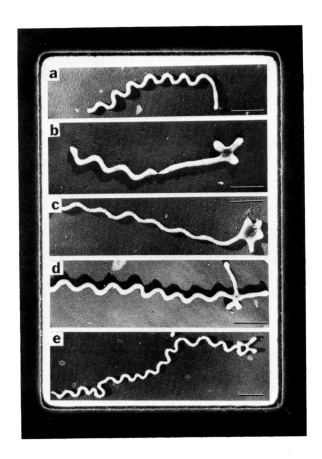

Fig. 6. Electron micrographs of shadow-cast Spiroplasma citri preparations, (a) a long unconstricted helical filament; (b) a helical portion connected to a straight filament with a branched structure; (c) a thin helical filament with a flattened branched end; (d, e) Long unconstricted helical filaments with branching at one end. Bars = 1 μm.

TABLE I.   Serological Grouping of Some *Spiroplasma citri*
Strains and Their Relationship to Some Other Spiroplasmas

Serological group	Subgroup	Spiroplasma strains
I	A	*Spiroplasma citri* - Maroc
	B	*S. citri* - C189
	C	Honeybee spiroplasma - BC_3 Honeybee spiroplasma - AS 576
	D	Tick spiroplasma - 277F
II		*Drosophila* sex ratio spiro- plasma
III		BNRI 23-6
IV		PPSI SR3
V		SMCA TP-2

B.   DIAGNOSIS AND ISOLATION

The symptoms of stubborn disease in affected field trees
are variable and often so obscure that positive identification
may require confirmation by experimental means.

1.   Tissue Grafting

Graft transmission, although influenced by the season of
grafting and the kind of tissue used as inoculum, has been
used as a fairly reliable method for field diagnosis (Calavan
and Christiansen, 1965; Calavan and Christiansen, 1966; Calavan *et al.*, 1968). Tests by Calavan and Christiansen (1966)
indicated that there are a large number of highly sensitive
citrus varieties but Marsh grapefruit, Sexton tangelo, and
Madam Vinous sweet orange were especially useful as indicators
of stubborn disease. The variability of responses obtained
from bud inoculations to the same clone of indicator plants
argues for the presence of strains which may also complicate
the use of indicator hosts as a definitive means of identification.

2. *Thin Sectioning*

Results of the electron microscope investigation published
by Doi *et al.* (1967) showing the presence of mycoplasma-like
bodies in the sieve tubes of plants with several yellows diseases stimulated the search for similar bodies as a diagnostic feature in numerous thin sections of plants infected with
diseases of questionable etiology, including stubborn disease
of citrus. There have subsequently been reports describing
the presence of typical mycoplasma-like bodies in sieve tubes
of stubborn disease-infected (Fig. 7) but not in healthy citrus (Lafleche and Bove, 1970; Igwegbe and Calavan, 1970).

3. *Antibiotic Susceptibility*

Igwegbe and Calavan (1970, 1973) also reported the recovery
of stubborn disease-infected citrus seedlings after holding
them in hydroponic solutions containing tetracycline but not
penicillin G. This type of evidence certainly can add to the
incrimination of an MLO as the disease agent of stubborn.

4. *Culturing*

The real breakthrough for disease diagnosis came in 1970
with the advent of spiroplasma cultivation in artificial media
(Saglio *et al.*, 1971; Fudl-Allah *et al.*, 1972, Calavan, 1979).
Although at first these cultured agents could be described
only as associated with citrus stubborn, fulfillment of Koch's
postulates soon followed (Daniels *et al.*, 1974; Rana *et al.*,
1975). The stubborn disease organism, *Spiroplasma citri*, has
been characterized in detail and its biological and physical
properties are well studied. Therefore, cultivation and subsequent serological identification provide more rapid and more
sensitive specific diagnostic tests than were available during
the early studies. In general the media that have been used
to culture *S. citri* are simple modifications of mycoplasma
media. The primary modification is the addition of sorbitol
to maintain osmolarity. It is also clear that serum is necessary for growth but can be replaced by PPLO serum fraction
or a combination of cholesterol, palmitic acid, and Tween 80
(Saglio *et al.*, 1977). Table II shows several media that have
been used for the cultivation of *S. citri*. It is apparent from
this table that individual isolates may adapt to medium variation. The range of this adaptability, however, has not been
fully studied (Chen and Davis, 1979). Just as no one medium

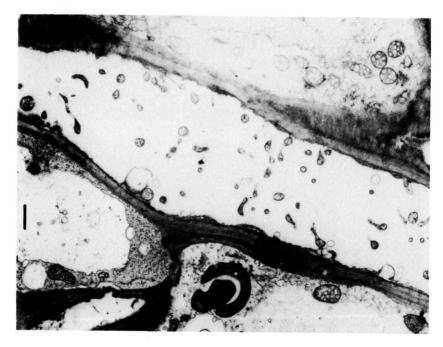

*Fig. 7. Mycoplasmalike bodies in a longitudinal section
through the sieve tube of a stubborn-infected orange leaf.
Bar = 1 μm.*

is apparently better than the other, there is no standardized
procedure for selection and preparation of tissue samples.
    Young leaves and shoots from greenhouse-grown sweet orange
seedlings showing leaf symptoms of stubborn were first used
for culture by Fudl-Allah et al. (1972) and Saglio et al.
(1971). Inoculum prepared from these tissues is usually sur-
face-sterilized in 0.5-1.0% sodium hyperchlorite, rinsed in
sterile water, and finely chopped with a razor blade in the
presence of culture media on the surface of a sterile plastic
petri dish. The liquid in which the tissue is triturated is
drawn through a 0.45 μm membrane filter. Aliquots of this
filtered liquid can then be added to tubes of liquid media or
seeded on agar media before incubating at 32ºC. If spiroplasma growth occurs, usually within 10-14 days, the liquid
cultures become very slightly turbid and the medium changes
from red to amber or yellow indicating a change of pH which
affects the phenol red in the medium. On agar, the organism
forms small, classical "fried-egg" colonies. These colonies
develop slowly reaching a maximum diameter of 0.1-0.2 mm in

TABLE II. Media Formulations Used for the *In Vitro* Cultivation of *Spiroplasma citri*

Ingredients	Medium designation				
	SMC[d]	---[e]	C-3[f]	M-1[g]	LD8[h]
PPLO broth base (gm/liter)	34	21	15	6.9	12
Tryptone (gm/liter)	10			13.3	
Peptone (gm/liter)				0.3	
Glucose (gm/liter)	1	1		0.3	4
Fructose (gm/liter)	1	1		0.3	
Sucrose (gm/liter)	10	1	160	3.3	60
Sorbitol (gm/liter)	70	50		23.3	
Inorganic salts[a]	No	No	No	No	Yes
Amino acids[b]	No	No	No	No	Yes
Organic acids[c]	No	No	No	No	Yes
HEPES buffer (gm/liter)					15
TC 199 [IX] (ml)			10		
CMRL 1066 [10X] (ml)			5		
Schneider's Drosophila medium [IX] (ml)			5	533	
Fresh yeast extract, 25% (ml)	100	100	100	33.3	
Yeastolate 2% (ml)					2
Fetal bovine serum (ml)				200	100
Horse serum (ml)	200	200	200		
Water (ml)	574	600	550	200	

[a] KCl, 0.4 g/l; KH_2PO_4, 0.3 g/l; $MgSO_4.7H_2O$, 0.2 g/l; $NaCl$, 1.4 g/l; Na_2HPO_4, 0.2 g/l; Na_2SO_3, 0.5 g/l.

[b] Arginine, 0.6 g/l; asparagine, 0.6 g/l; cystine, 0.4 g/l; glutamine, 0.6 g/l methionine, 0.4 g/l.

[c] In gm/liter: α-ketoglutarate, 0.4 g/l; pyruvate, 0.4 g/l.

[d] Saglio et al., 1971, 1973.   [e] Fudl-Allah et al., 1972.

[f] Chen and Liao 1975.   [g] Williamson and Whitcomb, 1975.

[h] Lee and Davis, 1978.

about 2 weeks (Fudl-Allah *et al.*, 1972). Aborted seeds of diseased fruits seem to be good sources for culturing *S. citri*.
The seeds can be aseptically removed from fruits and placed
in culture medium and incubated as above. Whole seeds, chopped seeds, and seed coats can all be used as sources for the
culture of *S. citri* (Fudl-Allah *et al.*, 1972).

Most consistent culture of *S. citri* from natural or experimentally infected citrus results from the trituration of
small pieces of albedo removed from the stem end of fruit,
followed by filtration and inoculation as described for leaves
and shoots. Optimum temperature for growth for all cultures
is 32°C (Saglio *et al.*, 1973). In addition, anaerobic conditions (5% carbon dioxide in nitrogen) favors the growth of
S. citri on solid media but has little effect on liquid cultures (Bove and Saillard, 1979).

The helical morphology and motility of cultured *S. citri*
may be examined and verified by light microscopy using darkfield optics. This can be a very useful means of evaluating
cultures to determine organism numbers and condition of culture. Cultures under optimum conditions have large numbers of
helical organisms with only 2-3 turns and very few long nonhelical filaments or clumps of filaments (Lee, 1977). Negatively stained spiroplasmas may also be examined by electron
microscopy. Since spiroplasma helicity is very dependent on
osmolarity of the surrounding medium, care must be taken in
the choice of stain and may best be preserved by fixation
with glutaraldehyde prior to staining (Bove and Saillard,
1979).

C. PATHOGENESIS

The symptoms of *S. citri* vary widely in different plant
species, however, in citrus they are expressed only at high
temperatures and consist of chlorosis, stunting, bushy growth
caused by shortened internodes and axillary proliferation,
and small misshapen fruit. Infected citrus trees may vary in
symptom severity but they rarely will die no matter how severe
the symptoms. The most striking symptom of *S. citri* occurs
in herbaceous plants like pea, bean, and periwinkle where it
causes a sudden wilting and death of the plant. Examination
of thin sections in the electron microscope revealed that
symptom severity in citrus is correlated with increased numbers of spiroplasmas in the sieve tubes (Daniels, 1979).
Likewise, in herbaceous plants expressing wilting symptoms
the sieve tubes contain large numbers of organisms. Daniels
(1979) has given as one explanation for the correlation between increased numbers and increased symptoms, that *S. citri*

growing in the phloem secretes a translocatable toxin. The
concentration of this toxin is thus dependent upon the number
of spiroplasmas present. Pursuing this idea, Daniels (1979)
has isolated and partially purified a rather unstable toxin
from culture fluids. Little information is available concerning the chemistry of this toxin which has been found to
be acidic in nature with a molecular weight of less than 400.
At present, insufficient information is available to make definitive statements regarding the role of the toxin in disease
development.

V. EPIDEMIOLOGY

A. *NATURAL SPREAD*

1. *Vectors*

    Natural spread of the stubborn agent was suspected for
years by research workers and growers. Trees surrounded by
apparently healthy trees as well as experimental seedling
trees inexplicably acquired stubborn disease. In addition,
diseased plants were found in nurseries under circumstances
that strongly suggested natural spread. As evidence accumulated that the stubborn agent was being spread naturally in
the field, an effort was made to discover a possible vector.
Early research directed toward finding a vector depended upon
feeding a variety of insect species on diseased plants and
then transferring them to indicator seedlings. A fortuitous
collection of *Scaphytopius nitridus* (DeLong) adults and nymphs
from citrus trees (Kaloostian and Pierce, 1972) made this
leafhopper a prime suspect as a vector of stubborn disease in
southern California. During the 1973 epidemic of stubborn
disease (Cartia and Calavan, 1974), Lee *et al.* (1973) cultured
S. citri from beet leafhoppers, *Circulifer tenellus* (Baker)
collected in a stubborn-disease field plot at the University
of California Moreno Farm, east of Riverside. Subsequently,
S. citri was also isolated from *S. nitridus, S. acutus delongi*
Young, and several other species of insects (Kaloostian *et al.*,
1979b; Oldfield and Kaloostian, 1979). This work demonstrated that at least the three named species of leafhoppers could
acquire *S. citri* naturally. Experimental leafhopper transmission studies which are discussed later in this section
(part C), also determined that in California these three leafhopper species are known to transmit *S. citri* (Calavan *et al.*,
1979; Kaloostian, *et al.*, 1979b; Oldfield and Kaloostian, 1979).

Vector studies done in Morocco using the enzyme-linked immunoabsorbent assay (ELISA) determined that 7 of 41 species of insects collected in stubborn-affected sweet orange orchards contained *S. citri* (Bove *et al.*, 1979b; Nhami *et al.*, 1980). Those insects determined by ELISA to harbor *S. citri* included one planthopper species, *Laodelphax striatellus* (Fallen), and six leafhopper species, *Neoaliturus (Circulifer) haematoceps* (Mulsant and Rey), *Exitianus capicola* (Stahl), *Euscelis alsius* (Ribaut), *Psammotettix striatus* (Linnaeus), *Toya propinqua* (Fieber), and *Recilia angustisectus* (Linnavouri). Some, or all, of these species may also prove to be vectors of *S. citri*.

Other laboratories interested in vector studies demonstrated that cultured *S. citri* injected into several species of leafhoppers could increase in the insects body (Whitcomb *et al.*, 1973; Daniels *et al.*, 1973). In addition, injected *Euscelis plebejus* transmitted *S. citri* to plants with about 4% efficiency and *Euscelidius variegatus* with up to 25% efficiency (Markham and Townsend, 1977). Bar-Joseph and Garnsey (1981) detected *S. citri* in *E. variegatus* 6 and 10 days after injection.

2. *Environmental Effects*

Temperature effects are the only environmental conditions that have been studied. There is very little natural spread of stubborn disease in citrus or periwinkle on the cool coastal plain of southern California (Calavan *et al.*, 1979; E. C. Calavan and R. L. Blue, unpublished data). We believe this to be due primarily to two factors: (1) the poor growth made by *S. citri* below 28°C and (2) the low populations of stubborn-disease vectors normally present in this cool area. High populations of *C. tenellus* are commonly found in warm or hot arid or semiarid regions.

3. *Variety and Age of Citrus Hosts*

Virtually all citrus varieties appear to be susceptible to *S. citri* when graft inoculated (Calavan and Christiansen, 1966) but in California natural infection occurs primarily in sweet orange, grapefruit, and tangelo. Navel orange and grapefruit are severely damaged and Valencia orange trees may suffer crop losses exceeding 50% (Calavan, 1969). We have observed some differences in susceptibility to stubborn based on tree age. Young trees seem much more susceptible. It is also easier to observe symptoms in younger trees. The stub-

born pathogen is unevenly distributed in the various parts of
infected citrus plants (Calavan and Gumpf, 1974) which compli-
cates observation of symptoms especially in large trees. Thus
some branches of trees may be infected with *S. citri* while
others remain free of the organism, making diagnostic and
transmission studies difficult.

4. Relationships with Non-citrus Hosts

Spiroplasma citri has been found in or transmitted to
plants in 18 dicotyledonous and one monocotyledonous family.
All of these hosts are herbaceous except those in the Rutac-
ceae and Rosaceae (Calavan and Oldfield, 1979; Oldfield and
Kaloostian, 1979; Oldfield and Calavan, 1981). Species in
which *S. citri* has been found to occur naturally are periwinkle, *Catharanthus roseus;* London rocket, *Sisymbrium irio;*
turnip, *Brassica rapa;* Chinese cabbage, *B. pekinensis;* pak-
choi, *B. chinensis;* short pod mustard, *B. geniculata;* black
mustard, *B. nigra;* a mustard of North African origin, *B.
tournefortii* and wild radish, *Raphanus sativus* (Granett *et
al.,* 1976; Calavan *et al.,* 1976; Oldfield *et al.,* 1977). Some
of these species are biennial plants that, like periwinkle,
may overwinter following fall infection.

In addition to the naturally infected species, *S. citri*
has been transmitted by *S. nitridus* to China aster, *Callistep-
hus chinensis;* Shasta daisy, *Chrysanthemum maximum;* red clover,
Trifolium pratense; cabbage, *B. oleracea capitata;* broccoli,
B. oleracea botrytis; clover, *T. repens;* sweet-william, *Dian-
thus barbatus;* hollyhock, *Alcea rosea* Bing cherry, *Prunus avium*
and onion, *Allium cepa,* to name a few (Calavan and Oldfield,
1979).

Of all the known hosts for *S. citri, Citrus* spp. and their
hybrids are the only ones known to survive infection by the
stubborn pathogen more than a few months under warm conditions except when concurrent infection with another mycoplasmalike pathogen exists, such as that mentioned by Oldfield
(1980).

5. Relationships to Nonsusceptible but Vector-Preferred
       Plants

Little or no definitive work has been done in this area.
Some field observations, however, have been made which merit
at least some comment here (E. C. Calavan and D. J. Gumpf,
unpublished results). Even though an insect propagation host
may not itself act as a reservoir for *S. citri*, it will most

certainly affect the level of spread of stubborn. The wide
host range of S. citri makes it widely available for transmission by large numbers of potential vectors that can arise
from nonhosts of S. citri. The organism is carried persistently as a result of S. citri multiplication in the leafhoppers, therefore S. citri can overwinter in the vector in
the mild climates of California, whether it is feeding on a
host for the stubborn agent or not. We have also tested and
observed that a leafhopper-attractive plant like sugarbeet
which is not a host for S. citri, can act as a trap plant
to protect young plantings of susceptible species from the
natural spread of stubborn disease. This subject is discussed
in Section VI, E.

In summary, existing evidence indicates that natural
spread of citrus stubborn disease in California, Arizona, Israel, Morocco and probably some other Mediterranean and mideastern countries is by cicadellids. A large reservoir of
S. citri exists in herbaceous hosts in extensive areas in
California and other western states. Presumably a portion of
northwestern Mexico is similarly infested. Recent reports
from Morocco (Bove et al., 1979b; Nhami et al., 1980) and
from Israel (Clark et al., 1978) show that S. citri is present in Madagascar periwinkle and in brassicaceous weeds in
southwestern and eastern Mediterranean regions. Stubborn disease now appears to have much greater potential in a variety
of crop and weed plants than it did a few years ago.

B. GRAFT TRANSMISSION AND PERPETUATION IN SCION

Citrus propagators inadvertently transmitted stubborn
disease into healthy scions by grafting them onto diseased
navel orange trees in unsuccessful attempts to improve fruit
quality and production of affected trees, (Fawcett et al.,
1944). Fawcett (1946) and Fawcett and Klotz (1948b) confirmed
the growers' findings and transmitted stubborn disease to a
seedling sweet orange rootstock from which the pathogen moved
into a healthy Washington navel orange scion grafted to the
rootstock after the original diseased scion died (Calavan,
1969). Tissue from this tree was used by Calavan and Christiansen (1961, 1966) to graft transmit the stubborn pathogen
to numerous citrus varieties and species. Tissue-graft indexing for stubborn disease was also developed by Calavan and
Christiansen (1965), who used Madam Vinous sweet orange seedlings as indicators. They determined that side grafts using
twig pieces more than 5 cm long were more effective than buds.

The distribution of the pathogen in sweet orange and grapefruit trees at various seasons was studied by Calavan et al.

(1968), who obtained transmission only from graft pieces containing phloem and the most consistent year-round transmission
from leaf patches excised from affected young leaves. This
finding helped Igwegbe and Calavan (1970) find *S. citri* in
sieve tubes.

The results of graft transmission attempts with stubborn
disease are highly variable, from 1 or 2% to nearly 100%
success, depending on inoculum variety, season, temperature,
and other factors. Carpenter and Calavan (1969) obtained
about 50% infection of plants inoculated from severely affected Valencia orange. Even side grafts sometimes provided very
low percentages of transmission (E. C. Calavan, D. W. Christiansen and R. L. Blue, unpublished results). Low percentages of transmission are, we suspect, due to the erratic distribution of *S. citri* in citrus phloem, where only low to
moderate populations have been observed by electron microscopy
or indicated by ELISA even under favorable conditions (Igwegbe
and Calavan, 1970; Lafleche and Bove, 1970; Saillard *et al.*,
1980). We found *S. citri* difficult or impossible to culture
from many parts of infected field-grown citrus trees, sometimes including the fruit, especially during the cool season.

Stubborn disease occurs in all commercial citrus areas of
California and Arizona and has been widely distributed in
northern Africa and mideastern countries by movement of infected budwood and trees and presumably by insect vectors. In
the 1940's, H. S. Fawcett (personal communication) found that
the trees in a navel orange orchard with an unusually high
amount of stubborn disease in eastern Los Angeles County had
been propagated from another orchard that also contained many
diseased trees.

Experiments involving perpetuation of the stubborn disease
pathogen produced variable results. Carpenter and Allen (1966)
noted erratic perpetuation in grapefruit. Calavan (1969) observed strong stubborn symptoms in 9 of 10 Washington navel
orange trees propagated from the C-189 stubborn tree; two of
the severely affected trees later improved. Subsequently,
E. C. Calavan and D. W. Christiansen (unpublished results) recorded the buds and branches used for propagations from a Valencia orange tree apparently uniformly affected by stubborn
disease; although many scions grew poorly for 1-3 years, most
eventually improved leaving only about 5% diseased after 5 years
in the field. *Spiroplasma citri* could be isolated only from the
conspicuously affected trees, several of which originated
from a single budstick. The reasons for the recovery phenomena noted in young navel and Valencia orange trees (Calavan,
1969) have not been investigated. Possibly *S. citri* is sometimes suppressed following propagation from diseased buds or
perhaps slow growth is attributable to the weakness of buds,

in which *S. citri* failed to move into new sieve tubes as the
buds grew into shoots.

Navel orange and some tangelo varieties perpetuate *S. citri* in most bud progeny, grapefruit perpetuates it less frequently, while Valencia orange and lemon may yield low percentages of infected buds and bud progeny (Carpenter and
Allen, 1966; Calavan, 1969 and unpublished results; Carpenter
and Calavan, 1969).

C. EXPERIMENTAL TRANSMISSION BY LEAFHOPPER

Experimental attempts to transmit stubborn disease by
using insects were started at Riverside about 30 years ago by
R. A. Flock, renewed in the late 1960's by R. C. Dickson,
E. F. Laird, Jr., and E. C. Calavan, and continued by G. H.
Kaloostian, G. N. Oldfield, J. W. Bowyer, and E. C. Calavan
in the 1970's. Many species of insects were fed on citrus
plants infected by one of several virulent forms of the stubborn pathogen, such as C-189, that had been maintained in
citrus tissue for at least 20 years. All results were negative or inconclusive. Sweet orange seedlings that had been
recently infected by vectors in the field became available in
1971 (Calavan *et al.*, 1974) for inoculum to tissue graft citrus plants which were subsequently used for acquisition feeding by potential insect vectors. Actual plant-to-plant transmission of *S. citri* by insect vectors was first obtained experimentally with *S. nitridus* from these graft-inoculated,
stubborn-affected citrus plants to healthy Madagascar periwinkle (Kaloostian *et al.*, 1975). The pathogen was later
transmitted with relative ease from periwinkle to periwinkle
and back to citrus by *S. nitridus* (Kaloostian *et al.*, 1979b).
Citrus-to-citrus transmissions were also accomplished with
C. tenellus (Oldfield *et al.*, 1976). Noninfective *C. tenellus*
fed on a pure culture of *S. citri* transmitted the pathogen to
a sweet orange seedling from which *S. citri* was twice reisolated (Rana *et al.*, 1975). Cells of *Spiroplasma citri* were
also visualized in *S. nitridus* (Russo *et al.*, 1976).

The existence of non-citrus hosts of the stubborn pathogen was strongly indicated in 1970 when sweet orange seedlings
grown far from other citrus developed stubborn disease (Calavan *et al.*, 1974) but transmission of *S. citri* to a non-citrus
plant was first reported by Daniels *et al.* (1973), who transmitted it to white clover with injected *Euscelis incisus*
(Kirschbaum) = *plebejus* (Fallen). Markham *et al.* (1974) used
injected *E. plebejus* to transmit *S. citri* to 2 of 49 small
citrus plants and reisolated the organism from the stubborn-
affected plants, thus completing Koch's postulates. Soon

thereafter, Kaloostian *et al.* (1975), by successive feeding of
leafhoppers on diseased citrus and healthy Madagascar periwinkle, transmitted *S. citri* to periwinkle plants, which later wilted and died. Allen (1975), in Arizona, and Granett
et al. (1976) in California reported the culture of *S. citri*
from diseased and dying, naturally infected periwinkle plants.
This discovery was followed by a rapid increase in the number
of species shown to be natural hosts or hosts susceptible to
inoculation (Calavan and Oldfield, 1979; Oldfield and Kaloostian, 1979; Oldfield and Calavan, 1981).

Oldfield and Kaloostian (1979) isolated *S. citri* from 7 of
17 groups of *C. tenellus* collected at Moreno in 1973 and 1974
but from only 1 of 76 groups of *S. nitridus* collected there
in 1974. Their attempts late in 1974 to transmit from Moreno
collections of *C. tenellus* (136 insects) to 4 periwinkle
plants and from *S. nitridus* (868 insects) to 28 plants failed.
In 1975, they transmitted *S. citri* to 4 of 6 periwinkles using
300 *C. tenellus* collected at Moreno in September-November but
53 *S. nitridus* collected during this period failed to transmit
to any of 4 plants.

Oldfield and Kaloostian (1979) also cultured *S. citri*
from wild *C. tenellus* collected from non-citrus hosts in areas
far from citrus. In 1975, *C. tenellus* from south central
Washington yielded *S. citri* in pure cultures. *Scaphytopius
nitridus* fed on these cultures, in the manner described by
Rana *et al.* (1975), transmitted the pathogen to 7 of 8 periwinkle and 3 of 4 sweet orange plants, producing symptoms indistinguishable from stubborn disease originating from other
sources. Six of 30 groups of *C. tenellus* collected in 1976 in
the Mojave Desert about 40 miles from commercial citrus transmitted *S. citri* directly to 4 of 13 periwinkle and 3 of 25
sweet orange plants. In February 1976, *C. tenellus* collected
from wild vegetation, mostly *Brassica tournefortii,* in the
low desert about 25 miles west of Blythe, California, transmitted *S. citri* directly to 2 of 2 periwinkles; another collection from radish in Riverside also transmitted to periwinkle. With the exception of collections from the warm desert
areas of southeastern California (Flock, 1977; Oldfield and
Kaloostian, 1979), cultures or transmissions were rarely obtained from wild leafhoppers collected before August or later
than November.

Spiroplasma was also cultured from *Ollarianus strictus*
but this species failed to transmit the stubborn pathogen.
Empoasca spp. and *Acertagallia* spp. also failed to transmit
(Oldfield and Kaloostian, 1979).

VI.   CONTROL OR PREVENTION

     Satisfactory control of citrus stubborn disease in suscep-
tible varieties has been achieved in California only in orchards established with disease-free trees in areas where
populations of the beet leafhopper appeared to be consistently low for several years after planting. We believe that any
abnormally low incidence of stubborn disease in highly susceptible varieties is related to a low degree of vector activity
or to inadequate inoculum of *S. citri* during the first 6 years
of tree growth. Chemical control efforts have been ineffective but disease-free nursery trees are useful in stubborn
disease prevention.

A.   *ANTIBIOTICS*

     The suppressive effect of tetracycline-HCl on stubborn
symptoms in small sweet orange seedlings led Igwegbe (1970)
to discover mycoplasmalike structures in sieve tubes of infected plants (Igwegbe and Calavan, 1970). Most treated plants
remained symptomless 8 months after treatment (Igwegbe and
Calavan, 1973). Later, small diseased Valencia orange bud-
lings were grown in solutions containing 50, 100, and 150 ppm
for 11 weeks, 250 ppm for 8 weeks, and 500 ppm for 8 days
before being planted in soil. Two of five plants treated at
50 ppm developed symptoms but none treated at a higher concentration showed symptoms within 20 months (Calavan, 1975). In
May 1973, the symptomless trees were planted in the field,
where they remain symptomless 8-½ years after treatment (Calavan and Blue, unpublished results).
     Oxytetracycline, 100-1000 ppm, injected by the method of
Nyland and Moller (1973) or sprayed, produced no beneficial
result in large stubborn-diseased trees at Riverside (E. C.
Calavan, C. N. Roistacher, and R. L. Blue, unpublished data).
Injections of field-grown stubborn-diseased sweet orange trees
near Bakersfield with tetracycline-HCl or with oxytetracycline
by gravity flow or pressure systems caused considerable wood
necrosis and formations of strap-shaped leaves but no improvement of growth or increase in yield (E. C. Calavan and
R. L. Blue, unpublished data).
     Although *S. citri* is highly sensitive to tetracyclines
in vitro (Saglio *et al.*, 1973; Bowyer and Calavan, 1974),
tetracycline injected into the xylem of large trees apparently
is not translocated to the sieve tubes in amounts sufficient
to inhibit the pathogen enough for affected trees to improve.
Prolonged root immersion, while effective, has no practical
value in the field.

B. CLEAN PROPAGATIVE MATERIAL

The use of propagative material free of stubborn disease
(clean stock) was recommended by Fawcett *et al.* (1944) and
often has been advised as a preventive measure (Carpenter,
1959; Chapot and Delucchi, 1964; Calavan and Carpenter, 1965;
Calavan, 1969, 1975; Calavan *et al.*, 1978, 1979; Wallace,
1978). In areas where considerable natural spread of *S. citri* occurs, clean stock has little value for prevention of
stubborn disease during epidemic years (Calavan *et al.*, 1979)
because young sweet orange, grapefruit, and tangelo trees are
especially susceptible to infection with *S. citri* from leafhopper vectors.

Propagations from diseased trees show highly variable
percentages of infection, from about 5% in Valencia orange and
Lisbon lemon to 90% or more in Washington navel orange (Calavan, 1969; E. C. Calavan, D. W. Christiansen, and R. L. Blue,
unpublished data). Thus, clean propagative material cannot,
by itself, assure prevention of stubborn disease in endemic
areas but must be supplemented by roguing and other cultural
practices.

Spiroplasma citri does not appear to be perpetuated in
seedlings grown from seeds of diseased trees (Calavan *et al.*,
1974).

C. QUARANTINE

The presence of stubborn disease in all commercial citrus
areas of California would make any quarantine measure against
infected plants and budwood ineffective here, particularly
since the pathogen is present in other plant hosts and, with
adequate leafhopper vector populations, can be transmitted
from citrus or other hosts to young citrus trees (Oldfield
et al., 1978; Calavan *et al.*, 1979; Kaloostian *et al.*, 1979b;
Oldfield and Kaloostian, 1979). Areas where stubborn disease
is not present in citrus should, of course, exclude stubborn-
infected propagative material.

Stubborn-free nursery trees can be produced in greenhouses
from clean propagative materials even when epidemic conditions
prevail in outdoor nurseries (Calavan, 1975). This reduces
tree exposure to possible infection during the highly susceptible period by about 2 years although the trees remain susceptible several years after being planted in the field (Pappo
and Bauman, 1969; Calavan *et al.*, 1976, 1979). During an
epidemic, a high percentage of nursery stock in the field can
be infected in one growing season (Cartia and Calavan, 1974;
Calavan *et al.*, 1979). Indoor production of nursery trees at

Riverside, California, facilitated establishment of a Valencia
orange orchard with only about 2% infection where the previous
orchard had more than 50% infected trees (E. C. Calavan, R. L.
Blue, and P. W. Moore, unpublished results).

D.   CONTROL OF NON-CITRUS HOSTS

    London rocket was the first naturally infected weed host
of *S. citri* identified in California (Calavan *et al.*, 1976;
Kaloostian *et al.*, 1976). The presence of *S. citri* in London
rocket and another brassicaceous weed, *Hirschfeldia incana*
(L.), collected in Israel in April and July, respectively, was
indicated by ELISA (Clark *et al.*, 1978). Non-citrus hosts,
particularly plantain and brassicaceous weeds and vegetables,
contribute to seasonal build-up of *S. citri*, which may be
transmitted to citrus by leafhopper vectors (Flock, 1977; Calavan *et al.*, 1979; Kaloostian *et al.*, 1979b; Oldfield and
Kaloostian, 1979). We have no experimental evidence that
controlling weed hosts of *S. citri* decreases the incidence of
stubborn disease in young citrus trees but clean cultivation
in and near citrus plantings has been suggested (Wallace, 1978)
and may be useful in some situations. However, in or near
citrus orchards the destruction of non-citrus host plants infested with vectors causes many of them to feed on citrus,
which might otherwise be largely avoided by beet leafhoppers.
Situations of this type may occur when cover crops are turned
under. R. Southwick (personal communication), who predicted
that mustard would prove to be a host of the stubborn disease
pathogen, observed that extensive spread of stubborn disease
occurred in southern California after several years of widespread use of mustard cover crops in young citrus orchards.

E.   TRAP PLANTS

    We have observed field plots of citrus seedlings and peri-
winkle plants designed to determine the effect of bordering
trap plants on the incidence of stubborn disease. Using sugarbeet which is leafhopper attractive and a nonhost for *S.
citri* as the trap plant, the incidence of stubborn disease can
be reduced. This reduction may be as much as 53% in periwinkle when compared to control plots without a surrounding
border of trap plants. We also observed that the application
of a systemic insecticide to the trap crop was of little additional value for protecting the plants in the test plots from
stubborn. Although the use of trap plants does not provide
complete protection from stubborn, the incidence was reduced

sufficiently to merit consideration of its use in a nursery
situation.

F. *VECTOR CONTROL*

In field trials also using citrus seedlings and periwinkle
plants, the effect of systemic insecticide application for
direct control of leafhopper vectors was evaluated. Two systemic insecticides, Temik and Di-Syston, evaluated in our
tests were of little value in reducing the incidence of stubborn even though leafhopper feeding tests clearly indicated
that the application rate used (10 lb/acre) at 6-week intervals
was sufficient to cause 70% insect mortality within a 4-hour
feeding period or 100% mortality within a 24-hour feeding period 40 days after treatment. Similar high percentages of
kill in leafhopper vectors fed on leaves of citrus and periwinkle plants treated with aldicarb (Temik) were obtained by
Kaloostian *et al.* (1979a) and E. C. Calavan, G. H. Kaloostian,
G. N. Oldfield, and R. L. Blue (unpublished results) from
leaves of treated field-grown periwinkle plants. However,
Temik failed to provide much protection for treated field-
grown periwinkles against natural infection with *S. citri*
(E. C. Calavan and R. L. Blue, unpublished results).
It appears from these studies also that the time required
for the transmission of the stubborn organism by infective
leafhoppers is much less than 24 hours. Therefore, it is unlikely that we will be able to kill the insect on the plant
before it transmits *S. citri*.

G. *ROGUING*

This likewise is a procedure which may affect the incidence of stubborn, however, there has been very little work
done in this area. We have been monitoring the spread of
stubborn in specific blocks of trees at different locations
in California. Trees found infected are removed immediately
following positive diagnosis. The pattern of spread in these
particular blocks of trees, however, is consistent with disease movement from outside the block. We have also initiated
a study to evaluate the effects of pruning out stubborn affected parts of trees and leaving the remaining healthy portions.
Insufficient progress has been made on this study at the present time to comment on any results.
In summary we consider the following measures to be useful in lowering the incidence of stubborn disease of citrus
or in minimizing losses from it. Neither the total nor rela-

tive values of these measures are known but the first two are of great value in locations where nursery trees might be exposed to epidemic conditions outdoors.

1. Use budwood from disease-free trees for all propagations.

2. Produce nursery trees in greenhouses or in areas having a very low incidence of stubborn disease.

3. Grow only the more tolerant varieties in areas where the incidence of stubborn disease is too high for profitable production of highly susceptible varieties such as navel oranges.

4. Avoid the use of cover crops that are highly susceptible to *S. citri* in orchards less than 7 years old in areas subject to high populations of stubborn-disease vectors.

5. Survey young orchards annually and remove trees that appear diseased or abnormally stunted.

6. Where nursery trees are produced in the open in high risk areas, use substantial borders of vector-attractice trap plants that are immune to *S. citri*. Treat the trap plants with systemic insecticide sufficient to give a rapid killing of the feeding vectors.

REFERENCES

Allen, R. M. (1975). Spiroplasma organism found in naturally infected periwinkle. *Citrograph 60,* 428-446.
Anonymous. (1970). Citrus production problems in the Near East and North Africa. *U. N. Devel. Program, FAO Rep. TA 2870,* 95 pp.
Bar-Joseph, M. and Garnsey, S. M. (1981). Enzyme-linked immunosorbent assay (ELISA): principles and applications for diagnosis of plant viruses. *In* "Plant Diseases and Vectors: Ecology and Epidemiology" (K. Maramorosch and K. F. Harris, eds.), pp. 35-39. Academic Press, New York.
Bové, J. M. and Saillard, C. (1979). Cell biology of spiroplasmas. *In* "The Mycoplasmas" (R. F. Whitcomb and J. G. Tully, eds.). Vol. III, pp. 83-153. Academic Press, New York.
Bové, J. M., Calavan, E. C., Capoor, S. P., Cortez, R. E., and Schwarz, R. E. (1974). Influence of temperature on symptoms of California stubborn, South Africa greening, India citrus decline, and Philippines leaf mottling diseases. *Proc. 6th Conf. Int. Organ. Citrus Virol., 1972,* pp. 12-15.

Bové, J. M., Moutous, G., Saillard, C., Fos, A., Bonfils, J., Vignault, J. C., Nhami, A., Abassi, M., Kabbage,K., Mouches, C., and Viennot-Bourgin, G. (1979a). Mise en evidence de *Spiroplasma citri* l'agent causal de la maladie du "stubborn" des agrumes dans 7 cicadelles du Maroc. *C. R. Acad. Sci. (Paris) Ser. D288,* 335-338.

Bové, J. M., Nhami, A., Saillard, C., Vignault, J. C., Mouches, C., Garnier, M., Moutous, G., Fos, A., Bonfils, J., Abassi, M., Kabbage, K., Hafidi, B., and Viennot-Bourgin, G. (1979b). Presence au Maroc de *Spiroplasma citri* l'agent causal de la maladie du "stubborn" des agrumes dans des pervenches *(Vinca rosea* L.)implantees en bordure d'orangeraies malades, et contamination probable du chiendent *(Cynodon* dactylon (L.) Pers.) par le spiroplasme. *C. R. Acad. Sci. Paris Ser. D288,* 399-402.

Bové, J. M., Saillard, C., Junca, P., Degarce-Dumas, J. R., Richard, B., Whitcomb, R. F., Williamson, D., and Tully, J. G. (1980). Guanosine plus cytosin content, hybridization percentages and EcoRI restriction enzyme profiles of spiroplasma DNA. *Abstr. 3rd Conf. Int. Organ. for Mycoplasmol.,* p. 2.

Bowyer, J. W. and Calavan, E. C. (1974). Antibiotic sensitivity in vitro of the mycoplasma-like organism associated with citrus stubborn disease. *Phytopathology 64,* 346-349.

Bruno, A. (1964). Citrus virus symptoms (psorosis, xyloporosis, exocortis, and stubborn disease) in Sardinia. *Nature (London) 202,* 932.

Calavan, E. C. (1968). A review of stubborn and greening disease of citrus. *Proc. 4th Conf. Int. Organ. Citrus Virol. 1967,* pp. 105-117.

Calavan, E. C. (1969). Investigations of stubborn disease in California: indexing, effects on growth and production, and evidence for virus strains. *Proc. 1st, Int. Citrus Symp., 1968, Vol. III,* pp. 1403-1412.

Calavan, E. C. (1975). The control of greening and stubborn, two mycoplasma-like diseases of citrus. *In* "Advances in Mycology and Plant Pathology" (S. P. Raychaudhuri, A. Varma, K. S. Bhargava, and B. S. Mehrotra, eds.), pp. 325-332. Prof. R. N. Tandon's Birthday Celebration Committee, New Delhi.

Calavan, E. C. (1976). Stubborn. *In* "Description and Illustration of Virus and Virus-like Disease of Citrus: a Collection of Color Slides" (J. Bove and R. Vogel, eds.), Vol. III, 11 p. + slides. Institut Francais de Recherches Fruitieres Outre-Mer, Paris.

Calavan, E. C. (1979). Symptoms of stubborn disease and the culture of *Spiroplasma citri. Proc. R. O. C.-United States Cooperative Science Seminar on Mycoplasma Diseases of plants, 1978,* pp. 67-73.

Calavan, E. C. and Carpenter, J. B. (1965). Stubborn disease
    of citrus retards and growth, impairs quality, and de-
    creases yields. *Calif. Citrogr. 50,* 86-87, 96, 98-99.
Calavan, E. C. and Christiansen, D. W. (1961). Stunting and
    chlorosis induced in young-line citrus plants by inocu-
    lations from navel orange trees having symptoms of stub-
    born disease. *Proc. 2nd, Conf. Int. Organ. Citrus Virol.,
    1960,* pp. 69-76.
Calavan, E. C. and Christiansen, D. W. (1962). Some effects
    of stubborn disease on bark, foliage, fruits, and size
    of citrus trees. *Phytopathology 52,* 727.
Calavan, E. C. and Christiansen, D. W. (1965). Rapid indexing
    for stubborn disease of citrus. *Phytopathology 55,* 1053.
Calavan, E. C. and Christiansen, D. W. (1966). Effects of
    stubborn disease on various varieties of citrus trees.
    Israel J. Bot. 15, 121-132.
Calavan, E. C. and Gumpf, D. J. (1974). Studies on citrus
    stubborn disease and its agent. *Coll. Inst. Nat. Sante
    Rech. Med. 33,* 181-186.
Calavan, E. C. and Oldfield, G. N. (1979). Symptomatology of
    spiroplasmal plant diseases. *In* "The Mycoplasmas" (R. F.
    Whitcomb and J. G. Tully, eds.), Vol. III, pp. 37-64.
    Academic Press, New York.
Calavan, E. C., Roistacher, C. N., and Christiansen, D. W.
    (1968). Distribution of stubborn disease virus within
    trees of *Citrus sinensis* and *C. paradisi* at different
    seasons. *Proc. 4th, Conf. Int. Organ. Citrus Virol.,
    1966,* pp. 145-153.
Calavan, E. C., Olson, E. O., and Christiansen, D. W. (1972).
    Transmission of the stubborn pathogen in citrus by leaf-
    piece grafts. *Proc. 5th, Conf. Int. Organ. Citrus Virol.
    1969,* pp. 11-14.
Calavan, E. C., Harjung, M. K., Fudl-Allah, A. E.-S. A., and
    Bowyer, J. W. (1974). Natural incidence of stubborn in
    field-grown citrus seedlings and budlings *Proc. 6th Conf.
    Int. Organ. Citrus Virol. 1972,* pp. 16-19.
Calavan, E. C., Kaloostian, G. H., Oldfield, G. N., and Blue,
    R. L. (1976). Stubborn disease found in London rocket.
    Citrograph 61, 389-390.
Calavan, E. C., Mather, S. M., and McEachern, E. H. (1978).
    Registration, certification, and indexing of citrus trees.
    In "The Citrus Industry" (W. Reuther, E. C. Calavan, and
    G. E. Carman, eds.), Vol. IV, pp. 185-222. University of
    Calfironia, Div. of Agricultural Sciences, Berkeley.
Calavan, E. C., Kaloostian, G. H., Oldfield, G. N., Nauer,
    E. M., and Gumpf, D. J. (1979). Natural spread of *Spiro-
    plasma citri* by insect vectors and its implications for
    control of stubborn disease of citrus. *Proc. Int. Soc.
    Citric., 1977,* pp. 900-902.

Carpenter, J. B. (1959). Present status of some investigations on stubborn disease of citrus in the United States. *In* "Citrus Virus Diseases" (J. M. Wallace, ed.), pp. 101-107. Univ. Calif. Div. Agric. Sci., Berkeley.

Carpenter, J. B. and Allen, R. M. (1966). Occurrence of stubborn disease symptoms in trees propagated from normal-appearing and diseased "Marsh" grapefruit. *Phytopathology 56,* 146.

Carpenter, J. B. and Calavan, E. C. (1969). Effects of stubborn virus on young Valencia orange trees. *Proc. 1st Int. Citrus Symp. 1968, Vol. III,* pp. 1505-1511.

Carpenter, J. B., Calavan, E. C., and Christiansen, D. W. (1965). Occurrence of excessive seed abortion in citrus fruits affected with stubborn disease. *Plant Dis. Rep. 49,* 668-672.

Cartia, G. and Calavan, E. C. (1974). Further evidence on the natural spread of citrus stubborn disease. *Riv. Patol. Veg. 10,* 219-224.

Cassin, J. (1965). Research on stubborn disease in Morocco. *Proc. 3rd Conf. Int. Organ. Citrus Virol. 1963,* pp. 204-206.

Chapot, H. (1959). First studies on the stubborn disease of citrus in some Mediterranean countries. *In* "Citrus Virus Diseases" (J. M. Wallace, ed.), pp. 109-117. Univ. Calif. Div. Agric. Sci., Berkeley.

Chapot, H. and Cassin, J. (1961). Maladies et troubles divers affectant les citrus au Maroc. *Al Awamia (Rabat) 1,* 107-142.

Chapot, H. and Delucchi, V. L. (1964). "Maladies, Troubles et Ravageurs des Agrumes au Maroc," 359 pp. Institut National de la Recherche Agronomique, Rabat.

Chapot, H., Cassin, J., and Larue, M. (1962). Nouvelles varieties d'agrumes atteintes par le stubborn. *Al Awamia (Rabat) 4,* 1-6.

Chen, T. A., and Davis, R. E. (1979). Cultivation of spiroplasmas. *In* "The Mycoplasmas" (R. F. Whitcomb and J. G. Tully, eds.), Vol. III, pp. 65-82. Academic Press, New York.

Chen, T. A. and Liao, C. H. (1975). Corn stunt spiroplasma: Isolation, cultivation and proof of pathogenicity. *Science 188,* 1015-1017.

Childs, J. F. L. and Carpenter, J. B. (1960). Observations on stubborn and other diseases of citrus in Morocco in 1959. *Plant Dis. Rep. 44,* 920-927.

Clark, M. F., Flegg, C. L., Bar-Joseph, M., and Rottem, S. (1978). The detection of *Spiroplasma citri* by enzyme-linked immunosorbent assay (ELISA). *Phytopathol. Z. 92,* 332-337.

Cochran, L. C. and Samadi, M. (1976). Distribution of stubborn disease in Iran. *Proc. 7th Conf. Int. Organ. Citrus Virol. 1975,* pp. 10-12.

Cole, R. M., Tully, J. G., Popkin, T. J., and Bove, J. M. (1973). Morphology, ultrastructure, and bacteriophage infection of the helical mycoplasma-like organism (*Spiroplasma citri* gen. nov., sp. nov.) cultured from "stubborn" disease of citrus. *J. Bacteriol. 115,* 367-386.

Daniels, M. J. (1979). Mechanisms of spiroplasma pathogenicity. *In* "The Mycoplasmas" (R. F. Whitcomb and J. G. Tully, eds.) Vol. III, pp. 209-227. Academic Press, New York.

Daniels, M. J., Markham, P. G., Meddins, B. M., Plaskitt, A. K., Townsend, R., and Bar-Joseph, M. (1973). Axenic culture of a plant pathogenic spiroplasma. *Nature 244,* 523-524.

Davis, R. E. and Lee, I.-M. (1980). Spiroplasmas: comparative properties and emerging taxonomic concepts: a proposal. *3rd Conf. Int. Organ. Mycoplasmol., p. 1. Abstr.*

Doi, Y., Teranaka, M., Yora, K., and Asuyama, H. (1967). Mycoplasma or PLT group-like microorganisms found in the phloem elements of plants infected with mulberry dwarf, potato witches' broom, aster yellows or Paulownia witches' broom. *Ann. Phytopathol. Soc. Jap. 33,* 259-266.

Fawcett, H. S. (1946). Stubborn disease of citrus, a virosis. *Phytopathology 36,* 675-677.

Fawcett, H. S. and Klotz, L. J. (1948a). Citrus diseases and their control. *In* "The Citrus Industry, Vol. II: Production of the Crop" (L. D. Batchelor and H. J. Webber, eds.), pp. 495-596. University of California Press, Berkeley, Los Angeles.

Fawcett, H. S. and Klotz, L.J. (1948b). Stubborn disease, one cause of nonbearing in navels. *Citrus Leaves 28,* 8-9.

Fawcett, H. S., Perry, J. C., and Johnston, J. C. (1944). The stubborn disease of citrus. *Calif. Citrogr. 29,* 146-147.

Flock, R. A. (1977). Citrus stubborn disease in relation to the beet leafhopper control program in Imperial County, California. *Proc. Amer. Phytopathol. Soc. 4,* 205.

Freeman, B. A., Sissenstein, R., McManus, T. T., Woodward, J. E., Lee, I. M., and Mudd, J. B. (1976). Lipid composition and lipid metabolism of *Spiroplasma citri. J. Bacteriol. 125,* 946-954.

Freundt, E. A. and Edward, D. G.ff. (1979). Classification and taxonomy. *In* "The Mycoplasmas" (M. F. Barile and S. Razin, eds.), Vol. I, pp. 1-41. Academic Press, New York.

Fudl-Allah, A. E.-S. A., Calavan, E. C., and Igwegbe, E.C.K. (1972). Culture of a mycoplasmalike organism associated with stubborn disease of citrus. *Phytopathology 62,* 729-731.

Granett, A. L., Blue, R. L., Harjung, M. K., Calavan, E. C.,
    and Gumpf, D.J. (1976). Occurrence of *Spiroplasma citri*
    in periwinkle in California. *Calif. Agric. 30(3),* 18-19.
Haas, A. R. C., Klotz, L. J., and Johnston, J. C. (1944).
    Acorn disease of oranges. *Calif. Citrogr. 29,* 168-169.
Hilgeman, R. H. (1961). Response of stubborn-infected trees
    to iron chelates. *Proc. 2nd Conf. Int. Organ. Citrus
    Virol. 1960,* pp. 84-92.
Igwegbe, E. C. K. and Calavan, E. C. (1970). Occurrence of
    mycoplasmalike bodies in phloem of stubborn-infected
    citrus seedlings. *Phytopathology 60,* 1525-1526.
Igwegbe, E. C. K. and Calavan, E. C. (1973). Effect of tetra-
    cycline antibiotics on symptom development of stubborn
    disease and infectious variegation of citrus seedlings.
    Phytopathology 63, 1044-1048.
Igwegbe, E. C. K., Calavan, E. C., and Fudl-Allah, A. E.-S.
    (1971). Inclusions in a mycoplasmalike organism associated
    with stubborn disease of citrus. *Phytopathology 61,*
    1321-1322.
Kaloostian, G. H. and Pierce, H. D. (1972). Note on *Scaphyto-
    pius nitridus* in California. *J. Econ. Entomol. 65,* 880.
Kaloostian, G. H., Oldfield, G. N., Pierce, H. D., Calavan,
    E. C., Granett, A. L., Rana, G. L., and Gumpf, D. J. (1975)
    Leafhopper - natural vector of citrus stubborn disease?
    Calif. Agric. 29(2), 14-15.
Kaloostian, G. H., Oldfield, G. N., Calavan, E. C., and Blue,
    R. L. (1976). Leafhopper transmits disease to weed host.
    Citrograph 61, 389-390.
Kaloostian, G. H., Oldfield, G. N., Gough, D., and Calavan,
    E. C. (1979a). Control of citrus stubborn vectors in the
    laboratory. *Citrograph 65,* 17-18, 25.
Kaloostian, G. H., Oldfield, G. N., Pierce, H. D., and Cala-
    van, E. C. (1979b). *Spiroplasma citri* and its transmission
    to citrus and other plants by leafhoppers. *In* "Leafhopper
    Vectors and Plant Disease Agents" (K. Maramorosch and
    K. F. Harris, eds.), pp. 447-450. Academic Press, New
    York.
Laflèche, D. and Bové, J. M. (1970). Mycoplasmes dans les
    agrumes atteints de "greening" de "stubborn" ou de mala-
    dies similaires. *Fruits 25,* 455-465.
Lee, I.-M. (1977). The *in vitro* life sycle, morphology, and
    ultrastructure of the citrus stubborn organism, *Spiroplas-
    ma citri.* Ph.D. Thesis. University of California, River-
    side.
Lee, I.-M. and Davis, R. E. (1978). Identification of some
    growth-promoting components in an enriched cell-free me-
    dium for cultivation of *Spiroplasma citri. Phytopathol.
    News 12,* 215.

Lee, I.-M. and Davis, R. E. (1980). DNA homology among di-
    verse spiroplasma strains representing several serological
    groups. *Can. J. Microbiol. 26,* in press.
Lee, I.-M., Cartia, G., Calavan, E. C., and Kaloostian, G. H.
    (1973). Citrus stubborn disease organism cultured from
    beet leafhopper. *Calif. Agr. 27(11),* 14-15.
Markham, P. G., and Townsend, R. (1977). Transmission of pro-
    karyotic plant pathogens by insect vectors. *John Innes
    Inst. Ann. Rept. 68,* 17-19.
Markham, P. G., Townsend, R., Bar-Joseph, M., Daniels, M. J.,
    Plaskitt, A., and Meddins, B. M. (1974). Spiroplasmas are
    the causal agents of citrus little-leaf disease. *Ann.
    Appl. Biol. 78,* 49-57.
Mouches, C., Menara, A., Tully, J. G., and Bove, J. M. (1980).
    Polyacrylamide gel analysis of spiroplasma proteins and
    its contribution to the taxonomy of spiroplasmas. *Abstr.
    3rd Conf. Int. Organ. Mycoplasmol. 1980,* p. 4.
Mudd, J. B., Ittig, M., Roy, B., Latrille, J., and Bove, J. M.
    (1977). Composition and enzyme activities of *Spiroplasma
    citri* membranes. *J. Bacteriol. 129,* 1250-1256.
Nhami, A., Bove, J. M., Bove, C., Monsion, M., Garnier, M.,
    Saillard, C., Moutous, G., and Fos, A. (1980). Natural
    transmission of *Spiroplasma citri* to periwinkle in Morocco.
    Proc. 8th Conf. Int. Organ. Citrus Virol. 1979, pp. 153-
    161.
Nour-Eldin, F. (1967). A tumor-inducing agent associated with
    citrus trees infected with safargali (stubborn) disease in
    the United Arab Republic. *Phytopathology 57,* 108-113.
Nyland, G. and Moller, W. J. (1973). Control of pear decline
    with a tetracycline. *Plant Dis. Rep. 57,* 634-637.
Oldfield, G. N. (1980). A virescence agent transmitted by
    Circulifer tenellus (Baker); aspects of its host range and
    association with *Spiroplasma citri. Abstr. 3rd Conf. Int.
    Organ. Mycoplasmol. 1980,* p. 46.
Oldfield, G. N. and Calavan, E. C. (1981). Stubborn disease in
    non-rutaceous plants. *In* "Description and Illustration of
    Virus and Virus-like Disease of Citrus: a Collection of
    Color Slides. (J. Bove and R. Vogel, eds.). (Rev. ed.).
    Institut Francais de Recherches Fruitieres Outre-Mer.
    Paris.
Oldfield, G. N. and Kaloostian, G. H. (1979). Vectors and host
    range of the citrus stubborn disease pathogen, *Spiroplas-
    ma citri. Proc. R. O. C.-United States Cooperative Science
    Seminar on Mycoplasma Diseases of Plants, 1978,* pp. 119-
    124.
Oldfield, G. N., Kaloostian, G. H., Pierce, H. D., Calavan,
    E. C., Granett, A. L., and Blue, R. L. (1976). Beet leaf-
    hopper transmits citrus stubborn disease. *Calif. Agr. 30
    (6),* 15.

Oldfield, G. N., Kaloostian, G. H., Pierce, H. D., Sullivan, D. A., Calavan, E. C., and Blue, R. L. (1977). New hosts of citrus stubborn disease. *Citrograph 62*, 309-312.

Oldfield, G. N., Kaloostian, G. H., Sullivan, D. A., Calavan, E. C., and Blue, R. L. (1978). Transmission of the citrus stubborn disease pathogen, *Spiroplasma citri,* to a monocotyledonous plant. *Plant Dis. Rep. 62,* 758-760.

Olson, E. O. (1969). Mottled-leaf symptom on index plants graft-inoculated from citrus trees showing various symptoms of stubborn disease. *Proc. 1st Int. Citrus Symp. 1968, Vol. 3,* pp. 1413-1420.

Olson, E. O. and Rogers, B. (1969). Effects of temperature on expression and transmission of stubborn disease of citrus. *Plant. Dis. Rep. 53,* 45-49.

Pappo, S. and Bauman, I. (1969). A survey of the present status of little-leaf (stubborn) disease in Israel. *Proc. 1st Int. Citrus Symp. 1968, Vol. III,* pp. 1439-1444.

Rahimian, H. and Gumpf, D. J. (1980). Deoxyribonucleic acid relationship between *Spiroplasma citri* and corn stunt spiroplasma. *Int. J. Syst. Bacteriol., 30,* 605-608.

Rana, G. L., Kaloostian, G. H., Oldfield, G. N., Granett, A. L., Calavan, E. C., Pierce, H. D., Lee, I.-M., and Gumpf, D. J. (1975). Acquisition of *Spiroplasma citri* through membranes by homopterous insects. *Phytopathology 65,* 1143-1145.

Reichert, I. (1930), Diseases new to citrus, found in Palestine. *Phytopathology 20,* 999-1002.

Reichert, I., and Perlberger, J. (1931). Little leaf disease of citrus trees and its causes. *Hadar 4,* 193-194.

Reuther, W., Nauer, E. M., and Roistacher, C. N. (1979). Some high temperature effects on citrus growth. *J. Amer. Hort. Soc. 104,* 353-356.

Russo, M., Rana, G. L., Granett, A. L., and Calavan, E. C. (1976). Visualization of *Spiroplasma citri* in the leafhopper *Scaphytopius nitridus* (DeLong). *Proc. 7th Conf. Int. Organ. Citrus Virol. 1975,* pp. 1-6.

Saglio, P., Lafleche, D., Bonissol, C., and Bove, J. M. (1971). Isolement et culture in vitro des mycoplasmes associees au "Stubborn" des agrumes et leur observation au microscope electronique. *C. R. Acad. Sci. (Paris) Ser. D272,* 1387-1390.

Saglio, P., L'hospital, M., Laflèche, D., Dupont, G., Bove, J. M., Tully, J. G., and Freundt, E. A. (1973). *Spiroplasma citri* gen. and sp. n.: a mycoplasma-like organism associated with "stubborn" disease of citrus. *Int. J. Syst. Bacteriol. 23,* 191-204.

Saillard, C., Garcia-Jurado, O., Bové, J. M., Vignault, J. C.,
    Moutous, G., Fos, A., Bonfils, J., Nhami, A., Vogel, R.,
    and Viennot-Bourgin, G. (1980). Application of ELISA to
    the detection of *Spiroplasma citri* in plants and insects.
    Proc. 8th Conf. Int. Organ. Citrus Virol. 1979, pp. 145-
    152.
Scaramuzzi, G. and Catara, A. (1968). Studies on sour orange
    stem-pitting in Sicily. *Proc. 4th Conf. Int. Organ.
    Citrus Virol. 1967,* pp. 201-205.
Schneider, H. (1966). South Africa's greening disease and
    Morocco's stubborn disease. *Calif. Citrogr. 51,* 299-305.
Smith, P. F. and Reuther, W. (1949). Observations on boron
    deficiency in citrus. *Proc. Fla. State Hort. Soc. 62,*
    31-37.
Vogel, R. and Bove, J. M. (1974). Studies of stubborn disease
    in Corsica. *Proc. 6th Conf. Int. Organ. Citrus Virol.
    1972,* pp. 23-25.
Wallace, J. M. (1978). Virus and virus-like diseases. *In*
    "The Citrus Industry" (W. Reuther, E. C. Calavan and
    G. E. Carman, eds.), Vol. IV, pp. 67-184. University
    of California Div. of Agri. Sci., Berkeley.
Whitcomb, R. F., Tully, J. G., Bove, J. M., and Saglio, P.
    (1973). Spiroplasmas and acholeplasmas: multiplication
    in insects. *Science 182,* 1251-1253.
Williamson, D. L. and Tully, J. G. (1980). Characterization
    of spiroplasmas by serology. *Abstr. 3rd Conf. Int. Organ.
    Mycoplasmol.,* p. 3.
Williamson, D. L. and Whitcomb, R. F. (1975). Plant mycoplas-
    mas: a cultivable spiroplasma causes corn stunt disease.
    Science 188, 1018-1020.
Zelcer, A., Bar-Joseph, M., and Loebenstein, G. (1971). Myco-
    plasma-like bodies associated with little-leaf disease
    of citrus. *Israel J. Agr. Res. 21,* 137-142.

PAULOWNIA WITCHES' BROOM DISEASE

Yoji Doi and Hidefumi Asuyama

Tokyo University, Tokyo, Japan

I. INTRODUCTION

Witches'-broom is a disease of striking and fantastic symptoms. Paulownia witches'-broom was the first case that was
demonstrated to be of mycoplasma-like organism (MLO) etiology
among the tree witches'-broom (Doi *et al.*, 1967). Later, a
number of other diseases were shown to be associated with MLO's,
namely, apple proliferation (Giannotti *et al.*, 1968; Marwitz
et al., 1973) and witches'-broom of ash (Hibben and Wolanski,
1971), willow (Holmes *et al.*, 1972), and jujube (*Zizyphus* spp.)
(Okuda, 1977; Nakamura *et al.*, 1977), respectively.
    Most MLO diseases of plants, including jujube witches'-
broom, are transmitted by leafhoppers, while paulownia witches'
broom is uncommon, being transmitted by a stink bug.

II. DISTRIBUTION AND LOSS

Paulownias (empress tree, princess tree) are fast-growing
deciduous trees belong to the Bignonia family (formerly classified to Scrophulariaceae), and are found in the temperate
zone of eastern Asia, namely, Japan,Korea, China or Taiwan.
The lumber of paulownia is light, soft, and durable. Therefore, it is an important and valuable material for manufac-
turing furnitures, musical instruments, footgears, and wood
crafts. Paulownia is propagated either by seed or by root-
cutting. It reaches maturity and is ready for cutting in 8-20
years. After the first cutting, a sprout arises from the
stump, which grows rapidly to full size in a few years.
    Witches'-broom of paulownia was first observed in about
1880 in the southern district of Japan and described in 1902
by Kawakami (1902). The disease then gradually spread north-

ward; and its occurrence in the central district was observed
by 1950 (Kondo, 1960), and in the northern district by 1960.
It is now widely distributed in Japan except Hokkaido. It was
also observed in Korea and China (Yoshii, 1931, 1967). Recently the disease was reported from Taiwan (Chang et al.,
1978; Ying, 1978).

In Japan and in Taiwan in heavily affected plantings, more
than 70% of the trees were infected (Chang et al., 1978; Kondo, 1960). In the affected trees growth is suppressed, which
is followed by a decline and usually premature death. Lumber
is of markedly poor quality. Witches'-broom is one of the
most destructive diseases of paulownia growing in eastern
Asia.

III.   SYMPTOMS

Witches'-broom is characterized by closely grouped clusters of fine slender twigs. In a healthy paulownia, the
growth of a new stem or new branch ceases before autumn, and
the axillary buds do not sprout in the current season. In
the case of affected paulownia, however, the primary axillary
buds on the new stem or new branch soon sprout and grow to
secondary shoots. The axillary buds produced on the secondary
shoots, in turn, then grow to the tertiary shoots, and thus
sprouting is repeated until late in the autumn. In this way,
the restless sprouting of the axillary buds and successive
branching during a prolonged period create a bush of slender
twigs, which looks like a broom (Tokushige, 1951) (Fig.1a,b,
d).

The disease is generally systemic. At first, the broom
of bushy proliferations appear on a few branches in late
spring, but later they arise throughout the crown. The proliferated axillary shoots are slender, brittle, and yellowish-
green in color. The leaves borne on the diseased shoots are
small, thin, often malformed, and faded to yellow green (Tokushige, 1951; Chang et al., 1978). The flower clusters,
which rarely develop, show different degrees of distortion,
and virescence, along with sterility (Nakamura, 1963)
(Fig. 1c). In the diseased shoots, necrosis occurs in the
phloem tissues, and cell arrangement in the woody cylinder
become irregular, especially around the vessels (Tokushige,
1951). Growth in the roots is greatly retarded; roots are
discolored and liable to decay. The affected saplings are
severely stunted and succumb within a few years (Tokushige,
1951).

Under certain circumstances, only localized infection
occurs on the affected trees. (Ito, 1971; Tahama, 1977). The

Fig. 1. (a-d). Symptoms of witches'-broom of paulownia
(P. tomentosa). (a) An affected tree, bearing many brooms in
the crown. (b) Incipient of witches'-broom: Development of
secondary shoots. (c) An affected flower cluster, showing
distortion and virescence. (d) Proliferated shoots on an
infected tree (left), and normal branching in a healthy tree
(right), photographed in winter.

affected tree develops bushy proliferations on a few branches,
but the proliferated shoots die together with the causal agent,
and the other branches remain healthy in the ensuing years.

IV. ETIOLOGY

     In 1902, a fungus, *Gloeosporium kawakamii,* was isolated
from the petioles and terminal buds of witches'-broom affected
paulownia, and was suggested as the causal agent of the disease (Kawakami, 1902). Later, however, the fungus was proved
to induce merely anthracnose, but not witches'-broom (Yoshii,
1931). Graft inoculation was then carried out to observe the
transmissibility of the disease. Healthy seedlings or stocks,
when grafted with scions from affected trees, became infected
and exhibited symptoms 2 months after grafting. Similarly,
healthy scions contracted the disease when grafted on diseased
stocks. Thus, paulownia witches'-broom disease was demonstrated to be both infectious and systemic. No visible organism
was found in the affected plants. Consequently, the causal
agent was presumed to be viral in nature (Tokushige and Yoshii,
1950; Tokushige, 1951; Yoshii, 1931, 1967; Hiromu Yoshii, 1950).
However, the particles of suspected virus have never been observed in the affected paulownia.
     In 1967, wall-less pleomorphic microorganisms resembling
mycoplasma were revealed in the phloem of witches'-broom-af-
fected paulownia by means of electron microscope. Mycoplasma-
like bodies are spherical to ellipsoidal in shape, surrounded
by a unit membrane, and packed with ribosome-like granular
particles and fibrillar nuclear materials. Mycoplasma-like
organisms (MLO's) are found solely in sieve tubes and phloem
parenchyma cells of affected plants, but not in healthy plants
(Doi *et al.,* 1967) (Fig. 2). In the affected plants, quantities of MLO are found in the young leaves and twigs of proliferated shoots and in the roots, with only a few in the old
leaves and stems, and almost none in the symptomless shoots on
partially affected trees. The quantity of MLO apparently corres-
ponds to the severity of symptoms. In winter, MLO disappear
in the diseased, defoliated shoots, and reappear in spring in
a new flush of growth, which later manifests symptoms. These
observations led to the conclusion that MLO is the causal
agent of the disease (Doi *et al.,* 1967; Doi and Okuda, 1973).
MLO etiology has been further supported by research in Taiwan,
using electron microscopy, graft transmission, and therapeutic treatments with tetracycline (Chang *et al.,* 1978; Ying,
1978). The bodies of the agent were shown to measure 230-380
X 360-1340 nm (Chang *et al.,* 1978). In Japan, efforts to
isolate the MLO from the affected paulownia in axenic culture

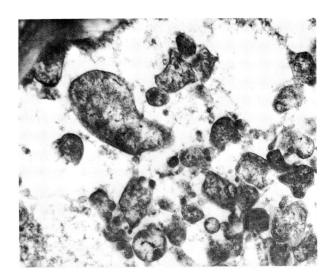

Fig. 2. Electron micrograph of Mycoplasma-like organism (MLO) in sieve tube of paulownia plant affected with witches'-broom. x 22,500.

have been unrewardeing (Okuda 1977). However, it is said that MLO or mycoplasma has been isolated in China from infected paulownia on PPLO medium, although pathogenicity of the isolate has not as yet been ascertained (Ming-Chi Wang *et al.*, 1978, personal communication).

V. TRANSMISSION

The disease is graft transmissible as was stated above. No evidence has been found to indicate its transmission through seed (Tokushige, 1951) or soil. An investigation in a paulownia field for 5 years, showed that new infections occurred every year at a rate of 10-15%, and spread nonconcentrically (Tokushige, 1952a; Furukawa and Ishii, 1961). In nature the disease was suspected to be spread by a vector which has only recently been identified.

In 1968, it was reported from Korea that tobacco leaf bug *(Cyrtopeltis tenuis)*, reared on the affected paulownia could transmit the causal agent to paulownia (La *et al.*, 1968; La, 1968), but this finding has not yet been confirmed either in Taiwan (Chang *et al.*, 1978) or Japan. Recently, brown-marmorated stink bug *(Halyomorpha mista)* has been shown to transmit the disease (Shiozawa *et al.*, 1978, 1979) (Fig. 3). The stink bug became infective by 10 days acquisition feeding on

Fig. 3. Brown-marmorated stink bug (Halyomorpha mista Uhler), the vector of paulownia witches'-broom disease. (Shiozawa et al., 1978.)

the diseased plant, and, following 30 days incubation, could infect the paulownia seedling with 5-7 days inoculation feeding. Symptoms of the disease appeared 30-40 days after the inoculation (Fig. 4a). Electron microscopy revealed numerous MLO bodies in the infected plants and in the infective stink bugs. Since the brown-marmorated stink bug is widely distributed in Japan and is found frequently infesting on paulownia (Nakamura and Kobayashi, 1961; Shiozawa *et al.*, 1978), it is believed to be responsible for the transmission of the disease.

VI. HOST RANGE

In eastern Asia, *Paulownia tomentosa, P. kawakamii, P. taiwaniana, P. fortunei, P. coreana* and *P. fargesii* are cultivated. All these species are known to be susceptible to witches'-broom disease (Chang *et al.*, 1978; Furukawa and Ishii, 1961; Kondo, 1960; Tokushige, 1951). In nature, paulownias seem to be only hosts. As a result of artificial inoculation, tobacco leaf bug was reported to transmit the disease from paulownia to calendula and morning glory, but not to tomato, carrot, aster, zinnia, cosmos, soybean, and plantago (La *et al.*, 1968; La, 1968). This undoubtedly requires additional investigation. Another experiment has shown that brown-marmorated stink bug has transmitted the disease to periwinkle *(Vinca rosea)*. Infected periwinkle manifested yellows symptoms and was colonized with MLO (Shiozawa *et al.*, 1979). (Fig. 4b,c).

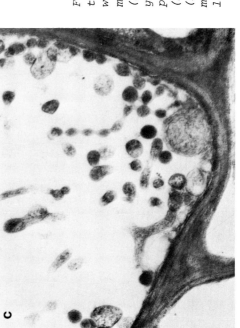

Fig.4(a-c). Transmission of the witches'-broom agent by
the insect vector. (a)Paulownia seedling infected with
witches' broom by inoculation with infective brown-marmorated stink bugs (left) and healthy control (right)
(Shiozawa et al., 1978). (b) Periwinkle plant showing
yellows and witches'-broom symptoms; transmitted from
paulownia by stink bugs 30 days after inoculation
(Shiozawa et al., 1978). (c) Mycoplasma-like organism
(MLO) in sieve tube of affected periwinkle plant transmitted from paulownia by the stink bugs (Shiozawa etal.,
1978). x 30,000.

VII. PHYSIOLOGY OF THE INFECTED PLANT

The physiological and biochemical changes brought about by
witches'-broom in paulownia were studied two decades ago by
comparison with the healthy plant, and based on the premise
that the disease was of a viral nature (Tokushige, 1950, 1952b,
c, 1955a-j.) As was expected from yellows symptoms, there was
a marked decrease in chlorophyll content and photosynthetic
activity in the leaves, while a significant increase in respiration rate occurred. Thus, an imbalance between photosynthesis and respiration arose. On the other hand, accumulation
of starch and soluble sugars in the leaves was evident, and
the loss of carbohydrates during the night was much lower. It
was concluded that translocation of sugars from the leaves was
inhibited and this might cause starvation in the diseased
shoots or the entire plant. Premature death of witches'-
broom-affected saplings might be, at least in part, attributable to starvation. An unusually high content of chlorogenic acid was present and higher activity of polyphenol oxidase was demonstrated. This may be connected with quick
browning of the sap from diseased leaves, but the role of
polyphenol is unknown.

Although MLO was proved as the causal agent, nutritional
physiology of the MLO itself has not yet been studied. Therefore, the action of its metabolites, particylarly the mechanism of stimulation of axillary buds, remains unsolved. However, since the MLO exclusively colonizes the phloem tissues,
the function of the phloem must be adversely influenced.

VIII. CONTROL

The following are suggested for control of the disease,
though practical measures have not yet been established: (1)
Disease-free seedlings should be selected for planting. Seedlings raised in the infected area, particularly those propagated by root-cuttings, are unsafe for use. (2) Roguing out
diseased individual trees from the field should be practiced
as soon as they are detected. The affected tree serves as a
source of infection. (3) Cultivated species of paulownia exhibiting immunity or resistance to the disease have not been
found; Paulownia tomentosa, P. fargesii, and P. kawakamii are
equally susceptible. (4) The affected saplings can be cured
by trunk injection with solutions of tetracycline antibiotics.
In Taiwan, 500 ppm tetracycline hydrochloride solution (2 li-
ters per tree) was injected through 3 holes drilled into the
trunk, in spring or in summer. The treated saplings of 2-3
years old were completely or partially cured, but recovery

was insignificant in those 4 years old (Ying, 1978). (5) Eradication of the insect vectors by the use of insecticides is
thought to be impractical for preventing natural spread in the
field plantings, but may be helpful for control in the nursery.

REFERENCES

Chang, Y. C., Su, H. J., and Wu, R. Y. (1978). Preliminary
    study on paulownia witches'-broom in Taiwan. *Proc. R. O.
    C. U. S. Cooperative Semin. Mycoplasma Dis. Plants,* March
    27-31, 1978. pp. 127-137. Nat. Sci. Council, Taipe.
Doi, Y. and Okuda, S. (1973). Electron microscopic study of
    Mycoplasma-like organisms (MLO) in yellows-diseased host
    plants. Shokubutsu Byogai Kenkyu 8, 203-221.
Doi, Y., Teranaka, M., Yora, K., and Asuyama, H. (1967). My-
    coplasma- or PLT group-like microorganisms found in the
    phloem elements of plants infected with mulberry dwarf,
    potato witches'-broom, aster yellows, or paulownia
    witches'-broom. *Ann. Phytopathol. Soc. Jap. 33,* 259-266.
Furukawa, T. and Ishii, Y. (1961). The relation between the
    plantation and the "Witches'-broom" of the *Paulownia.
    J. Jap. Forest. Soc. 43,* 72-74. (In Japanese.)
Giannotti, J., Morvan, G., and Vago, C. (1968). Micro-organ-
    ismes de type mycoplasme dans les cellules liberiennes de
    Malus sylvestris L. atteint de la maladie des prolifera-
    tions. *C. R. Acad. Sci. D. 267,* 76-77.
Hibben, W. B. and Wolanski, B. (1971). Dodder transmission of
    a mycoplasma from ash trees with yellows-type symptoms.
    Phytopathology 61, 151-156.
Holmes, G. R., Hirumi, H., and Maramorosch, K. (1972). Witches'
    broom of willow: Salix yellows. *Phytopathology 62,* 826-
    828.
Ito, K. (1971). Paulownia witches'-broom. "Pathology of Fo-
    rest Trees," Vol. I, pp. 232-234. Norin Shuppan, Tokyo.
    (In Japanese).
Kawakami, T. (1902). "On the Hexenbesen of *Paulownia tomento-
    sa.*" (In Jap.) Shokabo, Tokyo.
Kondo, H. (1960). Actual investigation on the witches'-broom
    disease of paulownia. *Res. Rep. Ibaraki Prefecture
    Forest Exp. Sta. 2,* 1-50.
La, Y. J., Pyun, B. H., and Shim, K. J. (1968). Transmission
    of paulownia witches'-broom virus by tobacco leaf bug,
    Cyrtopeltis tenuis Reuter. *Korean J. Plant Pro. 5,* 1-8.
La, Y. J. (1968). Insect transmission of paulownia witches'-
    broom disease in Korea. *Korean Observer 8,* 55-64.

Marwitz, R., Petzold, H., and Kunze, L. (1973). Elektronenmik-
    roskopische Untersuchungen über das Vorkommen mykoplasmaähn-
    licher Organismen in triebsuchtkranken Apfelbäumen. *Phy-
    topathol. Z. 77,* 84-88.
Nakamura, K. (1963). Symptoms on the flower cluster of pau-
    lownia affected with witches'-broom. *Forest Pests (Tok-
    yo) 12,* 127. (In Jap.)
Nakamura, K. and Kobayashi, H. (1961). Hemipterous insects
    observed in the virus-diseased paulownia. *J. Jap. Forest.
    Soc. 43,* 146. (In Jap.)
Nakamura, S., Saito, N., and Gohara, S. (1977). Mycoplasma-
    like organisms found in jujube tree, *Zizyphus jujuba* Mill.
    var *inermis* Rehd., infected with jujube witches'-broom in
    China. *Ann. Phytopathol. Soc. Jap. 43,* (Abstr. Jap.)
Okuda, S. (1977). Studies on the causal agents of yellows or
    witches'-broom diseases of plants. *Spec. Bull. College
    Agr. Utsunomiya Univ., No. 32,* pp. 1-70.
Shiozawa, H., Yamashita, S., Doi, Y., Yora, K., and Asuyama,
    H. (1979). Trial of transmission of paulownia witches'-
    broom by two species of bug, brown marmorated stink bug
    and brown-winged green bug, observed on paulownia. *Ann.
    Phytopathol. Soc. Jap. 45,* 130-131. (Abstr. Jap.)
Shiozawa, H., Yamashita, S., Doi, Y., Yora, K., and Asuyama,
    H. (1979). Transmission of Mycoplasma-like organisms (MLO)
    by brown-marmorated stink bug to *Vinca rosea* from paulow-
    nia affected with witches'-broom, and MLO in the infec-
    tive insects. *Ann. Phytopathol. Soc. Jap. 45,* 556.
    (Abstr. Jap.).
Tahama, Y. (1977). Mycoplasma-like organism in relation to the
    symptoms of paulownia witches'-broom. *Ann. Phytopathol.
    Soc. Jap. 43,* 377. (Abstr. Jap.)
Tokushige, Y. (1950). Witches'-broom of *Paulownia tomentosa*
    L. (On the respiration of paulownia infected with witches'-
    broom. Part 1.) *Bull.Kyushu Univ. Forests 19,* 71-82.
Tokushige, Y. (1951). Witches'-broom of *Paulownia tomentosa*
    L. *J. Fac. Agr. Kyushu Univ. 10,* 45-67.
Tokushige, Y. (1952a). Witches'-broom of *Paulownia tomentosa*
    L. (On the occurrence of the witches'-broom affected
    trees in paulownia forest.) *J. Jap. Forest. Soc. 34,*
    4-7.
Tokushige, Y. (1952b). On the respiration of paulownia in-
    fected with witches'-broom. 2. *Sci. Bull. Fac. Agr. Kyu-
    shu Univ. 12,* 309-314.
Tokushige, Y. (1952c). On the respiration of paulownia infec-
    ted with witches'-broom. 3. *Sci. Bull. Fac. Agr. Kyushu
    Univ. 12,* 315-319.
Tokushige, Y. (1955a). On the excess accumulation of chloro-
    genic acid in paulownia leaves affected by witches'-

broom. *Proc. Assoc. Plant Prot. Kyushu(Japan)* 1, 32-35.

Tokushige, Y. (1955b). On the oxidase of paulownia affected by witches'-broom. *Sci. Bull. Fac. Agr. Kyushu Univ.* 15, 145-150.

Tokushige, Y. (1955c). On the amylase and the peroxidase of paulownia affected by witches'-broom. *Sci. Bull. Fac. Agr. Kyushu Univ.* 15, 287-290.

Tokushige, Y. (1955d). On the catalase of paulownia affected by witches'-broom. *Sci. Bull. Fac. Agr. Kyushu Univ.* 15, 291-296.

Tokushige, Y. (1955e). On the chlorophyll and photosynthesis of paulownia affected by witches'-broom. *Sci. Bull. Fac. Agr. Kyushu Univ.* 15, 297-302.

Tokushige, Y. (1955f). On the accumulation of soluble sugars in the leaves of paulownia tree affected by witches'-broom. *Sci. Bull. Fac. Agr. Kyushu Univ.* 15, 303-307.

Tokushige, Y. (1955g). On the inhibition of translocation of starch in paulownia leaves affected by witches'-broom. *Sci. Bull. Fac. Agr. Kyushu Univ.* 15, 309-312.

Tokushige, Y. (1955h). On the decrease of starch syntheses from sugar solution in the leaves of the paulownia tree affected by witches'-broom. *Sci. Bull. Fac. Agr. Kyushu Univ.* 15, 313-138.

Tokushige, Y. (1955i). On the inhibition of translocation of carbohydrate in paulownia tree affected by witches'-broom. *Sci. Bull. Fac. Agr. Kyushu Univ.* 15, 319-326.

Tokushige, Y. (1955j). On the decrease of carbohydrate and nitrogen content of shoots of paulownia tree affected by witches'-broom. *Sci. Bull. Fac. Agr. Kyushu Univ.* 25, 327-331.

Tokushige, Y. and Yoshii, H. (1950). Graft transmission of paulownia witches'-broom. *Ann. Phytopathol. Soc. Jap.* 14, 107. (Abstr. Jap.)

Wang, Ming-Chi *et al.* (1978). (Futan Univ., Shanghai, China) Personal communication.

Ying, S. L. (1978). Witches'-broom disease of paulownia *(P. taiwaniana* Hu et Chang) in Taiwan. *Proc. R.O.C. U.S. Cooperative Semin. Mycoplasma dis. Plants,* March 27-31, 1978. pp. 161-168. Nat. Sci. Council, Taipei.

Yoshii, Hazime (1931). Does *Gloeosporium kawakamii* cause the paulownia witches'-broom? *Ann. Phytopathol. Soc. Jap.* 2, 388. (Abstr. Jap.)

Yoshii, Hazime (1967). Paulownia witches'-broom. *In* "Manual of Virus Diseases of Cultivated Plants in Japan" (H. Asuyama and T. Iida, eds.), pp. 318-332. Assoc. Adv. Agr. Sci., Tokyo, Japan. (In Jap.)

Yoshii, Hiromu (1950). Studies on the "witches'-broom" of the paulownia (1) On the infection experiments of the "witches' broom" of the *Paulownia. J. Jap. Forest. Soc.* 32, 301-305.

MULBERRY DWARF: FIRST TREE MYCOPLASMA DISEASE

Takashi Ishijima and Tatsuji Ishiie

Kyushu Branch Station
The Sericultural Experiment Station
Ueki, Kumamoto 861-01, Japan

and

Pathology and Entomology Division
The Sericultural Experiment Station
Yatabe, Ibaragi 305, Japan

I. INTRODUCTION

Recently, highly improved artificial diets for silkworms have been developed and are used for young silkworm larvae by sericultural farmers in Japan, one of the greatest silk-producing countries in the world. Mulberry trees, however, are definitely of the utmost importance as only plant food for silkworms. Since ancient times, Mulberry plants have been one of the most important crops. In 1976, mulberry plantations in Japan covered an area of 143,000 ha. The sericultural industry is apparently expanding, especially in some Asian countries such as Korea and Thailand, as well as in the People's Republic of China. Consequently, mulberry plantations are also expanding in those countries.

Of the approximately fifty diseases affecting mulberry plants, mulberry dwarf (MD) disease has caused serious losses throughout the history of the mulberry production in Japan because of its severity, wide distribution, highly infectious nature, and stubborness in occurrence. Once established mycoplasma diseases of perennial plants are difficult to eradicate. The establishment of effective control measures against mycoplasma diseases will be indispensable for further development of not only Japan's, but also of the sericultural industry of other countries. Unfortunately, recent social conditions in highly industrialized countries have impaired the successful

control of mulberry dwarf thus promoting the occurrence of the
disease while supressing the implementation of control measures.

Since the beginning of the Meiji era (1868) and throughout
the long history of the disease, numerous research reports on
MD have been published including the results of investigations
carried out by the Government Committee for investigation of
MD disease (4). In addition, since 1967, when a causual
association with a mycoplasma was postulated, there have been
vigorous attempts by many researchers in Japan to culture this
agent of the disease. Thus MD might be one of the most intensively studied mycoplasma diseases of trees or shrubs. Until now most of the information was only available in Japanese, although a few English reviews have recently been published (27, 51, 83). Therefore a comprehensive introduction
and description of the MD in Japan will be presented in this
chapter.

II. HISTORICAL BACKGROUND

A. *OCCURRENCE OF MD DISEASE AND ECONOMIC LOSSES IN SERICUL-
TURE*

Since the Ansei period (1854-1860), Japanese silk products
have become very important export goods; the sericultural industry has become remarkably prosperous. As the industry expanded year by year, the area of mulberry plantation increased
rapidly, and techniques of mulberry cultivation markedly advanced.

An original record of MD disease may be traced back to the
Bunsei period (1818-1831) when the pruning of mulberry shoots
was originally adopted as the means of harvesting leaves (13).
The disease occurred first in Fukushima Prefecture in the
northeastern part of Japan, and was called "Karichijimi"
(dwarf after cutting). In the beginning of the Meiji era
(1868-1912), the disease spread to Tokyo and Aichi Prefecture, frequently causing serious outbreaks. Thereafter,
MD occurred in fields of many localities following the expansion of the sericultural industry, from the latter part of the
19th and early part of the 20th centuries, as summer-autumn
rearing (the second rearing in a year) became highly developed
around 1888. There were recurrent outbreaks of the disease
throughout most of forefront areas of silkworm rearing and the
yields of cocoons dropped to disastrous levels. Hori and Tanaka (17) pointed out that since there were frequent introductions of large numbers of mulberry saplings, it was possible that a few infected ones were present as mulberry fields

expanded. This was one of the causes of the severe outbreaks
in Aichi Prefecture.

In 1896, the Japanese Ministry of Agriculture and Commerce
surveyed the damage caused by MD throughout the country, and
consequently, in 1898, organized a committee for investigation of MD, consisting of several renowned scientists of that
period. The Committee worked actively until 1904, and produced voluminous reports with extensive descriptions of MD
disease (4). There were slight differences of opinion among
members of the Committee, reflected in the conclusions and
based on their research activities.

The area damaged due to MD extended over approximately
4000 hectares (2.6% of the total area of mulberry plantation)
in 1893, and it increased strikingly to 56,000 hectares (12.1%
of total area) by 1916 (13).

The development of sericultural techniques in the Taisho
period (1912-1925) made possible multiple rearings of silkworms during a single year by breeding new silkworm varieties.
Consequently, summer pruning of mulberry plants was devised.
However, this new pruning method further accelerated the occurrence of MD and was actually one of the main reasons for the
occurrence of disease (as will be explained later). Numerous
research projects were undertaken to find effective control
measures against MD, and to study the possible relationship
between disease occurrence and cultural practices, saplings,
training, and harvesting methods, soils and fertilizers, as
well as trying to find the causal agent. By 1931 more than
320 reports were cited in the collection of the references
dealing with MD disease (87).

After World War II, the Japanese sericultural industry was
revived. Unfortunately, however, around 1953, immediately
after the reconstruction of the industry, MD again became
prevalent. Since that time, for about 20 years, the disease
spread to most of the sericultural areas throughout Japan,
being especially destructive in the southwestern part of the
country. The amount of the damage accumulated throughout these
years can hardly be calculated. Large numbers of sericultural
farmers have been forced to abandon mulberry cultivation and
silkworm rearing and to convert to other crops or other in-
dustries. Needless to say, during these years, the parties
concerned worked seriously in order to cope with the disease
outbreaks. Research steadily progressed: the infectious nature of MD was confirmed and transmission routes in fields
were clarified, with the exception of an exact determination
of the causal agent. Doi et al. (11) and Ishiie et al. (26)
then made the startling discovery of mycoplasma-like organisms
(MLO) in MD disease. Furthermore, striking improvements were
frequently achieved when control of the disease was accurately

carried out according to a standard program (7, 46) or by other
methods, such as the eradication of vectors and the removal of
diseased mulberry trees (79, 122). Nevertheless, MD disease
is still chronically epidemic in many sericultural areas. Even
worse, it is broadening its range and is entering the main
areas of cocoon production in Japan. Currently many difficulties are encountered which frustrate the successful control
of the MD disease (51).

B. *ETIOLOGY*

1. *Physiological Etiology Hypothesis*

     With early outbreaks of MD in the middle of the Meiji pe-
roid, various factors were considered as possible causes of
MD. Bacteria, fungi, various environmental conditions, and
nutritional and physiological disorders were considered by
different workers. Shirai (100) suspected that MD was caused
by rotting at the cutting edge of mulberry shoots at the time
of grafting or pruning; Yamamoto (133) speculated that the
occurrence of root rot was related to dead water in the soil;
Shirai (101) discovered the occurrence of decomposition in
the heartwood of roots, due to the attack of microorganisms
that entered through wounds incurred by hoeing. Hori and Tanaka (17) linked the disease to parasitism in roots caused by
microorganisms belonging to Chytridiaceae. Shirai (102)
claimed that there was an oversupply of nitrogenous nutrients
in trees and Tanaka and Ichikawa (127) incriminated excessive
leakage of plant sap from the cut edge of shoots.
     In 1904, the Government Committee concluded that MD di-
sease is a physiological disorder induced by overcutting of
shoots and overpicking of leaves during the summer (4, 107).
This was mainly based on the following facts: (1) MD is neither
transmitted in a concentric manner, nor does it always occur
in trees neighboring diseased ones; (2) frequent recoveries
of trees slightly affected by the disease following the stop-
page of cutting shoots for harvesting; (3) frequent disappearance of the disease from trees grafted with diseased scions;
(4) no disease transmission to trees planted successively at
the place of diseased ones; (5) no infection of plants inoculated with various microorganisms isolated from diseased roots;
and so on. Moreover, the Committee hypothesized that mulberry
plants grow vigorously in May and June, but at those times do
not store substantial amounts of nutrients. In addition, if
all their shoots were cut, they might possibly produce only
small amounts of nutrients because of the lack of assimilating
organs, and nutrients would abundantly leak out through the
cutting edges together with plant sap. Simultaneously, roots

might lose the function to absorb nutrients from the soil, they would stop growing, would begin to decay under the influence of shoot cutting above ground. On the other hand, the reproductive ability of shoots is highly stimulated by the pruning of shoots as a number of new shoots begin to extend rapidly, the accumulated nutrients in the tree are gradually exhausted, and dwarfed shoots are produced due to their feeble growth ability. In addition, overproduction of oxidase causes the retention of synthesized starch weakening the assimilation activity in the affected leaves and causing yellowing and wrinkling of leaves due to poor development of fibrovascular bundles.

This hypothesis had long-lasting effects. Control measures were based on the idea that the disease is physiological in nature and attempts to prevent it were concentrated on cultural practices and improvement of training and harvesting. Repeated attempts at controlling MD brought no noticeable results. This trend lasted until the confirmation of the infectious nature of MD which occurred after World War II.

2. Viral Etiology

Ikata and Matsumoto (19) in 1931 had some doubts about the physiological nature of MD, based on the following: (1) Occasional occurrences of MD in arbor trees or seedlings without any pruning of branches, (2) a large regional difference in the occurrence of MD, and (3) similarity of symptoms with those of other plant species infected with plant viruses. They succeeded in inducing MD using as a vector the leafhopper, *Hishimonus sellatus* Uhler, but two other leafhopper species under the same experimental conditions did not induce the disease. They were first to suggest a viral etiology of MD. Akiya (1) supported this conclusion when he successfully transmitted the disease agent to healthy plants by grafting. Sakai reconfirmed the transmissibility of MD by leafhoppers and grafting (89, 91, 92). In addition, he pointed out that even a single leafhopper is capable of initiating the disease in young mulberry plants, and reported the numbers and percentage of individual infective leafhoppers inhabiting mulberry fields heavily infested with MD (89, 93). He also studied the morphology and ecology of the leafhopper (90). Kawai (60) strenghtened the viral etiology hypothesis by his discovery of X bodies in the tissues of leaves affected with MD.

The evidence favoring viral etiology, however, was not totally convincing, therefore, it could not induce workers to consider eradication of leafhoppers in the field to control the disease. Consequently, until the 1960's, parallel research

activities on MD were continued on physiological and viral
etiology of MD. In the meantime, a new species, *Hishimonoides
sellatiformis* Ishihara (20) was established as the second known
leafhopper vector (44). The infectious nature of MD was further confirmed by many investigators (21, 36, 82, 103, 109,
112, 113) and the infection routes of the disease in fields
were also epidemiologically clarified (39, 42, 67). However,
all attempts to purify the causative agent by virological
methods, or to characterize it morphologically and chemically
as a virus remained unsuccessful.

Thus a new control program was begun around 1965, giving
first priority to the eradication of the leafhopper vectors
and to the removal of diseased trees as infectious sources.

III. SOME FEATURES OF MULBERRY CULTIVATION RELATING TO THE
     OCCURRENCE OF MULBERRY DWARF DISEASE IN JAPAN

The intensive cultivation of mulberry plants is generally
restricted in the Orient and differs from the cultivation of
fruit or other trees. In particular, the occurrence of MD is
closely related to the methods of cultivation, as shown by
many investigators (88, 96). (The reference list is extensive and only a few are cited here.) A brief explanation of
some features of mulberry cultivation in Japan will be presented here.

Mulberry plants grow well in tropical and temperate regions, and are widely distributed in various areas of the
world. These perennial woody plants belong to Moraceae and
are in nature arborescent or shrublike. There are now a number of cultivars (cv) in Japan and most of these are classified into three mulberry species, *M. bombysis* Koidz, *M. alba*
L., and *M. latifolia Poiret*. At present most of these mulberry plants are cultivated in plantations, although they were
left in the wild and they grew to towering trees by farmyards,
along fields with other crops, or at road sides in ancient
times. In recent years, the standard number of trees planted
in mulberry plantations, though it fluctuates depending on
conditions, is roughly 6,000-20,000 trees per hectare.

Since leaves of mulberry plants are used to feed silkworms for cocoon production, harvesting of leaves is very important. At present there are various methods of harvesting.
Mulberry trees, in general, are pruned in different ways
according to the climate, geographical conditions, and forms
of silkworm rearing. Currently, in Japan, 70% of the mulberry
plantations are in the form of low-cut training (main trunk is
lower than 30 cm in height above ground surface), while 20%
is medium-cut (lower than 90 cm), and 10% is high-cut (higher
than 90 cm) or naturally formed.

Low- and medium-cut trainings of mulberry plants were rapidly developed during the period from the Meiji to the Taisho era (1870-1925). Consequently, plantations have changed remarkably through the adoptation of these cuttings for training, resulting in the present intensive form. The low-cut mulberry trees are convenient. Since trees are small in size farmers can care for them easily, harvesting leaves and preventing trees from injurious insects and diseases. Optimum growth occurs in a few years, so farmers can obtain a large number of leaves per unit area. On the other hand, this type of training is very harmful to the growth of trees, often causing MD infection and root diseases.

After the Taisho era (1912), an improvement in silkworm varieties has enabled the rearing of silkworms two to three times a year, that is, in spring and autumn, or in summer and autumn, or in spring, autumn, and late autumn. Consequently, much progress has been made in the methods of pruning of mulberry trees. Summer pruning for harvest in spring and autumn rearing seasons, and spring pruning for harvest in summer and autumn rearing seasons are now practised. The branch-rearing method caused the expansion of the pruning for harvesting "Joso" (mulberry leaves harvested with branches) for every rearing season within a year. The spring pruning and summer pruning of the plants are important for farmers, since they now use these methods to harvest mulburry leaves whenever they are needed. Thus, a plantation which was formerly used only for a spring rearing can now be utilized for two to three rearings a year.

In summer pruning, branches are cut after the harvest for spring rearing, usually in the end of May or the beginning of June. In spring pruning, they are cut before the start of the silkworm rearing, namely, before the mulberry sprouts. Both types of pruning present severe problems occurring from the standpoint of growth of mulberry trees. First, cutting of all branches is involved; this extreme trimming cannot be compared to the pruning of fruit trees or teas. Second, summer pruning means cutting of branches during the growing period, which has profound affect on trees. Both of these prunings are unfavorable to mulberry trees. Fortunately, however, the trees can survive because of their vigorous growth patterns, provided they are not attacked by the MD-causing agent. As compared to spring pruning, summer pruning has a more pronounced affect upon the growth of mulberry trees. Moreover, "Kabunaoshi" (cutting back of shoots) is an important part of summer pruning; all of the remaining shoots after harvesting for spring rearing are cut down at the same position. There are two methods of "Kenshiki" (fist style) and "Mukenshiki" (non fist style) depending on the position of cutting back.

In the former, the branches are cut at the base of the main
trunk and, in the latter, they are cut so as to leave some of
their basal parts on the trunk. Although it is very important
to be able to harvest a good crop of leaves, at the same time,
it is necessary to prune carefully, to avoid disturbing the
physiological functions of the trees. In fact, these culturing operations greatly disturb the natural growth of mulberry
trees. Therefore, MD disease is most prevalent and more serious in trees subjected to low-cut training and summer pruning. Furthermore, the life span of these trees and the number
of years for effective culturing are limited, even if they remain free from MD disease. It is estimated that this span is
about 15 to 20 years for commonly cultivated trees in most
districts of Japan. In this respect, the mulberry cultivation in Japan differs from that in European countries where
mulberry plants have reportedly become towering trees and are
harvested once a year for the spring rearing season of silkworms.

IV. THE MYCOPLASMA-LIKE ORGANISM ETIOLOGY OF MULBERRY DWARF DISEASE

In 1967, Doi *et al.* (11) demonstrated the presence of
plemorphic, mycoplasma-like bodies in the phloem of plants,
affected by four different yellows diseases, including MD.
Iishiie *et al.* (26) reported that tetracycline antibiotics
caused remission of disease in plants infected with MD. Based
on these findings, they proposed that the yellows disease
agents might be mycoplasma or psittacosis-lymphogranulema-like
organisms rather than viruses.

A. *ELECTRON MICROSCOPIC OBSERVATIONS*

Since 1967, further studies on the morphology and distribution of MLO in diseased mulberry plants and infective vectors have been reported. According to Doi and Okuda (12) and
Kawakita (65) the distribution of the mulberry dwarf mycoplasma-like organism (MDM) was restricted mainly to the sieve
tubes and the adjacent phloem parenchyma passing through sieve
pores and plasmodesmatas of phloem cells. Large quantities of
MDM were found in young stems, leaves, and roots but few were
found in older tissues. Most of the MDM were spherical to
irregularly ellipsoidal in shape and ranged from 50 to 800 nm
in diameter. Among them, the intermediate MDM of 250 to 500
nm in diameter were most frequently found. MDM had a two-
layered limiting membrane (unit membrane) of about 8 nm in
thickness, and the cells were filled with ribosomes (about

12 nm in diameter) and contained nuclear net strands. It was
found that small MDM in the sieve tubes were produced
by fission, budding, constriction, or disruption of large MDM,
and newly formed small MDM grew into larger ones. Phloem necrosis was observed in all diseased materials examined, and
electron-dense MDM were found filling the dead shrunken sieve
tubes. The shape and the size of MDM vary considerably depending on the growing conditions, and it was not possible to
morphologically distinguish various mycoplasma species. The
symptom severity of diseased materials was related to the
quantity of MDM detected in the materials. Disappearance of
MDM in dormant shoots and reappearance in young sprouts was
confirmed. However, in root tissues MDM could always be observed in winter, although their presence diminished after
January in Ichinose, the only cultivar examined. It is
thought that the phloem necrosis of diseased plants results
from plugging of sieve tubes with MDM. Onishi *et al.* (86)
also reported that MDM in the phloem tissues of diseased mulberry trees collected from plantations were seldom found in
early June, but increased gradually in number during the summer and fall seasons, and reached a peak in late fall. On
the other hand, Onishi *et al.* made the interesting observation
that nearly all symptom-free trees, not only in the fields in
which MD disease occurred sporadically, but also on an island
where the disease had not been observed at all, had small numbers of MDM in early June. They concluded that such trees already suffered from the invasion of MDM and the symptoms would
appear when the trees were subjected to physiologically unfavorable conditions. Yamada *et al.* (131) worked on the ultrastructural features of mycoplasma-like organisms (MLO) in
dwarf-infected mulberry trees, using scanning and transmission
electron microscopy (SEM and TEM); SEM revealed a three-dimensional fibrillar meshwork in diseased phloem tissue cells, but
not in healthy ones. The characteristic meshwork consisted
mainly of both straight and branching filaments among which
rounded bodies could be discerned. TEM of thin sections
showed that both rounded and filamentous organisms were bounded only by a unit membrane and that they seemed to contain
ribosome-like particles. The fibrillar meshwork structures
observed under SEM may correspond to the filamentous structures observed under TEM. No similar structures have been
found in vascular tissues of other yellows-type diseased
plants. An MLO colony in plant cells under SEM consisted of
the following three basic morphological forms: (1) rounded
forms, presumably main body, approximately 200 to 300 nm in
diameter, (2) elongated and irregular forms, and (3) filamentous forms, roughly 120 nm in width, emerging from the main
body. Moreover, by SEM, a better visualization of the spa-

tial arrangement and configurations of single organisms, as
well as colonies, were observed. These results with SEM showed that the MLO associated with MD disease are similar in gross
structural features to the cultured large colony-forming mycoplasmas described in the literature.

   Numerous MDM, morphologically similar to those found in
dwarf-diseased mulberry plants, were detected in the salivary
glands and abdomens of infective leafhopper vectors, *H. sellatus* and *H. sellatiformis* (62). In salivary glands, numerous
particles were in the cytoplasm at the periphery of glandular
cells. In abdomens, they were detected in large numbers in
the fatbody-like cells near dorsal epithelial tissues. In
H. sellatus, MDM's in salivary glands were observed only on
the 26th day from the start of feeding on diseased plants,
whereas in the brain, mesothoracic ganglion, and fatbody
tissues MDM's were observed on the 40th day.

B. CHEMOTHERAPY

   Suppression of MD disease by antibiotics supported the
tentative hypothesis that the causal agent of MD might be an
MLO. Since the breakthrough discovery by Ishiie *et al.* (26),
chemotherapy of MD has been extensively used for experimental
purposes, as well as for practical application to combat the
disease in the field.

   Ishijima (43) treated MD-infected trees with various kinds
of tetracyclines (TC), macrolide, and other antibiotics using
foliar sprays, immersion of roots, and immersion of the bases
of scions, cuttings, and shoots detached from trees. The
remarkable effectiveness of TC antibiotics was observed regardless of the methods of application, although the symptoms
in the plants treated redeveloped after cessation of the
treatment. None of the other antibiotics tested including
mitomycin C suppressed MD in plants. Permanent recovery from
the disease was frequently observed when cuttings and scions
prepared from diseased plants were treated with the various
TC antibiotics. Takahashi *et al.* (123) also reported that the
symptom development was markedly suppressed by oxytetracyline
(OTC) and only slightly by erythromycin, but not by actinomy-
cin D or mitomycin C. The effectiveness of OTC-HCl and OTC-
base against MD was superior to chlortetracycline (CTC) (118).
Diseased trees once immersed in 50 ppm OTC in a hydroponic
solution began to recover within 7 days, but the symptoms reappeared 3 weeks later. Some plants treated repeatedly, however, grew continuously with healthy leaves for 14 months,
suggesting complete recovery from MD. The diseased trees
treated with 500 and 1000 ppm OTC-HCl began to produce healthy
leaves within 4 days.

   Antibiotics were then applied to MD-infected trees in a
field (30). The remissive effect on MD symptoms was observed
for a time when 100 ppm TC-HCl solution was sprayed a total
of 15 times at an interval of 2 to 3 days after summer pruning.
At that time, no practical application of antibiotics to prevent MD was anticipated. On the other hand, Tahama (119)
pointed out that mulberry trees infected with the mild Kikuchi
strain of MD (116) responded to the foliar spray of OTC only
at high concentrations of 500 and 1000 ppm, but not at concentrations below 200 ppm. Trees infected with the severe Tamana
strain (116) when treated with antibiotics did not show any
suppressive effect on symptoms irrespective of treatment, even
at higher concentrations of 1660 and 2300 ppm when the solutions were sprayed only on leaves. However, treatments were
slightly effective when the foliage was sprayed with a solution containing 500 and 1000 ppm and the branches were in
addition immersed in double the amount of solution used for
foliar spray. La et al. (69) stated that field control of MD
disease in Korea with OTC-HCl was feasible by using Manget
injection units; remission of dwarf symptoms was obtained by
applying 30 ml of antibiotic solution per plant, but the effect was only temporary at this small dosage and a concentration of 2000 ppm. Kasumi et al. (59) reported that 10 applications of 500 ppm OTC to MD-diseased trees in a field provided better remission of symptoms than fewer applications at
lower concentrations. However, symptoms reappeared soon after
treatment ceased. When antibiotics were applied to plants
with mild symptoms the remissive effect was rapid but even on
such trees, symptoms reappeared soon after treatment was terminated. Among the antibiotics available commercially for
agricultural use, Agrimycin and Aureomycin were slightly effective for remission of the disease, while spiramycin had no
effect.
   The effect of tetracycline antibiotics on the insect transmission of MD was also studied (106); the transmission of MD
to test seedlings by infective leafhopper vectors was completely prevented when the seedlings were treated with 100 ppm TC
solution, by means of root dipping, just before inoculation.
When the seedlings were treated after the inoculation, the
latent period was markedly longer, but eventually transmission
occurred. Transmission rates of leafhoppers fed on mulberry
plants treated with TC before or after acquisition feeding on
diseased mulberry trees were not affected. The transmission
rates of leafhoppers which were injected with TC, were slightly lower. The results suggest that the inhibitory effect of
TC on the multiplication of the MD agent is weaker in the insect vector than in the plant.

The inhibitory effect of OTC on the development of MD
symptoms in shoot tip cultures was also studied (66, 123).
When dwarf-infected shoot tips were treated with 50 or 100
ppm OTC's and then cultured, the appearance of symptoms was
delayed 14-35 days as compared to nontreated controls. Also,
MDM were found first in the cultured specimen treated with
50 ppm on the 18th day after the treatment, but not in the
specimen treated with 100 ppm on the 35th day. The results
suggest this method could be used in a simple way for the
screening of antibiotics against MD.

C. CULTIVATION ATTEMPTS

Since the publication of the first reports by Doi et al.
(11) and Ishiie et al. (26) extensive attempts at culturing
MLO have been made in Japan. The Japanese Ministry of Agriculture, Forestry and Fisheries organized a project team on
"Culturing and Clarification of Pathogenicity of Plant Parasitic Mycoplasma-like Organisms," and actively promoted the
work on various aspects of yellows diseases such as rice
yellow dwarf, Cryptotaenia japonica witches' broom, gentian
witches' broom, celery yellows, western X of peach (WXM),
and MD. The results were published as a voluminous report in
1978 (95).
Among the achievements, Nasu et al. (76, 77) reported the
culturing of MLO in insect organs. These authors developed
the "A-1 lobe method" (75) as a new bioassay. The four an-
terior lobes of salivary glands of the leafhopper Colladonus
montanus infected with WXM and filled with numerous MLO's
were aseptically excised, and cultured for a defined period in
a new glass culture ring. Ultrastructural changes of WXM in
the cultured lobes were electron microscopically observed to
discover the effects of the media components and of cultural
conditions such as, osmotic pressure and pH. Subsequently,
the infectivity of the cultured MLO was bioassayed by injecting homogenized lobes into healthy leafhoppers and confining
the insects to healthy seedlings for feeding tests (104).
Nasu et al. first noticed that the morphological integrity
and the infectivity of WXM were best retained in the primary
cultures prepared with insect tissue culture media or plant
cell suspension culture medium supplemented with 0.2 to 0.9 M
sucrose to raise the osmotic pressure. The ingredients of
media have been highly improved. Thus new specific media
have been created for MLO's, WXM, and MDM, namely, SM-1, SM-2,
and SM-3. Following some success with primary culturing or
subculturing of WXM in these media, the same method was applied to the culturing of MDM. As a result, the cultured MDM

proved pathogenic in the primary culture with AcTC medium (10) and Grace's insect tissue culture medium (16). Infectivity was retained in the first subculture with AcTC and in subculture with plant cell suspension medium under respective supplementation of sucrose. MDM was also cultured in the first specific medium for MLO, SM-1 medium (Table I), showing 6.6-22.0% of the infectivity rate in the subcultures. The ultrastructural morphology of MDM was retained in organs maintained in these media. Takahashi *et al.* (123) observed the same effects on the morphology of MDM in the SM-3 medium and the cultured MDM manifested slight infectivity on the 14th day after the start of the culturing. MDM's maintained in the A-1 lobe of the vector *H. seratiformis* for 15 days were shown in Fig. 1 (123).

TABLE I. Ingredients of SM-1 Medium[a]

Inorganic salts		Vitamins	
$MgSO_4$	360 mg	Ascorbic acid	100 mg
$Ca(NO_3)_2$	200 mg	Biotin	10 mg
NaCl	500 mg	Folic acid	10 mg
KNO_3	80 mg	Inositol	50 mg
KCl	800 mg	Pyridoxine	10 mg
NaH_2PO_4	270 mg	Thiamine-HCl	10 mg
$Fe(SO_4)_3$	25 mg	Carbon sources	
$MnSO_4$	45 mg	Glucose	4000 mg
$ZnSO_4$	15 mg	Sucrose	307,800 mg
H_3BO_3	15 mg	Others	
Amino acids		Yeastorate (Difco)	50,000 mg
L-Alanine	225 mg	Lactalbumin	
L-Asparagine	350 mg	hydrolyzate	
L-Cysteine	22 mg	(Difco)	10,000 mg
L-Glutamine	600 mg	Glycine buffer	
L-Leucine	75 mg	glycine	7,500 mg
L-Lysine	625 mg	$MgCl^2$	6,090 mg
L-Methionine	50 mg	Water	1,000 ml
DL-Serine	1100 mg		
l-Tryptophan	100 mg		

[a]*From Nasu et al. (77).*

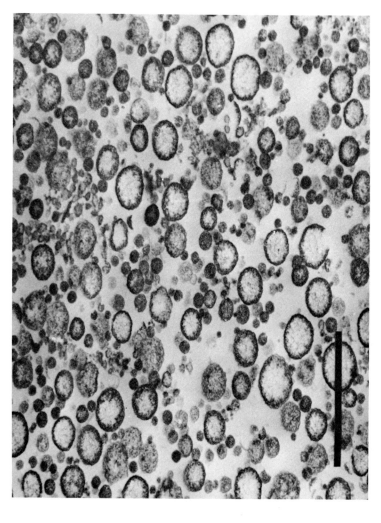

Fig. 1. MDM in A-1 lobe of *Hishimonoides sellatiformis* cultured for 15 days in SM-3 medium (Takahashi et al. (123)).

Different media for zoopathogenic mycoplasmas and phytopathogenic spiroplasmas and several modifications were repeatedly employed to isolate and culture MDM from shoots and
leafhopper vectors infected with MDM. No evidence of growth
was obtained after 15 days incubation and several blind passages when homogenized infected plants or leafhoppers or cut
pieces of infected plants were placed directly on agar, in
fluid cultures or in diphasic media, and even when media were
inoculated with extracts of infected leafhoppers which were
confirmed to be infectious immediately after maceration (48,
52, 123).

Ishijima (50, 52) assayed the infectivity of extracts or
cultures of macerated leafhopper vectors infected with MD in
a variety of balanced salt solutions (BSS) and in other
artificial media using the injection method. Very little and
unstable infectivity in five different BSS's, two tissue culture media, and glycine buffer was detected immediately after
maceration. The infectivity in glycine buffer was not improved by supplementation with different concentrations of
sucrose and horse serum or fetal bovine serum. Use of three
different media for mycoplasma or spiroplasmas combined with
Chen's medium (8), resulted in a slightly higher rate of infectivity but only a short time after maceration. A low rate
of infectivity was detected in three insect tissue culture
media on the day of maceration. In Grace's insect TC medium
(16) containing 10% horse serum and different concentrations
of sucrose, infectivity was occasionally retained for long
periods. In one experiment the initial infectivity was lost
on the 1st and 4th days. Afterward it began to increase from
the 10th or 15th day and was retained on the 20th and 25th
days, but was not detected after subculture. The results
suggested that long-term maintenance or limited multiplication
of MDM from infected leafhoppers might have occurred in Grace's
insect TC medium.

D. *MDM IN CULTURED TISSUES OF PLANTS AFFECTED BY THE DISEASE*

Morphological changes of MDM in cultured tissues from
young mulberry shoot segments of a weed "Kanamugura," *Humulus
japonicus* Sieb. et Zucc. infected with MD were electron microscopically studied in order to ascertain the suitability
of media for the growth of MDM (66, 123). Murashige-Skoog's
(MS) medium (74) and its modification (81) were tested. The
modified MS medium enriched with 2,4-D was suitable for the
formation of calluses from shoot segments, while MS medium
enriched with kinetin was not. The callus formation was better in the dark than in the light. MS medium enriched with

kinetin was best suited for bud development, while the medium
with 2,4-D was suitable for root formation from callus.   When
callus was cultured on MS medium enriched with kinetin neither
roots nor MDM developed.   The addition of 2,4-D to this medium induced abundant rooting from the callus and favored the
appearance of spherical, cylindrical, and doughnut-like
shaped MDM, although the number decreased in callus tissues
after the subculture.   Thus, the appearance of MDM in the cultured tissues was generally accompanied by rooting from callus,
viz., the development of vascular bundle system in tissues.
On the other hand, in the case of "Kanamugura," MS medium supplemented with 2,4-D and kinetin was most suitable than MS
medium alone for callus formation from shoot tips infected
with MD.   The callus was more friable than mulberry callus and
it produced neither buds nor roots on the MS media supplemented with 2,4-D and kinetin or with 2,4-D alone.   In the MS
medium enriched with IAA or potato extract, many roots developed
from "Kanamugura" tips, although formation of callus did not
occur on these media.   No MDM developed in the callus, nor did
the roots of "Kanamugura" form on the media mentioned above.
When small mulberry shoot tips were cultured on MS medium
supplemented with 6-benzyladenine and then transferred to the
new medium with IAA after leaves developed, the tips developed
into complete plants with leaves and roots.   The plants showed
typical symptoms of MD from the time of the root initiation
stage.   The number of MDM in the cultured plant increased
considerably at the termination of culturing compared to
the original tips, which showed that MDM multiplied during
the culturing.   For the isolation of MDM from diseased mulberry plants, a liquid MS medium with yeastolate, glucose,
and lactalbumin was used as a basic medium for the subsequent
culturing of the chopped tissues under germfree conditions.
In addition the medium was modified by adding several chemicals, such as, auxin, β-sitosterol, and mevalolactone.   The
tissues cultured on MS medium were chopped and put in the
liquid media, and then electron microscopically observed on
the 30th day after the incubation at 28°C in the dark.   MDM
particles in the tissues collapsed during cultivation in the
medium with 2,4-D or in sterilized water.   Normal appearing
organisms with nuclear or netlike structure, but without any
evidence of multiplication, were maintained in the medium
containing β-sitosterol.   Moreover, in the medium supplemented
with both ergosterol and DL-mevalolactone, the particles seemed
to have multiplied.

V.   GEOGRAPHIC DISTRIBUTION

Only limited information is available about the worldwide
geographic distribution of MD. There had been some discussion
in early research on MD in Japan on the identity or similarity of the diseases reported from Italy (Falcetto or Idropi-
sia) and France (Hydropisie or Plethore) (80), although no
clear conclusion could be drawn. From the limited available
references, it appears that the MD disease or similar diseases might occur in several countries.

In recent years (71), Korea has been facing serious problems due to frequent outbreaks of MD in mulberry fields. The
disease is mainly transmitted by *H. sellatus* and possibly by
H. sellatiformis, as the presence and distribution of the
latter already has been reported in Korea (70).

Possibly three types of MD or similar diseases is known in
the People's Republic of China (2, 14, 128, 129): mosaic-type
dwarf, common dwarf, and yellow dwarf. The mosaic-type dwarf
might be caused by a virus, whereas the other two are reportedly diseases which induce composite-type symptoms caused
by the double infection with a MLO and a virus (128, 129).
These are transmissible by grafting and by two species of
leafhopper vectors which are possibly identical with or similar to *H. sellatus* and *H. sellatiformis.* Severe outbreaks
of these diseases recently occurred in the south-central district of China. The most serious damage is caused by the
common dwarf. The rate of infected mulberry reached 80% within 3 or 4 years after new planting (2).

In West Georgia in the Soviet Union, a similar or identical disease is called "curly little leaf" or "curly dwarf."
Recently this disease became widespread. Transmission occurs
by grafting and by the vector *H. sellatus* (5, 6, 15).

In Thailand in recent years the Japanse-style methods of
mulberry cultivation have been introduced and highly developed.
So far, MD disease has not been found in Thailand. However
one species of mulberry cultivated in that country was shown
to be very susceptible to MD in Japan (94).

In Japan, MD is wiedspread in fields extending almost all
over the country, except in areas of the northern Prefectures,
and it has been most prevalent in the southwestern part since
early days of the disease occurrence. Curiously enough, there
has been no MD reported in the Hachijo island and the south-
west islands, including Ryukyu, in which the species, *M. kagayamae* Koidz or *M. australis* Poiret grow wild or as cultivated
plants. These plants are highly susceptible to MD and widely used as indicator test plants for studies on MD in Japan.

VI. SYMPTOMS

Symptoms of MD disease have many characteristics in common
with those induced by other yellows-type agents.

A. *EXTERNAL SYMPTOMS*

The first visible MD symptom after infection is curling
of a top leaf. Then, at the inception of the disease, the
affected leaves become somewhat reduced in size; there are no
serrations and lobings and the whole length of a leaf margin
becomes shorter. The surface of a leaf is markedly wrinkled
and the leaf bends downward. Branches of a tree affected
show weak growth, shortening of the whole length and of the
internodes, and irregular phyllotoxy. As the disease progresses, accessory buds and axillary buds sprout early on the
top, or whole branches become affected, producing groups of
many small branches that form a witches' broom growth. Leaves
are increasingly reduced in size and turn slightly yellow in
color. No flowers and syncarps are generally found on an
affected tree. In the last stage of the disease, the tree
becomes extremely small, producing a bunch of poor twigs, like
a broom, the leaves become extremely shrunken and yellow, and
often produce necrosis in susceptible varieties. In severely
attacked trees all twigs are killed within 2 or 3 years.

Mulberry plants seriously affected by MD have an unbalanc-
ed root system; the main root tends to be bent to one side
and markedly reduced in size. Moreover, rootlets and root
hairs become underdeveloped in number and length, and these
roots are often killed as the disease advances (78, 82, 84,
132).

B. *ANATOMICAL SYMPTOMS*

Affected leaves are markedly reduced in size, as well as
in thickness, and become yellowish because of reduced chlorophyll in assimilating tissues. The fibrovascular bundles are
stunted in the roots, twigs, and leaves. The vessels are thin
and reduced in number, their walls being thin and incompletely
lignified, the ramification of veins becomes reduced and discontinued in places, causing retention of synthesized starch
produced in the mesophyll. The pith is much thicker and the
xylem is thinner in affected twigs, as compared to healthy
ones. In roots the xylem is reduced and the cortical part
and pith are considerably thicker (3, 72). The pith of shoots
or roots is thicker in a diseased tree, and, consequently,

their density is less than in a healthy one (132). Latex
vessels of affected trees are underdeveloped thus, less latex
is produced (72).

C. SYMPTOM CATEGORIES

Tahama (116) found that the symptoms of MD in the field
of Taimei town, Kumamoto Prefecture were more severe than that
in the fields of Kikuchi city, Kumamoto Pre. At Taimei, wrinkled leaves appeared before the cutting back of mulberry
shoots in June (summer pruning), whereas at Kikuchi they were
recognized on the new shoots after pruning. These differences
in the symptom expression were not induced merely by environmental conditions of the two localities. Consequently, it
was concluded that the causal agent of MD was divided into
two strains: the Tamana severe strain and the Kikuchi mild
strain.

Ishiie and Matsuno (31) also reported two types of MD symptoms. Each group of naturally infected leafhopper vectors
collected from fields at two different localities in Tokyo
induced, respectively, extreme dwarfishness type (S type) and
non-dwarfishness type with small and somewhat curly leaves
(E type). Moreover, the groups of leafhoppers which were fed
separately on mulberry trees showing each type of symptom produced generally the same type of symptoms on the inoculated
test plants as on the original source. In addition, the diseased scions with one or the other type of symptom induced
the respective type of symptoms on grafted stocks. Thus it
was concluded that the types of MD symptoms were due to the
physiological differences of the causal agent.

D. SOME FACTORS INFLUENCING THE EXPRESSION OF MD SYMPTOMS

Factors such as low or high temperature, heat, and light
intensity profoundly influence the expression of MD symptoms
and the activity of the causal agent.

MD-diseased scions collected around December were completely cured by hot water treatment for 15 to 20 minutes at
50°C or for 2 to 5 minutes at 55°C long after having been
grafted onto healthy stocks and planted in the field (125).
The causal agent in the diseased scion wood collected during
the mulberry growing season became inactivated by dry heat
treatment for 2.5 days at 45°C or 1 day at 50°C and for 4
hours at 50°C or 40 minutes at 55°C in hot water (115). Diseased saplings were also cured by dipping their root area
in hot water for 15 minutes at 50°C or 30 minutes at 45°C
before planting (126).

Diseased trees with healthy appearing leaves under natural
conditions in April developed clear symptoms in 1-2 weeks when
placed at 30°C.   However, the severe symptoms on the trees
disappeared entirely from newly emerging leaves in 1-2 weeks
when they were kept at 37°C (111).

The symptoms expressed by the Kikuchi mild strain were
mild at high light intensity (viz. 25,000, 12,000, and 6.500
lux, with illumination for 8 hours followed by darkness for
16 hours at 25°C) and became severe at low intensity (viz.
1500 and 3000 lux) with the demarcation point being between
4000 and 5000 lux (120).   The symptoms are unchanged at 8-hour
light, 16-hour light, and 4-hour light of 1500, 3000, and
4000 lux.   At 24-hour light, the symptoms subsided slightly
at 1500 lux, two-thirds of the trees lost their symptoms at
3000 lux, and almost all of the diseased trees recovered at
4000 lux (121).

VII.   PHYSIOLOGY OF DISEASED MULBERRY TREES

Mulberry plants infected with MD generally undergo the
following physiological changes (3, 73): (1) low root pressure, (2) weak transpiration, (3) retention of assimilatory
starch in the mesophyll of leaves as a result of inactive
saccharification of starch grains and its translocation, and
(4) marked decrease in quantity of reserve starch.   Among
these changes, the most characteristic phenomenon is that
assimilatory starch, which was produced in leaves, was not
translocated through petioles, and remained in leaves due to
the defective conductive tissues in diseased leaves (73, 85).
These facts were reconfirmed in recent years (22).

The dry weight of all leaves on diseased shoots was about
40% of those on healthy ones and the respective weight of an
individual fresh leaf or dry leaf on diseased shoots averaged
below one-fifth or one-sixth of that of healthy shoots. The
respective amount of crude lipid and crude fiber was remarkably lower in diseased leaves than in healthy ones, and the
amount of non-protein nitrogen of diseased leaves was also
about two-thirds of healthy ones.   The amount of extract without nitrogen components (starch and sugar, etc.) was higher
in diseased leaves than in healthy ones (97).

The physiological changes in the tissues of mulberry
leaves at an early stage following MD infection were compared
with those of healthy trees growing under similar conditions
(130): (1) The respective specific gravity of tissue powder
of leaves without petioles from the upper, middle, and lower
parts of shoots was greater in the diseased trees, and the
value of the refractive index and osmotic pressure in the

aqueous solution, which was extracted from each of those lower leaves, was higher in diseased trees except for the material from the middle part of the shoot. (2) The respective
value of specific gravity in each part of one leaf, namely,
mesophyll, leaflet, petiole, was always higher in diseased
trees, whereas that of bark and xylem was lower in diseased
shoots, and the osmotic pressure of the extract from tissue
powder of mesophyll, excluding that from other parts, was
higher in the diseased material. (3) The specific gravity of
the tissue powder of roots (both main roots and lateral roots)
was lower in the diseased material. The translocation at the
petioles of the pigment (aqueous solution of safranin) to
leaves through a shoot was slower in diseased trees. (4) The
water content of healthy and diseased leaf blades and petioles
of leaves at the upper and middle parts of the shoots did not
differ. However, that in the blade at the middle part was
somewhat lower in diseased material. (5) The plant growth
substances were found abundantly in diseased leaves, but none
were present at the growing point, although they existed abundantly in healthy shoots. (6) Total acidity of roots decreased
as the disease advanced.

Diastase in diseased leaves was up to four times that
found in healthy ones (98). Moreover, there was generally more
oxidase and peroxidase in diseased leaves than in healthy
ones, with as much as 4-5 times the amount in healthy leaves
(107, 108). The highest activity of peroxidase was found in
the cells adjacent to discolored brown ones in very seriously
affected leaves; the backside of veins in those leaves had
turned brown, thus differing from those of healthy leaves
(23).

Nucleic acid concentrations in mulberry leaves affected by
MD were almost the same throughout different leaf ages. Therefore, in general, the top leaves of infected trees contained
lower concentrations of nucleic acid than healthy ones. In
healthy leaves the concentrations were high in the young leaves
at the top (124).

Using agarose gel electrophoresis (54), proteins in the
immature very young leaves from healthy trees were fractiona-
ted into three components; A, B, and C in cultivar Nezumigae-
shi and into A and C in cultivar Ichinose. In mature leaves
of both cultivars, however, the proteins formed only one
fraction, A. On the other hand, the proteins in immature
leaves from diseased trees revealed a marked decrease in the
concentrations of B and C components with advanced disease.
In mature leaves from diseased trees the pattern of the proteins was quite comparable to that from healthy ones.

The extract from leaf tissues, in the presence of Na_2CO_3,
contained two fluorescent components, detected by paper chro-

matography. These components inactivated MDM. Separated
mixtures of these components showed inhibitory action when the
infectivity was assayed by the insect injection method. Chemical analysis revealed that one component had a great similarity to chlorogenic acid. The other component has not been
characterized (123).

VIII. HOST RANGE

     MD has a limited host range. One known host is a weed,
"Kunamugura" (*Humulus japonicus* Sieb et Zucc.), which is an
annual plant growing on weedy areas near mulberry fields.
Hishimonus sellatus feeds readily and can be bred on this
weed. "Kanamugura" plants inoculated with MD by means of
leafhoppers showed the following symptoms: top leaves become
smaller in size and somewhat rounded followed by yellowing of
the leaf edges. A large number of MLO were found in the phloem tissue of the "Kanamugura" plants, and mulberry seedlings
expressed typical dwarf symptoms upon back-inoculation from
the diseased plants (63, 64). Another known host is the paper
mulberry, *Broussonetia kajinoki,* infected by means of *H. sellatus* (117) and by grafting (53). Recently, Shiomi *et al.* (99)
succeeded in experimentally transmitting the MD agent by infective leafhoppers to four species of plants out of 30 species
belonging to 12 families tested: white clover, red clover,
Chinese milk vetch, and *Vinca rosea,* although the back-inoculation was unsuccessful. MLO were also found in these diseased
plants.

IX. TRANSMISSION

     The MD disease agent is transmissible by grafting and by
leafhopper vectors, but not by sap inoculation, seed, or soil.

A. TRANSMISSION BY GRAFTING

     Although the specific phenomenon of graft transmission of
MD has been noted for many years, only recently has graft
transmission been properly understood. Hori and Tanaka (17)
in 1893, noticed, first, the fact that the disease was very
easily transmissible when healthy scions were grafted onto
diseased stocks. On the contrary, no transmission occurred
and the diseased scions recovered fully from the disease
when the scions were grafted onto healthy stocks. This seemed
to indicate that the causal agent of MD existed in the soil.
The authors proposed an etiological theory that a microorgan-

ism belonging to soil-inhabiting Chytridiaceae might cause MD
via graft transmission. The phenomemon of the recovery of
diseased scions was known for many years from studies of the
MD disease (18, 96, 126, 132). The following facts have been
clarified in recent years.

Scions obtained from diseased plants during the period
from May to November are highly infectious to healthy stocks
when routine grafting methods are used. However, symptoms of
MD disease do not occur on the scions or stocks when diseased
scions are obtained from fields during winter months or when
diseased scions have been stored for a long time at low temperatures before grafting onto healthy stocks. On the other
hand, diseased rooted stocks transmit the MD agent to healthy
scions at a high rate, even in the winter (21, 33, 51, 103,
110). The length of the dormant period or low temperatures
seems to strongly influence the inactivation of the agent (33).
It was shown that MDM in affected shoots can overwinter at
temperatures above 10°-15°C and at a depth of 150 cm below
ground surface (51). Under natural conditions, MDM may overwinter in some parts of the roots. Affected trees produce
apparently normal leaves at the beginning of sprouting, and
symptoms seem to appear gradually in new leaves, suggesting
that MDM may move from the roots to shoots with rising air
temperatures. Disappearance of MDM in dormant shoots and
reappearance in young sprouts was electron microscopically
confirmed (12, 65) at the same time as other mycoplasma disease agents of trees or shrubs such as paulownia witches'
broom (12) and blueberry stunt (9).

Grafting was carried out between dwarf- and mosaic-exhi-
biting plants, as well as plants with ring spot, yellowing,
or filiform symptoms. Dwarf symptoms with indistinct yellowing were observed on both stocks and scions when yellowing
seedling stocks were joined through grafts with dwarfed scions.
The dwarfing appeared systemically, giving rise later to typical filiform leaves among dwarfed leaves, in the grafting
combination of dwarf and filiform (25). Transmission of the
causal agent of the disease to healthy stocks by cleft-graft-
ing of the respective diseased shoots was obtained more easily
and more rapidly in MD than that of mosaic under the same conditions. The latent period was less than a month in the former and at least 2 months in the latter instance (24).

B. TRANSMISSION BY VECTORS

Needless to say, the most important route of infection in
natural epidemics of MD disease is transmission by insects.
Two types of leafhoppers are known as vectors: the rhombic-

marked leafhopper, *Hishimonus sellatus* Uhler, (Fig. 5), and
the false rhombic-marked leafhopper, *Hishimonoides sellatiformis* Ishihara (Fig. 4). The former is widely distributed
in mulberry fields throughout Japan (44), Korea (70), in Georgia, USSR (15), and possibly in mainland China (2), while the
latter was discovered in 1965 (20) and has been recorded at present only in Northern areas of Japan (44) and Korea (70). *Hishimonus sellatus* is said to be a possible vector of jujube witches'
broom in Korea (70).

Ikata and Matsumoto first described the transmission by
an insect (19) and recently some of the details of transmission by vectors have been studied (28, 29, 44, 51, 67, 103,
109, 114, 117).

1. *Acquisition and Infection Feeding*

The minimum acquisition feeding time of *H. sellatus* and
H. sellatiformis was found to be 3 hours (28) and definitely
less than 1 day (51), respectively. If the time of feeding
by either species becomes longer, the rate of acquisition becomes higher. Most of individuals of *H. sellatus* acquire
MDM from diseased plants in a feeding period of 10 to 20 days,
although not all acquire the disease even after feeding for
more than 35 days. Almost all individuals of *H. sellatiformis* can acquire MDM by feeding for more than 12 days (40, 51).

Infective individuals of *H. sellatus* can transmit MDM to
healthy mulberry seedlings during a minimum of 3 hours feeding
(29, 51). Transmission is considerable following 2 or more
days of inoculation feeding. Sixteen percent of *H. sellatiformis* individuals can transmit 1 hour and about 70% during 1
day of inoculation feeding (44, 51).

2. *Acquisition by Injection*

Both leafhopper vectors are able to transmit MDM when the
insects are injected with an extract made from infective insects. When both leafhopper species were injected with a 1:10
dilution of juice from infective leafhoppers, the rate of
transmission was about 40% with *H. sellatiformis* and 24-27%
with *H. sellatus*. However, the vectors were unable to transmit MDM when healthy leafhoppers were injected with extracts
made from the diseased leaves, ground with reducing agents,
enzyme inhibitors, or inactivator absorbent in buffer solution
or concentrated by differential centrifugation (105).

3. Incubation Period and Retention Period of MDM in Insects

Incubation periods of MDM in *H. sellatus* and *H. sellatiformis* are about 24 to 69 days (29, 35, 38, 103) and 18 to 35 days (38, 51, 94), respectively. Most individuals require incubation periods of about 30 and 26 days at 25°C, respectively. Therefore, the transmission of the disease agent in mulberry fields is actually by adults of both leafhopper species. After the incubation period, the leafhoppers usually remain infective over an extended period, almost until their death. Under experimental conditions some individuals retained the agent more than 100 days (34, 38). No evidence has been found for transovarial passage of MDM in *H. sellatus* (29, 103, 117).

4. Effect of Temperature during Incubation Period in Leafhoppers on Transmission of MDM

After feeding on diseased plants for 6 days, leafhoppers of both species were held at different temperatures; at 34°C there was no transmission by either species, since almost all leafhoppers died prematurely before completing their nymphal stages; little transmission occurred at 30°C. There was a higher rate at 28°C in *H. sellatiformis,* and the best transmission by both species occurred at 20°C to 25°C, although there was a considerable prolongation of their life span and of the incubation period of the agent in the insects (35, 38).

On the basis of results obtained so far, *H. sellatiformis* showed a higher transmission efficiency than *H. sellatus* in terms of shorter acquisition feeding and incubation period and since there is uninterrupted transmission without intermission until death (40). On the other hand, *H. sellatus* has a larger population and a wider distribution throughout the country. Consequently, it plays a more important role in the epidemiology of MD disease than *H. sellatiformis.*

5. Time of Vector's Transmission in Mulberry Fields

In the first generation of *H. sellatus* (three generations emerge generally in a year), there may be no (or only a small percentage of) infective leafhoppers in the field. Initial acquisition of MDM and its transmission are effected by the second generation of leafhoppers, around July to August, inhabiting any fields where a serious occurrence of the disease has been induced by summer pruning of mulberry trees. After this, in heavily infected fields, the percentage of infective individuals may be extremely high, with figures sometimes reaching 20-50% (32, 37, 42, 67, 68, 89, 93); this is particularly the case in fields where almost all healthy leaves have been removed for rearing silkworms and where diseased leaves and branches have been carelessly left on the trees. The different degrees of the MD prevalence in respective different habitats of *H. sellatus* do not relate to its ability for transmitting MDM when it is sufficiently fed on diseased mulberry (61, 103).

6. *Physiology of Infective Leafhopper Vectors*

The total soluble proteins of adult *H. sellatus* fed on diseased mulberry trees revealed a decrease or a disappearance of some protein fractions compared with those of the leafhoppers fed on healthy trees. Occurrence of the individuals showing hypoproteinemia among leafhoppers fed on diseased material was 71.1% in the female and 62.1% in the male, and the hypoproteinemia appeared markedly at more than 11-20 days after feeding on diseased sap (55). All of those leafhoppers were electron microscopically confirmed to have a number of MDM in their salivary glands and able to cause MD in healthy mulberry plants by feeding inoculation. Most of the leafhoppers that had no hypoproteinemia after feeding on the diseased trees lacked MDM in their salivary glands. Similar observations were made in leafhoppers collected from mulberry fields infested with MD (56). The rank reciprocity between the rate of leafhoppers with hypoproteinemia and that of infective leafhoppers followed by feeding inoculation to test plants was significant with r_s= 9.19 within 1% aberration. Therefore, it may be possible, to a certain degree, to estimate the rate of infective leafhoppers inhabiting mulberry fields by studying electrophoretic patterns of the vector's body protein (57).

Histopathological research revealed the high frequency of Feulgen-positive parts in the cytoplasm of salivary gland cells in adult *H. sellatus* infected with MD, but not in healthy ones. The cells containing Feulgen-positive parts were deformed and vacuolated. The multiplication of MDM was confirmed electron microscopically in the cytoplasm of salivary gland cells. Fewer Feulgen-positive parts were observed in Malpighian and fat body cells. These results suggested that the Feulgen-positive part in the cytoplasm coincided with the multiplication of MDM (58).

X.   RELATIONSHIP BETWEEN MDM AND HOST PLANTS

A.   INCUBATION PERIOD IN MULBERRY PLANTS

Under suitable greenhouse conditions, 10-20 days may be
required for the appearance of the initial symptoms in young
test plants of *Morus australis* or *Morus kagayamae* (29, 44, 94);
however, the appearance of symptoms in cultivated mulberry
plants is much more complicated. In the field, the appearance
of symptoms is related to the time of infection, methods and
amount of harvest, rate of plant growth, degree of susceptibility, and other factors. Under average field conditions,
about 1 month may normally be required for the appearance of
symptoms on mulberry plants inoculated by three infective
leafhoppers in June or July, although there may be a few plants
in which symptoms appear after a longer period, 300-350 days
(41, 47). Later inoculation is usually followed by the appearance of symptoms after summer pruning in the following
year, after a long incubation period ot about 300 days or more.
Inoculation made very late in the season (such as in October)
is much less effective. One might assume that MDM is inactivated at low temperatures during the winter before it can
reach the sites where it is capable of overwintering. The
inoculation period tends to become shorter with increases in
the amount and frequency of harvesting and with frequency and
degree of cutting of branches.

B.   RECOVERY

Under field conditions, recovery from the disease or a
reduction in the severity of the symptoms is sometimes very
conspicuous in plants in which symptoms have been mild or moderate, but is rarely seen in plants with severe symptoms.
This is especially noticeable after harvest or cutting is
halted, when mulberry trees are permitted to grow naturally
(73), and it is very frequent after spring pruning (51, 117)
when the influence of the disease on the growth of the trees
is reduced, compared with summer pruning. Cases of apparent
recovery may be so complete that it is difficult or impossible
to detect symptoms for a long time, even though, in some
plants, the symptoms reappear following harvesting during
early autumn. One explanation might be that recovery after
spring pruning may be due to the difference in optimal temperatures for the growth of plants and the multiplication of
MDM, that is, the plants can grow and sprout while the agent

is still latent in the roots because of the low temperatures.
Subsequently, good growth of the plant may provide a high degree of protection against attack by MDM after air temperatures rise. The recovery rarely occurs in plants pruned in
summer. In addition, the recovery is frequently observed in
the diseased plants in the field after they are transplanted
to other places in winter, and even in diseased saplings that
showed symptoms in nurseries, after they are planted in fields
in early spring (53). Thus, the phenomena of recovery is rather common in MD-diseased trees. The clarification of its
mechanism should be very important for practical control of
the disease.

XI. EPIDEMIOLOGY

     It was concluded that mulberry fields may suffer outbreaks
of MD disease within about 4-5 years after initial planting,
if the fields are adequately provided with two major factors -
the existence of nearby infected source-plants and leafhopper
vectors (39, 42). A high incidence of MD disease generally
occurs in large mulberry fields on level land along rivers or
near the seaside, under conditions that favor transmission.
In recent years, however, MD began to expand to scattered
fields in mountainous areas, as a result of the enlarged area
of diseased mulberry plantings.
     One epidemiological investigation was made on (1) newly
planted mulberry fields located at a distance of 2 km or more
from established fields which have experienced epidemics of
MD; (2) newly planted fields about 0.5 km distant from already
established fields; (3) newly planted fields adjacent to established fields; and (4) the established fields. The results
obtained (42) are described below.

A. ANNUAL OCCURRENCE OF THE DISEASE

     In newly planted fields at larger or smaller distances
from established ones, 0.1 to 0.2% of initial occurrence of
the disease was observed by the beginning of August, but there
was no increase thereafter. This type of occurrence, generally low in frequency, seems to be induced by mulberry saplings
already infected in the nursery (36). On the contrary, in
the fields adjacent to already established ones, following
the small initial occurrence of the disease by the beginning
of August, the incidence of the disease rapidly increased,
affecting about 30% of the trees by the middle of October.
These high frequencies of MD seemed to be due to transmissio
by *H. sellatus*. On the other hand, in already established

fields, after summer pruning at the beginning of June, a rapid
increase of the disease occurred and the incidence attained
50-60% of the trees by the beginning of August. This seemed to
be caused by the infection during the preceding year or even
earlier, judging by the time of appearance of infective leafhoppers and the known incubation period in plants. Furthermore,
the incidence of the disease increased to 60 or 80%, with mild
symptoms at the tips of shoots, by the middle of October. This
increase seemed to be caused by infection during the current
season. However, the incidence noted in newly planted fields
decreased in the second year to about one-third of that in
the first year. This was assumed to be due to recovery after
spring pruning, as already indicated, which is usually done in
the second year after planting. In any case, the newly planted fields adjacent to established ones suffered outbreaks of
about 45-70% of the disease rate only 3 years after planting,
in contrast with the situation in new, isolated fields.

B. SEVERITY OF THE SYMPTOMS AND THEIR COURSE

    In the first year of planting, all diseased trees showed
mild symptoms, which appeared mainly on the tips of shoots.
In the second and third year, 23 and 44% presented symptoms
of more advanced, moderate, and severe conditions, respectively. Thus the state of the disease becomes worse each year
after planting. All diseased plants showed a very complicated course of symptom progression following observations over
a 3-year period. The most frequently encountered course was:
mild (1st year), moderate or severe (2nd year), moderate or
severe (3rd year), under cultural conditions of summer pruning
in the second and third years after planting. Infrequently,
the course of mild-healthy-severe or of mild-severe-healthy,
was also observed. About 50% or more of the trees which had
been healthy in each of the preceding years presented moderate
or severe symptoms. This suggested that the symptoms in infected trees which had latent infections because of the time
of infection or some cultural conditions developed serious
symptoms because of the summer pruning in the following year.

C. HABITAT AND INFECTIVITY OF THE VECTOR

    Although the population of leafhoppers in isolated fields
was comparatively small, it showed a tendency to increase
with the passage of years. No infective leafhoppers were detected in any of these fields, which corresponded to a lack
of occurrence of MD. On the other hand, in the fields adjacent to established plantings, the population soon rose

after the trees had sprouted. Infective leafhoppers were detected from the beginning of August in the newly planted as
well as in the established fields. The percentage of infective leafhoppers increased during the period of observation,
corresponding with a rapid increase in the incidence of the
disease in the newly planted fields at the end of August. In
the second year and thereafter, large numbers of leafhoppers
were found in the newly planted fields and many of them were
found to be infective.

XII. CONTROL

     The improvement of cultural practices had been the main
approach for many years, based on the hypothesis that MD had
a physiological etiology. However, this approach was not totally effective and serious outbreaks continued. There are
no practical methods of curing the affected trees with chemicals; some antibiotics known to suppress the development of
symptoms may, at present, be inapplicable under field conditions.
     In order to prevent the spread of MD disease from diseased
trees, the extermination of the leafhopper vectors and removal
of infected plants is of paramount importance in early epidemics. However, there are some problems: (1) restricted uses
of insecticides in mulberry fields because of their toxic effects on silkworms; (2) the increasing popularity of multiple
rearings of silkworms 6-8 times a year that would be deleter-
iously affected by chemicals; and (3) the arduous work and
other difficulties in removing diseased plants from fields,
especially during the growing seasons. A standard practical
control program has been proposed; the basic details are given below. The considerable effect of this program, when it
was almost fully implemented by farmers, accounted for its
efficiency in areas infested with MD (7, 45, 46).

A. EXTERMINATION OF VECTORS

     Under a standard rearing of silkworms 3-4 times a year
the following course of extermination is used:
(1) About 1 week before the mulberry plants sprout, usually
in April, spray a solution of PAP emulsion over all the stems
to kill hybernating eggs scattered under the epidermis of
shoots. (2) At the end of May or the beginning of June, after summer pruning, apply PAP dust or emulsion or similar
type of chemical to kill the first and second generations of
the leafhopper. One more spraying is desirable after about
10 days. (3) For the third spraying use DEP + MTMC (**or** mecar-

bam + DEP) dusts at the beginning or middle of July to kill
not only the leafhopper but also other harmful insects. (4)
The fourth spraying (as well as the third) is very important
in order to decrease the highest population of infective leafhoppers of the second and third generations; the fourth spraying is rather difficult because early autumn rearing is continuously followed by later autumn rearings of silkworms.
Therefore, chemical control with DDVP, DEP, or vamidothion,
that are of shorter duration or which are selectively toxic
to the leafhopper, should be performed (49). A mixed application of the former two chemicals with vamidothion is highly
recommended. For multiple rearings of silkworms, the modification of the times and frequencies of spraying, and the solution of appropriate chemicals are very important.

B. REMOVAL OF DISEASED TREES

The removal of affected trees to eliminate sources of
inoculum as early as possible after the appearance of symptoms
is highly recommended. If this is difficult, cutting off all
diseased branches, covering the whole plant after cutting with
a used vinyl fertilizer bag, or at least killing diseased
plants by herbicides or kerosene is recommended.

C. CULTIVATION METHODS

Affected fields should be exclusively used for summer-
autumn rearing, thus avoiding summer pruning. The methods of
training and harvesting should be improved, using high-out
training and following a 2-year rotation of mulberry harvest
with one alternate year of pruning.

D. INTRODUCTION OF RESISTANT VARIETIES

Mulberry varieties immune to MD have not yet been developed. Among the three main species grown in Japanese silkworm-
producing areas, the "Yamaguwa" type (*Morus bombycis* Koidz.)
has a tendency to be susceptible to MD, the "Karayamaguwa"
type (*M. alba* Linn.) is intermediate, while the "Roso" type
(M. *latifolia* Poiret) seems to be more resistant. Among these
types, Oshimaso, Syukakuichi, and Akameroso may be considered
as being relatively resistant cultivars.

XIII.   CONCLUSION

As indicated above, MD is one of the most destructive diseases of mulberry trees. As a consequence, the successful
control of the disease is urgently needed. Any further increase of disease outbreaks may result in the decline or even
destruction of the Japanese sericultural industry, which at
the present time is already confronted with many other difficulties. Therefore, it is absolutely necessary to make
every effort to overcome the MD disease. In addition, in view
of the serious nature of the MD disease, great care must be
taken in the future when promoting sericulture in Asian or
other countries. It has been found that a *Morus* species from
Thailand is also very susceptible to MD in Japan (94).

For the establishment of future efficient control measures
against MD, the following may be suggested as the most important steps: establishment of simplified and reliable methods
of culturing MDM on artifical media and cloning of the microorganisms; breeding of immune or more resistent cultivars with
good quality and high yield of leaves for silkworm rearing;
invention of more effective chemicals and their application
methods for curing of MD-infected mulberry and other trees.

REFERENCES

1.   Akiya, K. (1931). *J. Sericult. Sci. Jap. 2,* 199-200 (In
     Japanese.)
2.   An editorial committee for Chekiang plant protection handbook (1978). "Handbook on the Control of Diseases and
     Pests of Mulberry Tree," p. 201, Agr. Publ. Co., Peking.
     (In Chinese.)
3.   Aso, K. (1903). "Report of Investigation on Mulberry
     Dwarf Disease," Vol. 7, pp. 579-626. Bur. Agr., Min.
     Agr. Comm., Japan (In Japanese.)
4.   Bureau of Agriculture, Ministry of Agriculture and Commerce. (1897-1903). "Report of Investigation on Mulberry
     Dwarf Disease," Vol. 1, p. 204; Vol. 2, p. 256; Vol. 3,
     p. 154; Vol. 4, p. 277; Vol. 5, p. 590; Vol. 6, p. 589;
     Vol. 7, p. 688. Bur. Agr., Min. Agr. Comm., Japan. (In
     Japanese.)
5.   Chaduneli, M. D. (1973). *Tr. Nauch.-Issled. Inst. Zashch.
     Rast. Gruz. SSR 25,* 152-154. *(Rev. Plant Pathol. 54,*
     3413.)
6.   Chaduneli, M. D. and Chkheidze, N. Z. (1974). *Tr. Nauch.-
     Issled. Inst. Zashch. Rast. Grez SSR 26,* 129-130. *(Rev.
     Plant Pathol. 55,* 2813).

7. Chanoki, N. *et al.* (1978). "Investigation on the Control of Mulberry Dwarf Disease - Effect of Eradication of Leafhopper Vector on the Disease Occurrence in Mulberry Fields," p. 41. Miyazaki Agr. Expt. Sta. (In Japanese.)
8. Chen, T. A. and Granados, R. R. (1970). *Science 167*, 1633-36.
9. Chen, T. A. (1971). *Phytopathology 61*, 233-236.
10. Chiu, R. J. and Black, L. M. (1967). *Nature (London) 215*, 1076-1078.
11. Doi, Y. *et al.* (1967). *Ann. Phytopathol. Soc. Jap. 33*, 259-266. (Engl. sum.)
12. Doi, Y. and Okuda, S. (1973). *Shokubutsu Byogai Kenkyu 8*, 203-221. (Engl. sum.)
13. Endo, Y. (1935). *In* "The History of the Japanese Sericultural Industry" (Dainihon Sanshikai, ed.), Vol. 4, pp. 73-86. Dainihon Sanshikai, Tokyo (In Japanese).
14. Faan, H. C. *et al.* (1964). *Acta Phytopathol. Sinica 7*, 151-156. *(Rev. Appl. Mycol. 44, 416.)*
15. Giogadze, D. G. and Tulashvili, N. D. (1973). *Tr. Nauch. Issled. Inst. Zashch. Rast. Gruz. SSR. 24*, 209-213. *(Rev. Plant. Pathol. 53, 4527)*
16. Grace, T. D. C. (1962). *Nature (London) 195*, 788-789.
17. Hori, S. and Tanaka, E. (1893). *Dainihon Sanshikaiho 8*, 20-29. (In Japanese.)
18. Ichikawa, E. (1901-1902). *In* "Report of Investigation on Mulberry Dwarf Disease," Vol. 5, pp. 329-344; Vol. 6, 377-389. Bur. Agr., Min. Agr. Comm., Japan (In Japanese.)
19. Ikata, S. and Matsumoto, S. (1931). *Ann. Phytopathol. Soc. Jap. 13*, 440. (In Japanese).
20. Ishihara, T. (1965). *Jap. J. Appl. Entomol. Zool. 9*, 19-22.
21. Ishiie, T. *et al.* (1960). *J. Sericult. Sci. Jap. 29*, 319-326. (Engl. sum.)
22. Ishiie, T. *et al.* (1960). *Sansi-Kenkyu 33*, 33-38. (In Japanese.)
23. Ishiie, T. *et al.* (1960). *Sansi-Kenkyu 33*, 39-42 (In Japanese.)
24. Ishiie, T. *et al.* (1964). *Sansi-Kenkyu 53*, 19-28. (Engl. sum.)
25. Ishiie, T. *et al.* (1965). *J. Sericult. Sci. Jap. 34*, 115-120 (Engl. sum.)
26. Ishiie, T. *et al.* (1967). *Ann. Phytopathol. Soc. Jap. 33*, 267-275. (Engl. sum.)
27. Ishiie, T. (1970). *Jap. Agr. Res. Quart. 5*, 48-53.
28. Ishiie, T. and Matsuno, M. (1971). *Ann. Phytopathol. Soc. Jap. 37*, 136-140. (Engl. sum.)

29. Ishiie, T. and Matsuno, M. (1971). *Ann. Phytopathol. Soc. Jap. 37,* 141-146. (Engl. sum.)
30. Ishiie, T. *et al.* (1971). *J. Sericult. Sci. Jap. 40,* 395-398. (Engl. sum.)
31. Ishiie, T. and Matsuno, M. (1971). *Sansi-Kenkyu 78,* 65-75. (Engl. sum.)
32. Ishijima, T. (1963). *Ann. Phytopathol. Soc. Jap. 28,* 292 (In Japanese.)
33. Ishijima, T. (1965). *Ann. Phytopathol. Soc. Jap. 30,* 275. (In Japanese.)
34. Ishijima, T. (1966). *J. Sericult. Sci. Jap. 35,* 198 (In Japanese.)
35. Ishijima, T. (1966). *Ann. Meet. Sericult. Sci. Kyushu 41,* 10. (In Japanese.)
36. Ishijima, T. (1967). *Proc. Assoc. Plant. Protection Kyushu 12,* 127-139. (In Japanese.)
37. Ishijima, T. (1967). *Ann. Phytopathol. Soc. Jap. 34,* 204 (In Japanese.)
38. Ishijima, T. (1967). *Ann. Phytopathol. Soc. Jap. 33,* 102-103. (In Japanese.)
39. Ishijima, T. (1968). *Bull. Sericult. Exp. Sta. 23,* 81-101. (Engl. sum.)
40. Ishijima, T. (1969). *Ann. Meet. Sericult. Sci. Jap. 39,* 8. (In Japanese.)
41. Ishijima, T. (1969). *Proc. Assoc. Plant Protection Kyushu 15,* 198 *(In Japanese.)*
42. Ishijima, T. (1969). *Bull. Sericult. Expt. Sta. 23,* 411-440. (Engl. sum.)
43. Ishijima, T. (1969). *Ann. Phytopathol. Soc. Jap. 35,* 132. (In Japanese.)
44. Ishijima, T. (1971). *J. Sericult. Sci. Jap. 40,* 136-140. (Engl. sum.)
45. Ishijima, T. (1973). *Proc. Sericult. Sci. Kyushu 4,* 35. (In Japanese.)
46. Ishijima, T. (1974). *Proc. Sericult. Sci. Kyushu 5,* 33. (In Japanese.)
47. Ishijima, T. (1975). *Proc. Sericult. Sci. Kyushu 6,* 30. (In Japanese.)
48. Ishijima, T. (1976). *Proc. Sericult. Sci. Kyushu 7,* 48 (In Japanese.)
49. Ishijima, T. (1977). *Proc. Sericult. Sci. Kyushu 8,* 35. (In Japanese.)
50. Ishijima, T. (1977). *Proc. Sericult. Sci. Kyushu 8,* 34. (In Japanese.)
51. Ishijima, T. (1978). *In* "Plant Disease Due to Mycoplasmalike Organism," pp. 104-115. Compiled by FFTC. FFTC, Taipei.
52. Ishijima, T. (1978). *In* "Studies on the Culturing and Clarification of the Pathogenicity of Plant Parasitic

Mycoplasma-like Organisms," pp. 108-119. Sec. Agr. For. Fish. Res. Council, MAFF, Japan. (In Japanese.)

53. Ishijima, T. Unpublished data.
54. Ishizaka, T. (1971). *J. Sericult. Sci. Jap. 40*, 391-394. (Engl. sum.)
55. Ishizaka, T. (1972). *J. Sericult. Sci. Jap. 41*, 465-469. (Engl. sum.)
56. Ishizaka, T. (1976). *J. Sericult. Sci. Jap. 45*, 431-436. (Engl. sum.)
57. Ishizaka, T. (1980). *J. Sericult. Sci. Jap. 49*, 367-368. (In Japanese.)
58. Ishizaka, T. (1980). *J. Sericult. Sci. Jap. 49*, 342-346.
59. Kasumi, I. (1972). *Bull. Tottori Sericult. Expt. Sta. 22*, 1-63 (In Japanese.)
60. Kawai, I. (1939). *Ann. Phytopathol. Soc. Jap. 9*, 16-21. (Engl. sum.)
61. Kawakita, H. and Ishiie, T. (1969). *Sansi-Kenkyu 72*, 49-53 (In Japanese.)
62. Kawakita, H. and Ishiie, T. (1970). *J. Sericult. Sci. Jap. 39*, 413-419. (Engl. sum.)
63. Kawakita, H. and Ishiie, T. (1970). *Sansi-Kenkyu 75*, 38-46. (In Japanese.)
64. Kawakita, H. and Ishiie, T. (1976). *Bull. Sericult. Expt. Sta. 26*, 443-454. (Engl. sum.)
65. Kawakita, H. (1977). *J. Sericult. Sci. Jap. 46*, 404-410. (Engl. sum.)
66. Kawakita, H. (1978). *J. Sericult. Sci. Jap. 47*, 465-471. (Engl. sum.)
67. Kojima, A. and Takahashi, A. (1976). *Bull. Gunma Sericult. Expt. Sta. 49*, 1-8. (In Japanese.)
68. Kojima, A. and Takahashi, A. (1979). *Bull. Gunma Sericult. Expt. Sta. 52*, 51-56. (In Japanese.)
69. La, Y. J. *et al.* (1975). *Sericult. J. Korea 17*, 175.
70. La, Y. J. (1978). *In* "Plant Disease Due to Mycoplasma-like Organisms," pp. 84-90. Compiled by FFTC. FFTC, Taipei.
71. La, Y. J. (1978). Personal communication.
72. Miyoshi, M. (1900). *In* "Reports of Investigation on Mulberry Dwarf Disease," Vol. 4, pp. 188-241. Bur. Agr., Min. Agr. Comm., Japan. (In Japanese.)
73. Miyoshi, M. (1901-1903). *In* "Reports of Investigation on Mulberry Dwarf Disease," Vol. 5, pp. 465-571; Vol. 6, pp. 461-564; Vol. 7, pp. 627-628. Bur. Agr., Min. Agr. Japan. (In Japanese.)
    ige, T. and Skoog, F. (1962). *Physiol. Plant 15*,
    et al. (1974). *Jap. J. Appl. Entomol. Zool. 9*,

76. Nasu, S. *et al.* (1978). *In* "Studies on the Culturing and
    Clarification of the Pathogenicity of Plant Parasitic
    Mycoplasma-like Organisms," pp. 11-21. Sec. Agr. For.
    Fish. Res. Council, MAFF, Japan. (In Japanese.)
77. Nasu, S. *et al.* (1978). *In* "Studies on the Culturing and
    Clarification of the Pathogenicity of Plant Parasitic
    Mycoplasma-like Organisms," pp. 22-107. Sec. Agr. For.
    Fish. Res. Council, MAFF, Japan. (In Japanese.)
78. Nishikawa, M. (1929). "The Latest Views on Mulberry Dwarf
    Disease," p. 240. Meibundo, Tokyo. (In Japanese.)
79. Nishio, A. *et al.* (1972). *Bull. Tokushima Sericult. Expt.
    Sta. 13*, 1-84. (Engl. sum.)
80. Nomura, H. (1899). *In* "Reports of Investigation on Mul-
    berry Dwarf Disease," Vol. 3, pp. 145-154. Bur. Agr., Min.
    Agr. Comm., Japan. (In Japanese.)
81. Oka, S. and Ohyama, K. (1973). *J. Sericult. Sci. Jap. 42*,
    317-324. (Engl. sum.)
82. Okabe, K. (1955). *Bull. Gunma Sericult. Expt. Sta. 30*,
    28-33. (In Japanese.)
83. Okada, T. (1978). *In* "Plant Diseases Due to Mycoplasma-
    like Organisms," pp. 29-41. Compiled by FFTC. FFTC,
    Taipei.
84. Omori, J. (1900). *In* "Reports of Investigation on Mulberry
    Dwarf Disease," Vol. 4, pp. 110-120. Bur. Agr., Min.
    Agr. Comm. Japan. (In Japanese.)
85. Omori, J. (1903). *In* "Reports of Investigation on Mulberry
    Dwarf Disease," Vol. 7, pp. 527-547. Bur. Agr., Min.
    Agr. Comm. Japan. (In Japanese.)
86. Onishi, M. *et al.* (1971). *Bull. Fac. Text. Sci. Kyoto
    Univ. Ind. Art. Text. Fibers 6*, 101-107. (Engl. sum.)
87. Saitama Sericult. Experiment Station (1932). *Tech. Bull.
    Saitama Sericult. Expt. Sta. 10*, 1-110. (In Japanese.)
88. Saitama Sericult. Experiment Station (1936). *Tech. Bull.
    Saitama Sericult. Expt. Sta. 18*, 1-113. (In Japanese.)
89. Sakai, S. (1935). *Rept. Sericult. Assoc. Jap. 44*, 39-52.
    (In Japanese.)
90. Sakai, S. (1937). *Bull. Nagano Sericult. Expt. Sta. 39*,
    1-68. (In Japanese.)
91. Sakai, S. (1937). *Agr. Hort. 12*, 2683-2688. (In Japanese.)
92. Sakai, S. (1937). *Bull. Nagano Sericult. Expt. Sta. 39*,
    1-14. (In Japanese.)
93. Sakai, S. (1937). *J. Moricult. Sci. 9*, 8-13. (In Japanese.)
94. Sato, M. and Takahashi, K. (1976). *Sansi-Kenkyu 98*, 68-73.
    (Engl. sum.)
95. Secretariat Agriculture Forestry, Fisheries Research Coun-
    cil. (1978). "Studies on the Culturing and Clarification
    of the Pathogenicity of Plant Parasitic Mycoplasma-like
    Organisms," p. 295. Sec. Agr. For. Fish. Res. Council,
    MAFF. Japan. (In Japanese.)

96. Sericult. Experiment Station (1929). *Tech. Bull. Seri- cult. Expt. Sta. 38,* 1-100. (In Japanese.)
97. Sericult. Training Institute (1899). *In* "Reports of in- vestigation on Mulberry Dwarf Disease," Vol. 3, pp. 33-37. Bur. Agr., Min. Agr. Comm., Japana (In Japanese.)
98. Shibata, K. (1903). *In* "Reports of Investigation on Mul- berry Dwarf Disease," Vol. 7, pp. 435-526. Bur. Agr., Min. Agr. Comm., Japan. (In Japanese.)
99. Shiomi, T. *et al.* (1979). *Ann. Phytopathol. Soc. Jap. 45,* 87. (In Jap.)
100. Shirai, K. (1890). *J. Sericult. Ind. Jap. 32,* 1-2. (In Japanese.)
101. Shirai, K. (1892). *Rept. Sericult. Assoc. Jap. 2,* 12-19. (In Japanese.)
102. Shirai, K. (1893). *J. Sericult. Ind. Jap. 57,* 1-16. (In Japanese.)
103. Sugiura, M. (1963). *Spec. Rept. Lab. Plant Pathol. Univ. Kyushu 2,* 1-98. (In Japanese.)
104. Sugiura, M. *et al.* (1978). *In* "Plant Diseases Due to Mycoplasma-like Organisms," pp. 166-174. Compiled by FFTC. FFTC, Taipei.
105. Suto, Y. and Ishiie, T. (1970). *J. Sericult. Sci. Jap. 39,* 451-457. (Engl. sum.)
106. Suto, Y. and Ishiie, T. (1971). *Sansi-Kenkyu 79,* 63-70. (Engl. sum.)
107. Suzuki, U. (1900). *Bull. Coll. Agr. Tokyo Imp. Univ. 4,* 167-226.
108. Suzuki, U. (1901). *In* "Reports of Investigation on Mul- berry Dwarf Disease," Vol. 5, pp. 401-441. Bur. Agr. Min. Agr. Comm., Japan. (In Japanese.)
109. Tahama, Y. (1960). *Bull. Kumamoto Sericult. Expt. Sta. 7,* 1-20. (In Japanese.)
110. Tahama, Y. (1961). *Ann. Phytopathol. Soc. Jap. 26,* 165-169. (Engl. sum.)
111. Tahama, Y. (1963). *Ann. Phytopathol. Soc. Jap. 28,* 195-197. (Engl. sum.)
112. Tahama, Y. (1963). *Ann. Phytopathol. Soc. Jap. 28,* 49-52. (Engl. sum.)
113. Tahama, Y. (1964). *Ann. Phytopathol. Soc. Jap. 29,* 185-188. (Engl. sum.)
114. Tahama, Y. (1964). *J. Sericult. Sci. Jap. 33,* 167-170. (Engl. sum.)
115. Tahama, Y. (1964). *Ann. Phytopathol. Soc. Jap. 29,* 39-42. (Engl. sum.)
116. Tahama, Y. (1967). *Ann. Phytopathol. Soc. Jap. 33,* 156-161. (Engl. sum.)
117. Tahama, Y. (1968). *Bull. Kumamoto Sericult. Expt. Sta. 40,* 5-184. (Engl. sum.)

118.   Tahama, Y. *et al.* (1972). *Bull. Hiroshima Agr. Coll. 4,* 198-207. (Engl. sum.)

119.   Tahama, Y. (1973). *Bull, Hiroshima Agr. Coll. 4,* 283-293. (Engl. sum.)

120.   Tahama, Y. (1971). *Bull. Hiroshima Agr. Coll. 4,* 97-105. (Engl. sum.)

121.   Tahama, Y. (1974). *Bull. Hiroshima Agr. Coll. 5,* 23-33. (Engl. sum.)

122.   Takahashi, A. and Kojima, S. (1979). *Bull. Gunma Sericult. Expt. Sta. 52,* 51-56. (In Japanese.)

123.   Takahashi, K. *et al.* (1978). *In* "Studies on the Culturing and Clarification of the Pathogenicity of Plant Parasitic Mycoplasma-like Organisms," pp. 120-167. Sec. Agr. For. Fish. Res. Council, MAFF., Japan. (In Japanese.)

124.   Takamura, Y. *et al.* (1960). *Phytopathol. Z. 38,* 123-128.

125.   Takeichi, D. (1938). *Sanshikaiho 556,* 30-34. (In Japanese.)

126.   Takeichi, D. (1938). *Sanshikaiho 557,* 50-54. (In Japanese.)

127.   Tanaka, E. and Ichikawa, J. (1894). *Bull. Sericult. Ach. Sericult. Expt. Sta. 10,* 262-275. (In Japanese.)

128.   Virus Research Group, Shanghai Institute Biochemistry (1974). *Sci. Sinica 17,* 421-427 (*Rev. Plant Pathol. 54,* 3414).

129.   Virus Research Group, Shanghai Institute Biochemistry (1974). *Sci. Sinica 17,* 428-434 (*Rev. Plant Pathol. 54,* 3415).

130.   Yahiro, M. and Saeki, Y. (1956). *J. Sericult. Sci. Jap. 26,* 5-12. (Engl. sum.)

131.   Yamada, Y. *et al.* (1978). *Ann. Phytopathol. Soc. Jap. 44.* 35-46. (Engl. sum.)

132.   Yamagata Sericultural Experiment Station. (1937). *Bull. Yamagata Sericult. Expt. Sta. 2,* 1-20. (In Japanese.)

133.   Yamamoto, K. (1891). *J. Sericult. Ind. Jap. (37),* 7-11. (In Japanese.)

LETHAL YELLOWING DISEASE
OF COCONUT AND OTHER PALMS

Karl Maramorosch and Peter Hunt

Rutgers University, New Brunswick, New Jersey

and

L.P.T.I. Sub-Station, Solok, Sumatera Barat, Indonesia

I. INTRODUCTION

The importance of coconut palms is often underestimated,
especially in industrial nations, because the acreage under
production is not well known and a high proportion of the
products are consumed locally. It has been estimated that
about 70% of the area, totalling 7 million hectares, lies in
Asia, followed by 10% in the Caribbean Islands, Central and
South America, close to 10% in the Pacific Islands, and more
than 5% in Africa. Also important is the fact that coconut
palms often grow in areas that are unsuitable for other crops.
The individual holdings range from very small gardens to very
large plantations. In India, several million people, mainly
subsistence farmers with holdings of less than 1 ha, own more
than 90% of the total coconut growing areas.
    It is often overlooked that in southeast Asia, as well as
in India, Sri Lanka, and the Pacific Islands, more than 50%
of the coconuts are consumed fresh as food. Therefore, the
coconut palm is often called the "tree of life." In certain
small atolls of Oceania, coconut palms provide all major
necessities for the inhabitants: water from the nuts is the
only source of drinkable beverage, the oil is used for cook-
ing, the meat is consumed fresh, the wood and fronds are used
for the construction of houses, and the palms are the only
source of shade. In areas where coconut palms are essential
for the survival of the inhabitants, coconut palm epiphytotics have a tremendous impact which goes far beyond the destruction of a food source or export item.

185

The coconut palm is unusual among plants in that it is
afflicted by a large number of lethal diseases. Determination
of the causal agents of these diseases has proved exceptional-
ly difficult and for many it is still unknown. However, a
bizarre collection of plant pathogens have been implicated.
Thus, all evidence points to a mycoplasma-like agent as the
cause of lethal yellowing. Red ring disease in the east Caribbean region is caused by a nematode, which invades all parts
of the palm (Fenwick, 1963). Cedros wilt in Trinidad (Waters,
1978; Griffith *et al.*, 1979) and hartrot in Surinam (Parthasarathy *et al.*, 1976; Parthasarathy and Van Slobbe, 1978) are
associated with *Phytomonas,* a phloem-restricted protozoan
flagellate. Rickettsia-like bacteria have been observed in
coconut palms affected by a newly described disease in Tanzania (Steiner *et al.*, 1977). Lethal bole rot in East Africa
is believed to be caused by a fungus (Bock *et al.*, 1970).
Kerala wilt, a serious disease in southern India has been extensively studied; no pathogen has been confirmed, although
a virus is suspected and the occasional presence of virus-
like particles has been detected in leaves (Maramorosch and
Kondo, 1977). Recent evidence (Randles *et al.*, 1977, 1979)
implicates a viroid as the cause of cadang-cadang, a coconut
disease which has devastated coconut production in parts of
the Philippines. Bud rot is usually ascribed to the fungus
Phytophthora palmivora. A mycoplasma-like organism has been
seen in the phloem of young palms with the newly named disease,
coconut stem necrosis, in Peninsular Malaysia and Sumatra
(Turner *et al.*, 1978). Finally, the causes of Malaysian wilt
(Sharples, 1928) and of diseases in Sarawak (Turner, 1963),
the Natuna Islands of Indonesia (Sitepu, 1979), the New Heb-
rides (Calvez, 1979) and Ivory Coast (Quillec *et al.*, 1978;
Julia, 1979) all remain obscure.

Cadang-cadang, Kerala wilt, and lethal yellowing (LY) have
each had an enormous impact in the areas they affect and most
of the other diseases referred to above are locally or regionally of great concern. They all pose threats to coconut production internationally and should, therefore, be a major
consideration for those involved in the exchange of germ plasm
and the genetic improvement of coconuts (Van der Graaff, 1979).

It is significant that, in addition to the ultimate death
of the palm, all of the above diseases have many symptoms in
common. These include, in various combinations: yellowing,
browning and wilting of leaves; necrosis and foreshortening
of the spear leaf; nutfall; rachilla necrosis in unopened and
opening inflorescences; root necrosis. By contrast, the integrity of the growing point is maintained until very late
in the symptom sequence (excepting bud rot). This emphasizes
the dangers of relying on symptomatology alone for disease

diagnosis and stresses the urgent need to establish Koch's postulates for the putative disease agents.

The monocotyledonous nature of the coconut (no cambium present) with its solitary vegetative growing point perhaps dictates that the palm can respond with only a limited range of symptom expression to internal pathological disturbance. Certainly those features, combined with the long growth cycle of the coconut palm and our inability to vegetatively propagate it, make it an extremely difficult plant for pathological experimentation and have contributed to our poor understanding of so many of the diseases which attack it.

The disease we now call lethal yellowing (LY) has been recorded in the Americas since 1834 under a variety of names including bud rot, West-end bud rot, bronze leaf wilt, and "unknown disease" (Grylls and Hunt, 1971a). In West Africa, the names Kaincope disease, Cape St. Paul wilt, Kribi disease, and Awka disease are given locally to a disease which appears identical with LY using all currently applicable tests and which we regard as synonymous with LY in this chapter. With only limited information presently available, it is premature to judge if a condition recently described (Steiner, 1978) from Tanzania in East Africa is also LY disease.

In recent years a number of reviews have listed the pertinent literature on LY (Grylls and Hunt, 1971a; Heinze *et al.*, 1972; Sherman and Maramorosch, 1977). Since 1973, the International Council on Lethal Yellowing (ICLY) (Fisher, 1973), an informal organization for all those interested in the disease, has met in alternate years, most recently in Fort Lauderdale, Florida in August, 1979. The meeting serves as a forum for the exchange of technical information, assesses recent research effort, and attempts to identify and coordinate areas for future research. Proceedings of the meetings are published, usually in mimeographed form.

II. HISTORY OF LETHAL YELLOWING

The earliest description of LY is from the Cayman Islands in 1834 (Maramorosch, 1964) and was followed by a more complete description of the disease at Montego Bay, Jamaica in 1891 (Fawcet, 1891). Shortly thereafter, Cuba, the Bahamas, Haiti and, more recently, the Dominican Republic and Florida reported outbreaks. In Ghana, at Cape St. Paul, the disease was reported in 1932 and its first observation in Togo, near the village of Kaincope, came at about the same time (Maramorosch, 1964). The outbreaks in Nigeria in 1917 and 1951 in the Awka-Onitsha area and the recent incidence of the "Kribi" disease in Cameroun demonstrate the further spread on the West Coast of Africa.

In 1955, LY struck at the island of Key West, Florida,
killing 75% or more of the coconut palms in that area in the
following 13 years before it spontaneously disappeared. Key
Largo had an outbreak in 1969 and the disease reached the mainland when the Miami area was hit in 1971. In the first 3 years,
respectively, of the latter outbreak 1200, 20,000, and over
100,000 palms were affected and by the end of 1977 the figure
was estimated at 375,000 (McCoy, 1976a; Gwin, 1978). Present
accumulated losses in Florida are estimated at 50-60% of the
total 1 million palms and the disease occurs as far north as
Martin County and has appeared on the West (Gulf) Coast.

In Jamaica, the disease moved suddenly in 1961 from the
western part of the island where it had been confined for 70
years and spread rapidly, killing an estimated 100,000 palms
in 1972. Twice as many palms were killed yearly by 1974
(Harries, 1974) and currently only a few small areas of the
island remain unaffected. In most plantations the commonly
grown Jamaica Tall variety has been completely eliminated.
The disease is now active in the Bahamas and isolated trees
have been reported dying from LY in Costa Rica and in Mexico.
Very little progress has been achieved in containment of the
disease despite increased research efforts in recent years.

The hypothesis that LY is caused by a virus [first proposed by Nutman and Roberts (1955) who were the first to give
the disease its present name] was abandoned when the constant
association with mycoplasma-like organisms (MLO) was established by Plavsic-Banjac et al. (1972). These findings have
since been confirmed by many others. Progress of research remained slow because the MLO could not be transmitted experimentally and the natural mode of spread had not been discovered. Early reports of successful mechanical transmission by
Price et al. (1968) have never been substantiated by other
workers. Palms grow slowly, are large and unwieldy, and it
is, therefore, difficult to conduct controlled experiments
with trees in areas where the disease spreads naturally.

The widespread introduction and establishment of exotic
palms as ornamentals into southern Florida has provided a
unique opportunity to investigate the host range of LY disease.
To date 26 species of palms are regarded as susceptible to LY
(Florida Dept. Agr., memo of Sept. 11, 1979 on LY quarantine).
In the absence of a technique for experimental transmission,
it is impossible to be assured that other palm species with
unusual symptoms are necessarily suffering from LY. However,
the occurrence of palms with lethal declines, confined to
areas where LY of coconuts is active, provides strong presumptive evidence for a common cause. Also, the necrotic symptoms
in inflorescences, spear leaves, and roots in these other
species are often similar to those in coconuts. The strongest

evidence comes from the demonstration (Parthasarathy, 1973,
1974; Thomas, 1979) of MLO in the phloem of most of the suspected species. Thus, in Florida, the following 26 palm species are listed as susceptible to lethal yellowing:

1. *Allagoptera arenaria* (Gomes) Kuntze
2. *Arenga engleri* Becc.
3. *Arikyryroba schizophylla* (Mart.) Bailey (Arikury palm)
4. *Borassus flabellifer* L. (Palmyra palm)
5. *Caryota mitis* (Cluster fish-tail palm)
6. *Chrysalidocarpus cabadae* H. E. Moore (Cabada palm)
7. *Cocos nucifera* L. (Coconut palm) - all varieties, including Malayan dwarf
8. *Corypha elata* Roxb. (Buri palm, Gebang palm)
9. *Dictyosperma album* (Bory) H. Wendl. & Drude (Hurricane or
   Princess palm)
10. *Gaussia attenuata* (O. F. Cook) Beccari (Puerto Rican
    Gaussia)
11. *Howeia belmoreana* (C. Moore & F. Muell.) Becc. (Sentry
    palm)
12. *Latania* spp. (all species)
13. *Livistona chinensis* (N. J. Jacquin) R. Br. ex Mart. (Chi-
    nese fan palm)
14. *Mascarena verschaffeltii* (Wendl.) Bailey (Spindle palm)
15. *Nannorrhops ritchiana* (W. Griffith) J. E. T. Aitchison
    (Mazari palm)
16. *Phoenix canariensis* Hort. ex Chab. (Canary Island date
    palm)
17. *Phoenix dactylifera* L. (Date palm)
18. *Phoenix reclinata* Jacq. (Senegal date palm)
19. *Phoenix sylvestris* (L.) Roxb. (Sylvester date palm)
20. *Pritchardia affinis* Becc. (Kona palm)
21. *Pritchardia pacifica* Seem & H. Wendl. (Fiji Island fan palm)
22. *Pritchardia thurstonii* F. Meull. & Drude
23. *Ravenea hildebrandti* Wendl. ex Bouche
24. *Trachycarpus fortunei* (Hook.) Wendl. (Chinese windmill
    palm)
25. *Veitchia merrillii* (Becc.) H. E. Moore (Christmas palm,
    Manila, or adonidia)
26. *Veitchia montgomeryana* H. E. Moore (Montgomery's palm)

Moore (1978) points out that LY attacks, without evident
pattern, genera in seven of the fifteen major groups in the
palm family. It has also been noted that only introduced
palms are susceptible and that, so far, none of the palms
native to South Florida has been affected. With the inclu-

sion of *Phoenix* spp. in the susceptible list and an almost un-
broken supply of palms across the southern United States, date
palm growers on the West Coast have expressed their concern.
In south Florida, the percentage losses of some palm species
have been very high, notably with *Phoenix* spp., *Veitchia merrillii,* and *Pritchardia* spp. (Howard and Collins, 1978). Deaths
of the popular Christmas palm, *V. merrillii,* may have exceed-
ed those of coconut palms; its smaller size and apparently
high susceptibility make it an attractive alternative palm to
the coconut for experiments in LY transmission. *Dietyosperma
album* and *Trachycarpus fortunei* also appear to be quite susceptible (Henry Donselman, personal comm.)

III. SYMPTOMATOLOGY

    The name "lethal yellowing" refers to the progressive dis-
coloration of fronds which is seen in those coconut varieties
commonly grown in the Caribbean region and in West Africa.
Discoloration of the pinnae progresses from lemon yellow,
through darker yellow to an orange-yellow over a period of 1
or 2 weeks and is soon followed by drying and browning as the
frond dies. The experience in Jamaica is that with most of
the coconut varieties introduced from southeast Asia and the
Pacific, the yellowing of the pinnae is far less marked or is
absent altogether, the color passing from green directly to
bronze and brown.
    The sequence of pinna discoloration can be variable but
usually progresses from tip to base of frond and upward
through the canopy from the oldest fronds toward the spear
leaf. Successive fronds discolor, die, and fall to the ground
at the rate of one to several per week (cf. natural senescence
of approximately one frond per month). In some varieties dead
fronds may persist, hanging down around the trunk as a skirt,
in which case the petiole usually buckles near the point of
attachment of its basal pinnae. Leaf symptoms are sometimes
more advanced on one side of the tree than the other. Once
yellowing has commenced, the process is not naturally revers-
ible. Dabek and Hunt (1976) showed that regreening could occur
if detached pinnae were irrigated with various hormones and
trace elements, suggesting that the disease-induced yellowing
was associated with a hormonal imbalance.
    It has been observed, more commonly in Florida than in
Jamaica, that a frond high in the canopy may be the first to
show yellowing in some of the pinnae. This so-called "flag-
leaf" may have as few as two or three but more usually 10-50
pinnae on one or both sides of the rachis brightly colored;
these stand out in sharp contrast with the dark green of ad-

jacent fronds. In many diseased palms the flag-leaf symptom
is not seen.

    At the center of the crown, the spear leaf ceases growth.
Dabek (1975) presents data which show that, in prebearing
palms, the spear leaf growth ceases, those pinnae most recently exposed, but still folded, positioned just above the sheathing leaf base of the next older leaf, show a brown dry papery
necrosis in irregularly shaped patches. This necrosis ex-
pands and later leads to the collapse of the spear leaf which
then hangs down from the crown. At this stage, or earlier,
dissection of the crown reveals a wet brown necrosis in the
previously white tissues of the pinnae and/or petiole at the
base of the spear. Adjacent and younger leaves in this zone,
between the growing point and the exposed parts, are also
usually affected. These necrotic tissues soon become invaded
by putrefying microorganisms, nematodes, and insect larvae.
The resultant soft rot extends downward toward the growing
point and has a powerfully nauseating odor.

    Another early diagnostic symptom of LY is the premature
dropping of nuts in all stages of their development. This is
quite distinct from the shedding or "self-pruning" of newly
set nuts which occurs in healthy palms only during the first
few weeks after inflorescence opening. Typically all nuts are
dropped over a period of 1 to 4 weeks. Abscission is between
the calyx and the paericarp, the calyx persisting on the inflorescence rachilla. Nuts dropped from diseased trees, if
sufficiently mature, will germinate normally; there is no evidence for disease transmission through the seed.

    A further early symptom is necrosis of the inflorescence.
This is most easily seen in the recently opened inflorescence.
Tips of a few or many of the rachillae are brown and pendant
and the affected tips soon turn black and wither. In moist
weather the necrotic portions rapidly become colonized by
secondary fungi. In more advanced disease, spathes may also
show an outward browning and may split abnormally with a resultant failure of the rachillae to separate from one another.

    The browning of rachillae and male flowers can usually be
detected in one or more of the unopened inflorescences if
these are dissected. The extent of the necrosis may vary from
the tip (less commonly a mid-position) of a single rachilla to
almost all of the tissues enclosed by the spathe. In mildly
affected spathes which are preparing to open, the male flowers
are often drier than normal and are more easily dislodged from
their rachillae.

    Any one of the four symptoms of nutfall, inflorescence
blackening, mature leaf discoloration (yellowing), or spear
leaf necrosis may herald the onset of LY. Usually within

1 month, and often far less, a diseased palm will be showing all four symptoms.

Following the upward progression of leaf discoloration and leaf fall, the support normally afforded by the sheathing leaf bases is removed and the trunk collapses, usually a short distance below the growing point, bringing with it the few remaining leaves. Even at this late stage in symptom sequence, the trunk and growing point sometimes appear white and normal. An average of 4 to 6 months elapses between the appearance of first symptoms and the death of the palm.

In prebearing palms, symptoms are similar but the interval before death is usually shorter. In such palms the obvious absence of nutfall and inflorescence symptoms makes the assured diagnosis of LY more difficult. Eden-Green (1978a) has noted in Jamaica that a bole rot often accompanies the disease in prebearing trees.

Complete or partial recovery from LY is extremely rare. It has been observed in Jamaica (Harries, 1978a) in the Malayan Dwarf variety and small number of tall x tall and tall x dwarf hybrids, mainly of Panama Tall parentage.

Coconut root tips are not easy to find except where they emerge from the base of the trunk. Early observers of LY-affected palms merely reported that some spontaneous necrosis of root tips often occurred. Recently, Eden-Green (1976, 1978b) induced vigorous growth of accessible roots by banking-up moist coir dust around the boles of over 100 apparently healthy palms growing in an area of active disease spread in Jamaica.

Roots were sampled at intervals and the appearance of abnormalities was related to any symptoms of LY in the crown. With a few exceptions, all the vigorously growing root tips became devitalized and necrotic over a short period of time coincident with the appearance of the first yellowing symptoms in the lower leaves. At least some of the older and firmer roots, established in the ground, remained apparently healthy until late in the symptom development.

IV.   CAUSAL AGENT

Since 1898, the cause of LY has been attributed to bacteria, salt damage, toxic materials, nutritional injury, and various viral agents (Grylls and Hunt, 1971a,b). All of these now appear to be wrong. In 1972, the first electron micrographs of ultrathin sections from diseased inflorescences were published (Plavsic-Banjac et al., 1972), revealing the presence of membrane-bound microorganisms in the sieve tube elements. The microorganisms were pleomorphic and some showed filamentous characteristics. They were described as mycoplasma-like organisms (MLO) and were found in diseased, but not in healthy, coconut palms. No virus-like particles or microorganisms other than MLO were detected in the ultrathin sections of the phloem, xylem, or parenchyma tissue. Whether MLO were the etiological agent or merely a natural coinfective organism could not be determined at that time (Plavsic-Banjac *et al.*, 1972). A note published shortly thereafter by Beakbane *et al.* (1972) confirmed the presence of MLO in diseased coconut palm material from Jamaica. Subsequently, others also observed MLO in association with LY from Jamaica (Heinze *et al.*, 1972), Florida (Parthasarathy, 1973, 1974), Togo (Dollet and Gianotti, 1976; Nienhaus and Steiner, 1976), and Cameroun (Dollet *et al.*, 1977). Further evidence supporting MLO etiology of the disease came from the results of tetracycline treatments (McCoy, 1972; Hunt *et al.*, 1974; Steiner, 1976a).

The isolation of a mycoplasma, presumed to be the causative agent of LY, has been reported by Giannotti *et al.*, (1975). Unfortunately this report, similar to earlier ones by the same author(s), cannot be evaluated because the rules concerning the deposition of the cultured microorganisms have not been fulfilled and no "plant mycoplasmas" were made available to any of the designated type culture collections (Maramorosch, 1972) or to other workers in the disease.

Waters and Hunt (1980) have used the LY-infected coconut palm as a model system for reconstructing the three-dimensional morphology of MLO *in vivo*. Using serial sections obtained from diseased roots they were able to show how the individual profiles of MLO seen in ultrathin sections belie the complexity of structure which these organisms can assume. Simple circular, elliptical, and dumbbell-shaped profiles were demonstrated to be parts of MLO having very varied morphologies. Filamentous, moniliform, cylindrical, and erythrocyte-shaped forms were recognized in addition to the less commonly found, approximately spherical forms. Simple binary fission was never observed. The sizes of complete organisms, particularly the filamentous forms, were appreciably greater than had previously been interpreted for plant-inhabiting MLO. Thomas (1979) has also given data on the distribution, sizes, and frequency of occurrence of MLO in diseased coconuts and in other lethal decline-affected palms in Florida.

Extensive electron microscopic studies of the distribution of MLO-containing sieve tubes among the tissues of infected coconut palms have also been made in Jamaica (Waters and Osborne, 1978; Henry Waters, personal comm.). These studies attempt to quantify the numbers of MLO profiles in individual sieve elements, the numbers of infected sieve tubes in the cross section of any vascular bundle, and the proportion of infected bundles in various selected sites in diseased palms.

The results are then analyzed in relation to the symptom sequence. MLO have not been detected in any palm prior to the
appearance of first symptoms. At that time, MLO make a sudden appearance more or less simultaneously in the newly formed
sieve tubes at several sites. These sites are all juvenile
tissues and are active sinks of phloem transport; they include
the spear leaf and leaves younger than it, developing inflorescences, root tips and the stem apex below the growing point.
It is significant that no MLO have been detected in any fully
expanded leaves with the exception of the flag-leaf (see Section III). The richest tissues for detecting MLO are above
the bud in the petioles and lower pinnae of the several leaves
lying inside the spear leaf and also behind the apices of
primary, secondary, or tertiary roots whose tips are being
devitalized.

These results have significance for workers attempting
insect transmission of LY. There are very few tissues containing MLO which are accessible to candidate vectors during
acquisition feeding and the MLO may only be present in such
tissues for a limited time. It is also tempting to speculate
on the occurrence of MLO in flag-leaves. Could this site
represent the site of primary inoculation from which MLO have
later been exported to other parts of the palm? If so, and
assuming a 9-12 month incubation period in bearing palms, the
present position of a flag-leaf implies that it was in the
approximate position of the spear leaf at the time of infection.

A further outcome of the results (Henry Waters, personal
comm.) is that, in absolute terms, the quantity of MLO present
is insignificantly small. Even in heavily infected tissues
where perhaps half of the sieve tubes may contain crowded MLO,
calculations show that the MLO occupy only about 0.1% of the
total volume. This may explain why attempts to locate MLO in
extracts of coconut tissues (or tissues from other MLO-infected plants) have been unsuccessful.

Eden-Green (1978c) and Eden-Green and Tully (1979), in
attempting to isolate the MLO of lethal yellowing, obtained
several different cultures of sterol-nonrequiring mycoplasmas
(Acholeplasma spp.). Serological and biochemical tests showed
that most of the isolates were closely related to established
species. A. axanthum, and A. oculi (Eden-Green and Tully,
1979; Townsend et al., 1980), which were previously known only
from animal sources. Their occurrence, apparently as saprophytes on the enclosed surfaces near the growing points of
diseased palms was unexpected and dramatically extends the
range of ecological niches in which these mollicutes are
known to occur. Similar isolations have been made from palms
affected by diseases other than LY and no plant pathogenecity

could be demonstrated (Eden-Green *et al.*, 1980). Although it would thus seem unlikely that these organisms are causally realted to LY, their occurrence on decaying plant surfaces should be interpreted with caution for those attempting the isolation of plant MLO and may explain the claims by Giannotti *et al.* (1975) that the LY agent has been cultured.

When coconut inflorescences, which are approaching their natural opening time, are cut across, they may be induced to exude a sugar-rich fluid which is believed to be pure sieve tube sap. This fluid, toddy, is collected in many countries and used for the production of alcoholic beverages. Toddy (and likewise, the water from immature nuts which has a somewhat similar constitution) has long been suggested as a suitable medium for the growth of the LY agent or other plant MLO (McCoy, 1976c). Also, if diseased palms could be induced to exude toddy freely, this might provide a source from which the MLO could be extracted and enriched by filtration or centrifugation. Unfortunately, extensive attempts in Jamaica and Florida to follow these promising lines have so far been unsuccessful. It has, however, been shown that toddy is a suitable base medium for the cultivation of some other mollicutes, including *Spiroplasma citri,* the corn stunt agent and some *Acholeplasma* spp. (McCoy, 1978a; Eden-Green, 1978c).

V. DISEASE SPREAD

Observations made earlier in Key West, Florida, where the spread of LY disease was easily observable because of the comparatively small area, led to the statement that the spread was most likely through the soil (Maramorosch, 1964). In Togo, trees were quite often destroyed on one side of the concrete highway along the coast, while trees on the opposite side remained healthy (Maramorosch, 1964). Steiner (1976b) observed that in Togo the disease spreads from single trees or small foci and moves outward like an "oil stain." Just three rows beyond the border of a focus, no disease was observed. From this observation it was concluded that the spread of the disease in Africa differed from that in the Carribean area, where a jump-spread pattern has long been apparent. The spread in Jamaica and Florida is not continuous from tree to tree and clearly suggests that the disease agent is air-borne.

The rate of long distance spread in Jamaica appeared to be slower than in Florida, perhaps because Jamaica's mountains provide a physical barrier to long-distance air-borne dispersal (McCoy, 1976a). Steiner (1976b) and McCoy (1976a) speculated that the respective vectors in Africa and Jamaica might differ, thus accounting for the variation in the pattern of spread. We feel that it would be premature to conclude that the MLO of LY is carried by an air-borne vector in Jamaica and Florida and by a nematode vector, root contact, or other soil-associated route in Togo. Until now, there has been no proved instance of a plant disease agent of the MLO type that can be transmitted by nematodes or aphids. In fact, all presently known vectors seem to be confined to leafhoppers, planthoppers, or psyllids.

It was reported by Carter and Suah (1964) that the monthly incidence of primary symptoms of LY in Jamaica remained virtually constant over an 18-month observation period. This could suggest that disease transmission occurs with uniform efficiency throughout the year. It could also mean that the incubation period is affected by seasonal variation and that transmission occurs over a discrete time period (Carter and Suah, 1964). An incubation period of 9-12 months is a likely possibility for bearing palms, according to recent observations in Jamaica.

No further work was reported on the incidence of spread until 1969 when Dabek (1975) undertook a systematic investigation of LY infection, incubation, and growth rates. Over a period of 2 years, young transportable palms were carried to diseased areas, exposed to LY by natural infection, and then returned to incubate in a disease-free area. Although on 28 palms became infected from over 400 which were exposed, these provided considerable new information. With these approximately 3-year-old palms, the most probable incubation period was between 111 and 191 days. Palms exposed in March and April had the longest incubation periods. Maramorosch et al. (1970) have pointed out that increased temperatures produce a delayed expression of symptoms in MLO infections. Since all evidence points to an MLO etiology of LY, this would explain why palms infected after the summer have the shortest incubation period, while those infected in the spring have the longest (Maramorosch, 1978).

Although the number of palms becoming diseased were too small to allow confident analyses, the results (Dabek, 1975) also suggest a seasonality in the infection process. The main period of infection was between January and August, with no palms exposed in October and November becoming infected. This suggests a seasonality in the numbers or the transmission efficiency of the presumed insect vector of LY disease.

The lowest apparent infection rates for LY in Florida have been observed when palms are adjacent to the ocean or salt water bay (McCoy, 1976a). This has not been observed in Jamaica.

No vector has yet been clearly identified for LY, but the fact that the field infection rate at the height of an epi- phytotic is only 4% per month allows for some speculation (Johnson and Eden-Green, 1978). They argue that very large numbers of insects may be required to bring about experimental disease transmission. The vector could be a rare insect which transmits the disease efficiently or, conversely, a common insect which transmits inefficiently (Dabek, 1975). Another plausible explanation is that the vector does not ordinarily feed on coconut palms and does so only when displaced from its normal food plants. In this connection it might be worthwhile to point out that in Togo, areas where coconut palms were dy- ing from the disease also frequently contained *Vinca rosea* plants with typical symptoms of yellows-type disease (Mara- morosch, 1964). At the time this observation was made (1963) it was not known that such diseases, as well as LY, are asso- ciated with MLO. By contrast, extensive searches in Florida and Jamaica to recognize yellows-type symptoms in plants other than palms, and which are restricted to LY-affected areas, have all failed.

There are 117 insect species and 17 mites known to inhabit coconut plams in Florida (Reinert, 1973). Working under the hypothesis that a leafhopper or planthopper might be the MLO vector, many workers in Florida and Jamaica have made exten- sive transmission tests with these insects over a period of many years. Until very recently, and despite the prodigious numbers of insects handled, these transmission tests had all been negative or the very occasional transmissions observed had not been repeatable (Johnson and Eden-Green, 1978; Tsai, 1975, 1978; Eden-Green, 1978d; Eden-Green and Schuiling, 19- 78). At the 1979 meetings of ICLY the preliminary reports of successful transmission of LY by the fulgorid *Haplaxius crudus* were optimistically received. Field-collected insects, re- leased at a rate of ca. 18,000 per large cage, resulted in the subsequent development of LY symptoms in seven test palms of *Veitchia merrillii* (Howard, 1979a). Confirmation of these findings of a natural vector (or one of the natural vectors) is eagerly awaited. A full account of the vector studies with LY is given elsewhere in this volume.

Harries (1978b) pointed out that one should not be compla- cent that LY disease will never reach certain areas of tropi- cal Asia or South America. At present, the disease is not threatening major coconut growing areas outside of the Carib- bean islands and West Africa. However, a recent report by Steiner (1978) indicates that the disease may also occur in East Africa in Tanzania.

VI.   CONTROL

     Two appraches have been used to control LY disease: the
planting of resistant varieties and antibiotic chemotherapy.
The use of insecticides to control possible vectors has had
only limited success in some trials (Howard, 1979b).
     Independently of a complete knowledge of the cause or mode
of spread of LY, the finding of resistance in coconut varieties would obviously alleviate the dangers of this devastating disease. During the past 30 years, trials in Jamaica have
unquestionably established the high resistance of the Malayan
Dwarf variety. According to the records of the Coconut Industry Board, Kingston, Jamaica, the most reliable source of
resistance is in the progeny of "Red Dwarfs" introduced into
Jamaica from Trinidad in 1940. Other subsequent introductions
of red, yellow or green color variants of the Malayan Dwarf
from elsewhere in the world have also proved highly resistant.
The mechanism of Malayan Dwarf resistance is not understood
and the resistance is not absolute. Under conditions of high
and maintained inoculum pressure, a small proportion of these
palms are lost to LY. The fact that Malayan Dwarf palms or
hybrids may very occassionally show symptoms of LY disease
and then partially or totally recover (Harries, 1974, 1978a)
suggests that a physiological or biochemical tolerance of infection is operating, rather than immunity or infection es-
cape.
     Seed nuts from worldwide sources have been collected,
mostly under FAO auspices, and brought to Jamaica during the
past 20 years. These have been rated for resistance to LY
by exposure to natural infection in a number of sites in the
island. High inoculum levels have been maintained by the re-
supply of susceptible palms as these were lost to disease.
Harries (1973), quite early in these trials, recognized three
broad categories of resistance. The highly resistant category
(less than 3% losses at that time) included only the Malayan
Dwarf variety. Intermediate resistance (33-50% losses) was
shown by many other varieties originating mainly from southeast Asia. High susceptibility (62-05% losses) was shown by
the Jamaica Tall variety and others originating mainly in the
Indian subcontinent and in parts of the Pacific.
     Coconuts can be classed into two broad groups on the basis
of their botanical characteristics. In a comprehensive review
of their evolution, Harries (1978c) suggests two main centers
of origin, one in India and Sri Lanka (from where the palms
of Africa and the Atlantic coasts of America have come), the
other in southeast Asia (from where the palms in many of the
Pacific islands and the Pacific coasts of America have come).
He further points out the marked differences in the resistance

to LY displayed by these groups of palms under the conditions
of test in Jamaica; those believed to originate in the Indian
subcontinent (to which group the Jamaica Tall and West African Tall varieties belong) are highly susceptible. Harries
has further cautioned on the potential for the spread of LY
disease. He calculates (Harries, 1979b) that if the strain
of LY presently active in Jamaica were to be established
worldwide, an estimated 62% of the world's population of coconut palms would be killed. He argues that this is no cause
for complacency, particularly in the light of the continuing
spread of LY in the Caribbean and West Africa. The patterns
of susceptibility to LY have recently been summarized and
commented on by Chiarappa (1979) and a worldwide strategy for
breeding coconut palms for resistance to LY and other diseases
proposed by Van der Graaf (1979).

The resistance of the Malayan Dwarf has been further confirmed in Florida, but some preliminary results from Ghana
(Johnson and Harries, 1976; Addison, 1978) are less encouraging and their apparent susceptibility there suggests the possibility of a different strain of the LY pathogen occurring
in West Africa.

Unfortunately, the Malayan Dwarf variety has a number of
agronomic and aesthetic disadvantages when it is compared with
the Tall varieties that is is replacing in those areas affected by LY disease. It is a more delicate palm, less tolerant
of poor soils, drought and frost injury; it is more susceptible to leaf spots; as an ornamental it is said to be less
graceful; and, for copra production, having a smaller nut size,
it requires more labor. These objections can be largely overcome by the production of F_1 hybrids between Malayan Dwarf and
selected Tall varieties while at the same time preserving sufficient LY resistance. The national programs in many countries
now embrace the mass production of hybrid nuts which are produced in isolated seed gardens of Malayan Dwarf mother trees
(Harries, 1976). Inflorescences are emasculated as they open
and female flowers are hand-pollinated using fresh or lyophil-
ized pollen or are open-pollinated by selected Tall trees interspersed among the Malayan Dwarfs. Thus, the "Maypan" hybrid (Malayan Dwarf crossed with the intermediate resistant
Panama Tall) has been successfully introduced in Jamaica and,
recently, in Florida, and Malayan Dwarf X West African Tall
hybrids have been introduced into many countries by IRHO from
Ivory Coast. The Maypan hybrid proved less resistant than the
Malayan Dwarf to LY when tested in mixed plantings with susceptible varieties in Jamaica; however, acceptably small
losses are expected under the lower inoculum levels which
should prevail in pure stands of these hybrids in plantations.

The second approach to LY control is by tetracycline antibiotic therapy. Initially, this was used as an experimental
tool to further implicate MLO as the causal agent of LY (Mc-
Coy, 1972; Hunt et al., 1974). It has, however, been used in
controlling LY in the decorative coconut palms in Florida (Mc-
Coy et al., 1976; Donselman, 1978; McCoy, 1978b) but is successful only where a strict injection regime is followed. Trunk
injections of tetracycline-HCl or oxytetracycline will often
cause total remission of symptoms in palms already affected by
LY provided they are in a very early stage of symptom development. However, this is not a cure; symptoms soon return after
treatment is withheld. Similar injections can also be used
prophylactically to protect apparently still healthy palms in
areas of active disease spread. Both approaches require repeated doses with a recommended frequency of three injections
per year.

The U. S. Environmental Protection Agency has granted re-
gistration for the use of oxytetracycline as an economic control measure for LY disease in Florida. Although antibiotic
residues were below detectable levels in sampled nuts (McCoy,
1976b), concern has been expressed about the possible use of
such antibiotics in areas where nuts are consumed fresh for
their water or jelly.

It should be pointed out that tetracycline injections
carried out on a large scale in Florida might result in the
formation of tetracycline-resistant MLO which could then be
transmitted by vectors to other plants and possibly cause extensive damage. In the western United States, pear decline
is now commercially controlled by tetracycline injections in
large orchards. The same objections apply to this application
as to the temporary remissions of LY being carried out in Florida. The above considerations, combined with the high cost
of an injection program, do not make the use of antibiotic
therapy an attractive proposition in commercial coconut plantations. To the best of our knowledge, antibiotics are only
applied at present to control LY for scientific experiments
and in ornamental palms. Any more widespread use is not recommended.

VII.   POSSIBLE RELATIONSHIPS BETWEEN LETHAL YELLOWING AND
       OTHER COCONUT DISEASES

In a survey of coconut diseases of uncertain etiology,
Maramorosch (1964) was the first to note the similarity in
symptomatology between LY and Malaysian wilt diseases. The
similarity is especially marked in many of those coconut
varieties of southeast Asian origin when they are exposed to

and affected by LY in the Caribbean. An MLO etiology for LY
now seems assured but no agent has yet been implicated for
Malaysian wilt.

Sporadic and local outbreaks of some other coconut diseases in southeast Asia are probably related to or identical
with Malaysian wilt, on the basis of their symptomatology and
epidemiology. Thus, outbreaks in Sarawak in 1957 (Johnston,
1960; Turner, 1963), in Marinduque, Philippines in 1963 and in
the Natuna islands, Pulau Bintan, and eastern Sumatra of Indonesia from 1976 (Sitepu, 1979) are probably all Malaysian wilt.
In emphasizing the similarity with LY, Hunt (1979) suggested
that if the Indonesian palms could be magically transported
to a resistance trial in Jamaica, they would unquestioningly
be recorded as affected by LY disease. From the very close
similarity of symptoms with both LY and hartrot (Parthasarathy
and Van Slobbe, 1978), it is reasonable to suppose that Malaysian wilt is also a disease in which phloem transport is dis-
turbed. However, all attempts to identify a potential pathogen have proved negative. Recent unpublished electron micrographs (Hunt and Jones) of diseased material from Sumatra
show some abnormalities in developing phloem. A few structures resembling MLO were observed but could not be identified unequivocally.

Turner *et al.* (1978) have recently given the name "coconut
stem necrosis" to a long-known disease, with different symptoms
from Malaysian wilt, which attacks seedlings and young coconut plants up to 3 years old in North Sumatra and Peninsular
Malaysia. Observations to date show this disease to be restricted to Malayan Dwarf and to its hybrids with African Tall,
including those introduced from Ivory Coast where a similar
disease is also known (Quillec *et al.,* 1978). Turner *et al.*
(1978) recognized an MLO which was scantily present in the
phloem of stem necrosis-affected, but not healthy, palms.
They postulate that the disease is endemic in the region and
is unlikely to reach epidemic proportions. The evidence
points to a high level of resistance in the native palm population with the exception of the Malayan Dwarf variety. No
attempts have been reported to obtain symptom remission with
tetracyclines and so corroborate an MLO etiology.

Noting the similarity in symptomatology between LY and a
number of these coconut diseases in southeast Asia, Chiarappa
(1979) has examined all the available evidence and put for-
ward a working hypothesis that LY originated in southeast
Asia. He based his argument largely on the high levels of
resistance to LY shown by coconut palms of southeast Asian
origin in trials in the Caribbean and on the occurrence of
endemic palm diseases in southeast Asia which have a suspected MLO etiology. The existence of differential varietal

responses to these diseases in various parts of the world further requires the supposition that a number of strains of the LY pathogen have evolved. Over the course of time, palms would come into equilibrium with the local strains of their pathogen and disease levels would be low, but exotic palms introduced into an area would be expected to have poor levels of resistance. Chiarappa's proposal while highly plausible is also highly speculative and emphasizes the need to establish the etiology of these ambiguous southeast Asian diseases. As has already been pointed out, symptomatology alone is a most unreliable criterion for disease diagnosis in coconuts. The demonstration of an MLO in arecanut palms with yellow leaf disease in Kerala, India (Nayar and Seliskar, 1978) and the occurrence in Sarawak of diseased arecanut palms at the same time that coconut palms were affected with Malaysian wilt symptoms (Turner, 1963) may also be relevant.

VIII. CONCLUSION

Plant disease agents know no political or geographical boundaries and may be expected to spread to all regions within their environmental adaptability. The proved susceptibility of the majority of coconut palms now planted in the world to the strain of LY presently active in the Caribbean is cause for major concern. It is not only coconut palms that are at risk. How many palms beyond the 26 species presently reported from Florida may also be susceptible to LY? The ever-growing rise in air transportation further increases the risk of accidental transfer of infectious vectors or plant material to countries or continents presently free of LY disease.

These considerations emphasize the need for continued research into LY. Many fundamental questions remain unanswered. Although few would deny the probability of an MLO etiology, the presumed agent has not been reproducibly cultured or have Koch's postulates been established. (These features are also shared by other plant MLO diseases, with the exception of those involving spiroplasmas.) *Haplaxius crudus* is a likely vector but other species may well be able to transmit the disease and might be more important in the field. Even with *H. crudus*, experimental transmissions are a daunting task. We are ignorant of the manner of vector transmission, the site(s) of possible infection and acqustion outside of the palm family. The lethal and highly destructive nature of the disease in all susceptible palm species argues in favor of there being an alternate host, in which the disease is not lethal, as a more permanent reservoir of infection.

Answers to some of the above problems would greatly assist in the formulation of strategies against LY worldwide and might shed light on the nature of the resistance shown by the Malayan Dwarf and other varieties.

The suggestion that the LY pathogen may exist in a number of strains and may be endemic in southeast Asia merits serious consideration. This would explain the high LY resistance of palms from this region and the occurrence there of sporadic and local outbreaks of diseases with symptoms resembling LY. Acceptance of this suggestion requires as a minimum that MLO be detected consistently in these diseased palms (no easy matter as electron microscopists working on LY in Africa will testify) and that symptom remission is obtained with tetracycline antibiotics.

While most unaffected countries remain totally unprepared for an attack by LY, the results of control in affected countries have been encouraging after an initial traumatic interlude. Although antibiotic chemotherapy can be successful, there are ecological and environmental considerations against its use and economic factors preclude its use in commerical plantings or for subsistence farming. We suggest that, at most, tetracyclines should be used only for experimental purposes or as a stop-gap measure for ornamental palms.

With present knowledge, satisfactory control of LY can only be achieved by the use of resistant varieties. The results from Jamaica have demonstrated the practicability of breeding for resistance to LY and the strategies involved have been clearly defined (Harries, 1973, 1976; Van der Graaf, 1979). The deliberate hybridization of coconuts for disease resistance using germ plasm from worldwide sources has long been neglected (Child, 1974). The severity of the attack by LY disease has done much to stimulate such breeding programs. However, resistance to LY is only one consideration since many other serious diseases of coconuts are regionally extremely important, e.g., kerala wilt, cadang-cadang, and Natuna disease. The suggestion by Van der Graaff (1979) that the progeny from national breeding programs should be reciprocally exchanged and exposed in areas where coconut diseases of known or uncertain etiology are active has much merit. Such an interchange, preferrably under the coordination of an international agency such as F.A.O., would establish the genetic vulnerability of existing lines to all of the major coconut diseases. At present, exotic palms are often prone to locally endemic diseases when introduced into new areas. The resistance of many coconut varieties to the strain of LY active in the Caribbean is well known and has been successfully exploited there. The aim should now be to combine and test this resistance with resistance to local diseases in national pro-

grams. Should LY continue to spread, as seems probable, suit-
ably resistant hybrid material could then be immediately made
available to replace palms lost to the disease. Although
much fundamental research is still urgently needed, a practical control is thus available to those threatened by this
devastating disease.

IX. SUMMARY

    The symptomatology of lethal yellowing, its distribution
in the Caribbean, North America, and West Africa and the mycoplasma-like agent of the disease are described in detail. Although best known on coconuts, many other palm species are
apparently susceptible and the disease presents a serious
threat to date and some ornamental palms and a potential
threat to other important palms. Lethal yellowing is briefly
compared with other lethal diseases known to affect coconut
palms and the evidence is examined for the suggestion that
lethal yellowing may be related to various endemic diseases
in southeast Asia.
    Recent evidence implicates the fulgorid, *Haplaxius crudus,*
as a natural vector of lethal yellowing disease. Tetracycline
chemotherapy causes a temporary remission in palms treated
early in symptom development and protects apparently healthy
palms in areas of active disease spread; it has been successful in preserving the decorative coconut and other palms in
some areas of Florida, but is not attractive for the control
of lethal yellowing on a plantation scale. In field trials
relying on natural infection, various levels of resistance to
lethal yellowing have been demonstrated by coconut varieties
arising from different centers of origin. The extremely high
resistance of the Malayan Dwarf variety has been successfully
exploited and its present widespread use as a female parent in
breeding programs with selected Tall varieties offers the best
long-term control of the disease in areas affected or threatened by lethal yellowing.

REFERENCES

Addison, E. A. (1978). The Cape St. Paul wilt disease in
    Ghana: The present position. *Proc. 3rd. Meeting ICLY,
    November, 1977, Univ. Fla.* p. 7. *Publ. No.* FL 78-2.
Beakbane, A. B., Slater, C. H. W., and Posnette, A. F. (1972).
    Mycoplasmas in the phloem of coconut, *Cocos nucifera* L.,
    with lethal yellowing disease. *J. Hort. Sci. 47,* 265.

Bock, K. R., Ivory, M. H., and Adams, B. R. (1970). Lethal
    bole rot disease of coconut in East Africa. *Ann. Appl.
    Biol. 90*, 293-302.
Calvez, C. (1979). Tolerance of the Local X Rennell hybrid to
    the New Hebrides disease. *5th Session FAO Techn. Working
    Party Coconut Production, Protection Processing; Manila,
    Philippines, December, 1979, Paper No. AGP: CNP/79/44.*
Carter, W. and Suah, J. R. R. (1964). Studies on the spread
    of lethal yellowing disease of the coconut palm. *FAO
    Plant Prot. Bull. 12*, 73-78.
Chiarappa, L. (1979). The probable origin of lethal yellowing
    and its co-identity with other lethal diseases of coconut.
    *5th Session FAO Techn. Working Party Coconut Production,
    Protection Processing, Manila, Philippines, December,
    1979, Paper No. AGP: CNP/79/4*, 12 pp.
Child, R. (1974). "Coconuts" 2nd ed., 335 pp. Longman, London.
Dabek, A. J. (1975). The incubation period, rate of transmis-
    sion and effect on growth of coconut lethal yellowing di-
    sease in Jamaica. *Phytopathol. Z. 84*, 1-9.
Dabek, A. J. and Hunt, P. (1976). Biochemistry of leaf sene-
    scence in coconut lethal yellowing, a disease associated
    with mycoplasma-like organisms. *Trop. Agr. (Trinidad) 53*,
    115-123.
Dollet, M. and Giannotti, J. (1976). Maladie de Kaincope: Pre-
    sence de mycoplasmes dans le phloeme des cocotiers mala-
    des. *Oleagineux 31*, 169-171.
Dollet, M. and Giannotti, J., Renard, J. L., and Gosh, S. K.
    (1977). Etude d'un jaunissement lethal des cocotiers au
    Cameroun: La maladie de Kribi. Observations d'organismes
    de type mycoplasmes. *Oleagineus 32*, 317-322.
Donselman, H. (1978). Present effort in preserving South Flo-
    rida's palms. *Proc. 3rd Meet. Intern. Council Lethal
    Yellowing, November 1977, Univ. Fla. Publ. No. FL 78-2*,
    p. 28.
Eden-Green, S. J. (1976). Root symptoms in coconut palms af-
    fected by lethal yellowing disease in Jamaica. *FAO Plant
    Prot. Bull. 24*, 119-122.
Eden-Green, S. J. (1978a). Bole rots in pre-bearing palms
    apparently affected by lethal yellowing disease. *FAO
    Plant Prot. Bull. 26*, 13-15.
Eden-Green, S. J. (1978b). Further studies on root symptoms
    of lethal yellowing. *Proc. 3rd Meet. Intern. Council
    Lethal Yellowing, November, 1977, Univ. Fla. Publ. No.
    FL 78-2*, p. 12.
Eden-Green, S. J. (1978c). Attempts to extract and culture
    mycoplasmas from coconut palms. *Proc. 3rd Meet. November
    1977, Univ. Fla. Publ. No. FL 78-2*, p. 20.

Eden-Green, S. J. (1978d). Lethal Yellowing transmission tests
    in Jamaica, 1975-1977. *Proc. 3rd Meet. Intern. Council
    Lethal Yellowing, November, 1977, Univ. Fla. Publ. No.
    FL 78-2*, pp. 22-23.
Eden-Green, S. J. (1979). Attempts to transmit lethal yellow-
    ing disease of coconuts in Jamaica by leafhoppers (Homop-
    tera: Cicadelloidea). *Trop. Agr. (Trinidad), 56,* 185-192.
Eden-Green, S. J. and Schuiling, M. (1978). Root acquisition
    feeding transmission tests with *Haplaxius* spp. and *Proar-
    na hilaris,* suspected vectors of lethal yellowing of co-
    conut palm in Jamaica. *Plant Dis. Rep. 62,* 625-627.
Eden-Green, S. J. and Tully, J. G. (1979). Isolation of *Acho-
    leplasma* spp. from coconut palms affected by lethal yel-
    lowing disease in Jamaica. *Curr. Microbiol. 2,* 311-316.
Eden-Green, S. J., Markham, P. G., and Townsend, R. (1980).
    Acholeplasmas and lethal yellowing disease II: Transmission
    experiments. *Proc. 4th Meet. Intern. Council Lethal Yel-
    lowing. Univ. Fla., mimeographed.*
Fawcett, W. (1891). Report on the coconut disease at Montego
    Bay. *Bull. Botan. Dept. Jamaica 23,* 2.
Fenwick, D. W. (1963). Recovery of *Rhadinaphelenchus cocophi-
    lus* (Cobb, 1919) Goodey, 1960 from coconut tissue. *J.
    Helminthol. 37,* 11-38.
Fisher, J. B. (1973). Report of the lethal yellowing sympo-
    sium at Fairchild Tropical Garden, Miami. *Principes 17,*
    151-159.
Giannotti, J., Arnaud, F., Dollet, M., Delattre, R. and de
    Taffin, G. (1975). Mise en culture de mycoplasmes a par-
    tir de racines et d'inflorescences de cocotiers atteints
    par la maladie de Kaincope. *Oleagineux 30,* 13-18.
Griffith, R., Sealy, L., and Ramharack, R. (1979). Present
    status of investigations on Cedros wilt of coconuts in
    Trinidad. *5th Session FAO Techn. Working Party Coconut
    Production, Protection Processing, Manila, Philippines,
    December, 1979. Paper No. AGP: CNP/79/29.*
Grylls, N. E. and Hunt, P. (1971a). A review of the study of
    the aetiology of coconut lethal yellowing disease. *Olea-
    gineux 26,* 311-315.
Grylls, N. E. and Hunt, P. (1971b). Studies on the aetiology
    of coconut lethal yellowing disease in Jamaica, by mecha-
    nical and bacterial inoculations and by insect vectors.
    Oleagineux 26, 543-549.
Gwin, G. H. (1978). Distribution and impact of lethal yellow-
    ing in Florida. *Proc. 3rd Meet. Intern. Council Lethal
    Yellowing, Univ. Fla. Publ. No. FL 78-2,* p. 5.
Harries, H. C. (1973). Selection and breeding of coconuts for
    resistance to diseases such as lethal yellowing. *Oleagi-
    neux 28,* 395-398.

Harries, H. C. (1974). Natural symptom remission of lethal yellowing disease of coconuts. *Trop. Agr. (Trinidad) 51,* 575-576.

Harries, H. C. (1976). Coconut hybridisation by the Policaps and Mascopol systems. *Principes 20,* 136-147.

Harries, H. C. (1978a). Natural symptom remission and total recovery after lethal yellowing infection. *Proc. 3rd Meet. Intern. Council Lethal Yellowing, November, 1977, Univ. Fla. Publ. No. FL 78-2,* pp. 13-14.

Harries, H. C. (1978b). Susceptbility to lethal yellowing disease in global perspective. *Proc. 3rd Meet. Intern. Council Lethal Yellowing, November, 1977, Univ. Fla. Publ. No. FL 78-2,* pp. 29-30.

Harries, H. C. (1978c). The evolution, dissemination and classification of *Cocos nucifera* L. *Botan. Rev. 44,* 265-320.

Heinze, K. G., Petzold, H., and Marwitz, R. (1972). Beitrag zur Aetiologie der todlichen Vergilbung der Kokospalme. *Phytopathol. Z. 74,* 230-237.

Heinze, K. G., Schuiling, M., and Romney, D. H. (1972). The possible cause of lethal yellowing disease of coconut. *FAO Plant Prot. Bull. 20,* 58-66.

Howard, F. W. (1979a). Lethal yellowing transmission experiments with *Haplaxius crudus. Abstr. 4th Meet. Intern. Council Lethal Yellowing August, 1979, Ft. Lauderdale, Fl.* (mimeogr.)

Howard, F. W. (1979b). Foliar insecticide applications reduce spread of palm lethal decline. *Abstr. 4th Meet. Intern. Council Lethal Yellowing August, 1979, Ft. Lauderdale, Fl.* (mimeogr.)

Howard, F. W. and Collins, M. E. (1978). "Palm Species Susceptible to Mycoplasma-like Organism-Associated Lethal Declines in Fairchild Tropical Garden, 1971-1977." Fort Lauderdale, Agric. Res. Center, Univ. Florida. 26 pp.

Hunt, P. (1979). A coconut disease of uncertain etiology in Indonesia. *Abstr. 4th Meet. Intern Council Lethal Yellowing, August, 1979, Ft. Lauderdale, Fla.* (mimeogr.).

Hunt, P., Dabek, A. J., and Schuiling, M. (1974). Remission of symptoms following tetracycline treatment of lethal yellowing-infected coconut palms. *Phytopathology 64,* 307-312.

Johnson, C. G. and Eden-Green, S. J. (1978). The search for a vector of lethal yellowing of coconuts in Jamaica: A reappraisal of the experiments from 1962 to 1971. *FAO Plant Prot. Bull. 26,* 137-149.

Johnson, C. G. and Harries, H. C. (1976). A survey of Cape St. Paul wilt of coconut in West Africa. *Ghana J. Agr. Sci. 9,* 125-129.

Johnston, A. (1960). "A Preliminary Plant Disease Survey in Sarawak." FAO, Rome (mimeogr.). Cited by Chiarappa (1979).

Julia, J. F. (1979). Isolation and identification of insects carrying juvenile diseases of the coconut and the oil palm in the Ivory Coast. *Oleagineux 34,* 391-393.

Maramorosch, K. (1964). "A Survey of Coconut Diseases of Unknown Etiology." FAO, Rome 39 pp.

Maramorosch, K. (1972). The enigma of mycoplasma in plants and insects. *Phytopathology 62,* 1230-1231.

Maramorosch, K. (1978). Amarelecimento letal de coqueiro: distribuicao, impacto e implicacoes mundiais. *Fitopatol. Brasil. 3,* 135-148.

Maramorosch, K., Granados, R. R., and Hirumi, H. (1970). Mycoplasma diseases of plants and insects. *Advan. Virus Res. 16,* 135-193.

Maramorosch, K. and Kondo, F. (1977). Electron microscopy of leaf sections from Kerala wilt diseased coconut palms. *J. Plantation Crops 5,* 20-22.

McCoy, R. E. (1972). Remission of lethal yellowing in coconut palm treated with tetracycline antibiotics. *Plant Dis. Rep. 56,* 1019-1021.

McCoy, R. E. (1976a). Comparative epidemiology of the lethal yellowing, Kainkope and cadang-cadang diseases of the coconut palm. *Plant Dis. Rep. 60,* 495-502.

McCoy, R. E. (1976b). Uptake, translocation and persistence of oxytetracycline in coconut palm. *Phytopathology 66,* 1039-1042.

McCoy, R. E. (1976c). Plant phloem sap: a potential mycoplasma growth medium. *Proc. Soc. Gen. Microbiol. 3,* 155.

McCoy, R. E. (1978a). Attempts to culture mycoplasmalike agents at Fort Lauderdale. *Proc. 3rd Meet. Intern. Council Lethal Yellowing, November, 1977, Univ. Fla. Publ. No. FL 78-2,* p. 20.

McCoy, R. E. (1978b). Recent advances in antibiotic treatment of lethal yellowing. *Proc. 3rd Meet. Intern. Council Lethal Yellowing November, 1977, Univ. Fla. Publ. No. FL 78-2,* p. 29.

McCoy, R. E., Carroll, V. J., Poucher, C. P., and Gwin, G. H. (1976). Field control of coconut lethal yellowing with oxytetracycline hydrochloride. *Phytopathology 66,* 1148-1150.

Moore, H. E. (1978). Comments on the host family of lethal yellowing and suggestions for replacing susceptible species. *Proc. 3rd Meet. Intern. Council Lethal Yellowing, November, 1977, Univ. Fla. Publ. No. FL 78-2,* p. 8.

Nayar, R., and Selikskar, C. E. (1978). Mycoplasma like organisms associated with yellow leaf disease of *Areca catechu Eur. J. Forest Pathol. 8,* 125-128.

Nienhaus, F. and Steiner, K. G. (1976). Mycoplasmalike organ-
    isms associated with Kaincope disease of coconut palms in
    Togo. *Plant Dis. Rep. 60*, 1000-1002.
Nutman, F. J. and Roberts, F. M. (1955). Lethal yellowing:
    The "Unknown disease" of coconut palms in Jamaica. *Emp.
    J. Exp. Agr. 23*, 257-267.
Parthasarathy, M. V. (1973). Mycoplasma-like organisms in the
    phloem of palms in Florida affected by lethal yellowing.
    Plant Dis. Rep. 57, 861-862.
Parthsarathy, M. V. (1974). Mycoplasma-like organisms asso-
    ciated with lethal yellowing disease of palms. *Phytopa-
    thology 64*, 667-674.
Parthasarathy, M. V. and Van Slobbe, W. G. (1978). Hartrot
    or fatal wilt of palms. I. Coconuts *(Cocos nucifera)*.
    Principes 22, 3-14.
Parthasarathy, M. V., Van Slobbe, W. G., and Soudant, C.
    (1976). Trypanosomatid flagellate in the phloem of di-
    seased coconut palms. *Science 192*, 1346-1348.
Plavsic-Banjac, B., Hunt, P., and Maramorosch, K. (1972).
    Mycoplasma-like bodies associated with lethal yellowing
    of coconut palms. *Phytopathology 62*, 298-299.
Price, W. C., Martinez, A. P., and Roberts, D. A. (1968). Re-
    production of the coconut lethal yellowing syndrome by
    mechanical inoculation of young seedlings. *Phytopatho-
    logy 58*, 593-596.
Quillec, G., Morin, J. P., Renard, J. L., and Mariau, D.
    (1978). Les maladies du cocotier dans le jeune age.
    Causes, methodes de lutte. *Oleagineux 33*, 495-501.
Randles, J. W., Boccardo, G., Retuerma, M. L., and Rillo, E. P.
    (1977). Transmission of the RNA species associated with
    cadang-cadang of coconut palm, and the insensitivity of
    the disease to antibiotics. *Phytopathology 67*, 1211-1216.
Randles, J. W., Boccardo, G., Imperial, J. S., Palukaitis, P.,
    and Hatta, T. (1979). Cadang-cadang: Evidence for a vi-
    roid etiology. *5th Session FAO Techn. Working Party Co-
    conut Production, Protection Processing, Manila, Philip-
    pines, December, 1979, Paper No. AGP: CNP/79/38.*
Reinert, J. A. (1973). Lethal yellowing of coconut - Vector
    studies and a possible control. In Fisher (1973), p. 155.
Sharples, A. (1928). Palm diseases in Malaya. *Malay Agr. J.
    16*, 313-360.
Sherman, K. E. and Maramorosch, K. (1977). Present status of
    lethal yellowing disease of the coconut palm *(Cocos nuci-
    fera)*. *J. Plantation Crops 5*, 75-83.
Sitepu, D. (1979). Coconut wilt in Natuna Islands. *5th Ses-
    sion FAO Techn. Working Party Coconut Production, Protec-
    tion Processing, Manila, Philippines, December, 1979,
    Paper No. AGP: CNP/79/35.*

Steiner, K. G. (1976a). Remission of symptoms following te-
    tracycline treatment of coconut palms affected with Kain-
    cope disease. *Plant Dis. Rep. 60,* 617-620.
Steiner, K. G. (1976b). Epidemiology of Kaincope disease of
    coconut palms in Togo. *Plant Dis. Rep. 60,* 613-617.
Steiner, K. G. (1978). Suspected lethal yellowing disease of
    coconut palms in Tanzania. *FAO Plant Prot. Bull. 26,*
    10-12.
Steiner, K. G., Nienhaus, F., and Marschall, K. J. (1977).
    Rickettsia-like organisms associated with a decline of
    coconut palms in Tanzania. *Z. Pflanzenkr. Pflanzenschutz
    84,* 345-351.
Thomas, D. L. (1979). Mycoplasma-like bodies associated with
    lethal declines of palms in Florida. *Phytopathology 69,*
    928-934.
Townsend, R., Eden-Green, S. J., Markham, P. G., Archer, D.,
    and Clark, M. F. (1980). Acholeplasmas and lethal yellow-
    ing disease III: Microbiological and serological studies.
    *Proc. 4th Meet. Intern. Council Lethal Yellowing. Univ.
    Fla., mimeographed.*
Tsai, J. H. (1975). Transmission studies of three suspected
    insect vectors of lethal yellowing of coconut palm.
    FAO Plant Prot. Bull. 23, 140-145.
Tsai, J. H. (1978). Vector studies in Florida. *Proc. 3rd
    Meet. Intern. Council Lethal Yellowing, November, 1977,
    Univ. Fla. Publ. No. FL 78-2,* p. 24.
Turner, G. J. (1963). Observations on a previously reported
    wilt of the coconut palm in Sarawak. *Trop. Agr. (Trini-
    dad) 40,* 115-120.
Turner, P. D., Jones, P. and Kenten, R. H. (1978). Coconut
    stem necrosis, a disease of hybrid and Malayan Dwarf co-
    conuts in North Sumatra and Peninsular Malaysia. *Perak
    Planters J. 1978.* 14 pp.
Van der Graaff, N. A. (1979). Breeding coconuts for resis-
    tance to major crop parasites. The need to understand
    and manage the pathosystem. *5th Session FAO Tech. Working
    Party Coconut Production, Protection Processing, Manila,
    Philippines, December, 1979, Paper No. AGP: CNP/79/5.*
Waters, H. (1978). A wilt disease of coconuts from Trinidad
    associated with *Phytomonas* sp., a sieve tube-restricted
    protozoan flagellate. *Ann. Appl. Biol. 90,* 293-302.
Waters, H. and Hunt, P. (1980). The *in vivo* three-dimensional
    form of a plant mycoplasma-like organism by the analysis
    of serial ultrathin sections. *J. Gen. Microbiol. 116,*
    111-131.
Waters, H. and Osborne, I. (1978). Preliminary studies upon
    lethal yellowing and the distribution of MLO in coconut
    palms. *Proc. 3rd Meet. Intern. Council Lethal Yellowing,
    November, 1977, Univ. Fla. Publ. No. FL 78-2,* p. 15.

TRANSMISSION OF LETHAL YELLOWING MYCOPLASMA BY
MYNDUS CRUDUS

James H. Tsai and Darryl L. Thomas

Agricultural Research and Education Center
University of Florida
Fort Lauderdale, Florida

I. INTRODUCTION

Coconut palm *(Cocos nucifera L.)* is one of the most valu-
able plants in the world. Of the 2800 spp. of Palmae, the coconut, oil, and date palms are economically important in the
world because they are the top three major plantation crops.
Over 25 million tons of coconuts and about 5 million tons of
copra are produced annually on 6.5 million hectares of land
with over 500 million palms throughout the tropics, and half
are in the Philippine islands. Sri Lanka, Malaysia, the West
Indies, Central America, and southern India are the centers for
the remainder of coconut production. The United States is the
largest consumer of coconut and coconut oil; nearly 400,000
tons of the coconut oil are used yearly by American industry.
Coconut palms are of great importance to millions of people in
the tropics. They provide sources of shelter, drink, fiber,
and clothing (Woodroof, 1970). In other parts of the tropics,
however, coconut palms have woven themselves more tightly into
the fabric of the life of local tribes and become symbolic of
life in several cultures. The name assigned to this beautiful
tree always has the same meaning, whatever the language — the
tree of life. In addition, coconut palms constitute one of the
important components of the tropical forest ecosystems of Africa and South America (Moore, 1973).
Virtually every part of the coconut tree has economic value. The fibrous husks are processed into rugs, rope, uphols-
tery, and many rubberized coir products. The leaves are woven
into screens, baskets, hats, novelty items, and roofing materi-

als. The shells of the nuts are used for making lamp stand, containers and other ornamentals, and the carbonized shell is an excellent source of activated charcoal. Coconut lumber is used for furnitures and building materials. The fine-grained outer wood, known as porcupine wood is used for carving and veneers. Apart from the above uses, various parts of coconut trees also provide important foodstuffs. The palm heart is considered a delicacy and may be eaten cooked or raw. Sap from inflorescences is used as a soft drink and can be brewed into wine. Fresh or dried coconut flesh is a nutritious food for man and livestock. Coconut milk is an important constituent of media for tissue culture of plants and it is a good source of extracting coconut honey. Copra contains 60-68% oil which is an important ingredient in soaps, shampoo, detergents and resins where its lauric and myristic acids aid in leathering both in hard and soft water, and its fatty alcohols enhance both detergency and biodegradability. The oil has special qualities and is also used in products as varied as aviation brake fluid, margarine, and confectionatires.

In 1873, an American Union army doctor named Horace Porter found coconut palms near the Miami area and named the post office he founded there "Coconut Grove." The original seeds may have come from the Caribbean islands by the ocean's current. In 1878, the famous shipwreck of the Providentia brought a load of coconut seed to Lake Worth, Florida, where the coconut palms quickly thrived in their new home. A sizable tract of coconut palms was planted by Henry Lum on Miami Beach. Ezra Osburn and Elnathan Field planted about 330,000 coconuts along a 65-mile stretch of beach property from Cape Florida to Jupiter Inlet. During the early 20th century, William Matheson turned Key Biscayne into a massive coconut plantation. The history of coconut cultivation in southern Florida is a succession of high hopes and low yields. However, coconut palms have thrived in south Florida's landscape. They are grown solely for their pervasive symbol of all the magic of the tropics. Along the sunny beaches and avenues, no other plants can create the same tropical effect as the coconut palms. It is only coconut palms which can provide its unique environment for both residents and tourists of south Florida. The tourism business in the areas has depended heavily on the aesthetic value of coconut palms. The survival of coconut palms has been threatened by a disease called lethal yellowing.

Lethal yellowing (LY) disease of coconut palms was first reported in Jamaica (Fawcett, 1891) and, by 1979, killed more than one-half of the 4.3 million coconut palms on that island. It has also been reported in the Caribbeans (De La Torre, 1906; Leach, 1946; Martinez, 1965; Mijailova, 1967; Maramorosch, 1964; Seal, 1928; Schieber and Hichez-Frias, 1970) and West Africa (Dring, 1963; Leather, 1959; Ollagnier and West-steijn, 1961; Steiner, 1976; Westwood, 1953). This devastating disease was reported in Key West, Florida in 1965 (Martinez, 1965) and has killed of 20,000 coconut palms. In 1971, it appeared on the mainland of Florida. In eight years LY has killed over 50% of the estimated one million coconut palms in Florida. The aesthetic value and environmental impact of the coconut palms in Florida (Fisher, 1975) cannot be measured in terms of dollars.

Evidence indicates that LY attacks not only coconut palms but also 25 other species representing at least 7 of the 15 major groups of palms (Moore, 1973). Mycoplasma-like organisms were found in these species of infected palms in the LY epidemic areas (Parthasarathy, 1974; Thomas, 1974, 1976, 1979). The majority of the coconut varieties from many parts of the world that were tested in Jamaica have shown low resistance to LY (Harries, 1971), thereby posing a great threat to every coconut-growing country (Harries, 1978).

II. ETIOLOGY, SYMPTOMATOLOGY, AND EPIDEMIOLOGY

Since 1945, a number of pathogens have been studied as possible causes of LY disease, including viruses, fungi, bacteria, and nematodes (Tsai, 1980). In 1972, Plavsic-Banjac et al. first reported the association of a mycoplasma-like organism (MLO) with infected but not healthy tissue of a coconut palm, which suggested the MLO etiology. This theory was further supported by researchers from other laboratories (Beakbane et al. 1972; Heinze, 1972; Parthasarathy, 1973; 1974; Thomas, 1974, 1976, 1979; Waters and Osborne, 1978). Additional evidence of a remission of symptoms following treatment with tetracycline antibiotics, but not penicillin, also suggests the MLO etiology (Hunt et al., 1974; McCoy, 1972).

The LY symptoms in fruit-bearing palms include (a) premature nut fall; (b) necrosis of inflorescences in opened and unopened spathes; (c) root rot; (d) a yellow flag-leaf first appearing in the middle of the brown followed by upward movement of discoloration; (e) the decay of the spear leaf and heart; (f) toppling of the crown. The symptoms in young coconut palms begin with yellow leaf or leaves and end with decay of spear leaf and heart. The symptoms on the other species of palms are similar to that of coconut palms, except that infested palms exhibited leaf browning instead of yellowing. Water soaking marks are typical along the pinnae. In Veitchia palms, the older leaves may appear yellow-brown in color, and break at the junction of the sheathing leaf-base

and the midrib. The spear leaves often remain on top of the
dead tree. In *Pritchardia* palms, the spear leaves tend to die
first while the other leaves only show slight yellowing or
discoloration. (Parthasarathy and Fisher, 1973).

The typical spread of LY first involves a local center of
infection on one or two trees. Later, other infected trees
appear in a random pattern surrounding the initial center of
infection to provide further local spread. The other type of
spread is a long-distance spread which is followed by the pattern of local spread. The distance or jump spread can be as
far as 70 km (Carter, 1966). This is indicative of spread by
an airborne vector (McCoy, 1976).

III. PREVIOUS STUDIES ON LETHAL YELLOWING VECTORS

For many years, LY research both in Jamaica and Florida
has been hindered by the difficulties in isolation and cultivation of an LY causal agent, a long incubation period of
pathogen in its hosts, inability to transmit or graft LY agents
to a palm, and the unknown biology and rearing techniques of
the test insects. However, in recent years some progress has
definitely been made on LY research. Evidence obtained from
electron microscopy and the remission of symptoms after antibiotic treatment strongly suggests the MLO association with
LY of palms. Since 1955, a number of suspected vectors have
been used in transmission trials by various workers both in
Jamaica and Florida. The species included in the studies were
members of Cicadellidae, Fulgoridae, Derbidae, Delphacidae,
Issidae, Aphidae, Coccoidea, Pseudococcidae, Aleyrodidae, Cixiidae, Cicadidae (Homoptera), Tingidae (Hemiptera) and Thy-
sanoptera. Other suspected vectors also included eriophyid
mites and nematodes. Numerous earlier studies were summarized
in the recent reviews (Johnson and Eden-Green, 1978; Tsai,
1980).

IV. APPARENT TRANSMISSION OF LETHAL YELLOWING MYCOPLASMA TO
     CARYOTA spp., *PHOENIX* spp., and *PRITCHARDIA* spp. BY
     MYNDUS CRUDUS

Epidemiological evidence suggests the airborne vectors and
the mycoplasmal etiology of LY has led the search for potential vectors among the *Auchenorrchynchus* homopterans, namely,
the leafhoppers (Cicadelloidea) and planthoppers (Fulgoroidea).
Earlier works show that *Myndus crudus* Van Duzee (Kramer, 1979)
is the only predominant homopteran consistently found on coconut and other palms (Tsai, 1980). Initial emphasis was

placed on studying and testing this insect as a natural vector
on the assumption that spread of LY is from palm to palm (Tsai,
1975, 1977, 1978; Tsai et al., 1976; Tsai and Kirsch, 1978).
More tests were conducted with this insect by collecting the
adult Myndus from LY epidemic areas and directly caging them
on potted palms. The species of palms also known to be susceptible to LY are included in Table I. Therefore, they were
used in this study.

        There were 13 tests palms including one P. remota, two P.
thurstonii, six P. eriostachya, three P. reclinata and one C.
maximus from the above tests which had shown LY symptoms. Of
the 13 palms showing disease symptoms, only one P. remota con-

TABLE I. Transmission Tests of Lethal Yellowing Mycoplasma
Using Myndus crudus from LY Epidemic Areas

Test plant	Plant height (ft)	plants tested	insects tested
Pritchardia remota	2	2	153
P. thurstonii	2	2	199
Livistona chinensis	2	2	168
Arikuryroba schizophylla	2,3	2	161
Arenga engleri	2	2	265
Chrysolidocarpus cabadae	2	1	322
Phoenix reclinata	2	4	247
P. dactylifera	3	2	247
Latania sp.	1	2	174
Cocos nucifera	5	1	114
Veitchia merrilli	3	1	158
Caryota maximus	2	2	253
C. mitis	6,10,12,12,15	5	216
Phoenix reclinata	1.5	1	
Pritchardia eriostachya	1.5	1	3025
P. reclinata	1.5	1	
P. thurstonii	1	1	3200
P. eriostachya	1	5	
P. remota	1	1	11,660
P. eriostachya	1	6	13,100
P. eriostachya	1	5	11,950
P. eriostachya	1	5	16,500

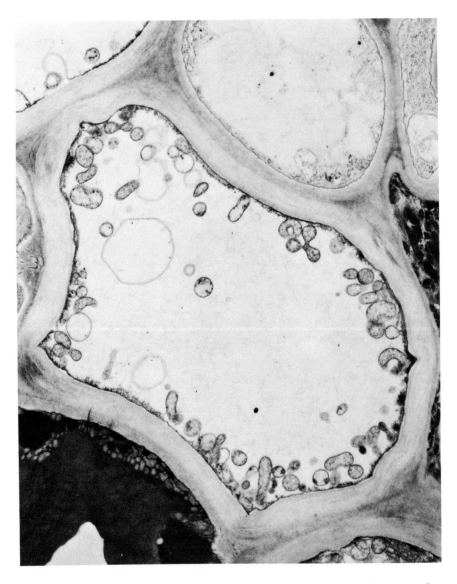

Fig. 1. Transverse section through a sieve element of Pritchardia remota containing MLO.

tained MLO in the thin section (Fig. 1). The rest yielded negative results in the EM examination.

Another experiment was performed by collecting *Myndus* a-dults from the field. They were caged over LY-infected leaves for 1-3 days acquisition access period (AAP) in the hope of

TABLE II. Attempts to Transmit LY Agent by *Myndus crudus*[a]

Source plant	Test plant	Height (ft)	Number of insects[b]
Veitchia merrillii	*Cocos nucifera*	7	3042/4345
V. merrillii	*C. nucifera*	8	2393/5025
V. merrillii	*C. nucifera*	8	1400/2200
V. merrillii	*C. nucifera*	9	3292/5150
C. nucifera			
V. merrillii	*C. nucifera*	8	3180/5720
C. nucifera			
V. merrillii	*C. nucifera*	8	2869/3860
C. nucifera			
C. nucifera	*Pritchardia remota*	3.5	2402/3400
V. merrillii	*P. thurstonii*	3.5	3194/3802
V. merrillii	*P. thurstonii*	2.5	3571/5320
V. merrillii	*P. thurstonii*	3	2389/4245
V. merrillii	*P. thurstonii*	20	1497/2600
V. merrillii	*P. woodfordi*	2	2074/3115

[a]*AAP = 1-3 days.*

[b]*No. of insects transferred after AAP/No. of insects at start of experiment.*

increasing the numbers of infective insects. At the end of the AAP, the survivors were recovered and caged on potted palms growing indoors and on one 20-ft tall *P. thurstonii* which was grown outdoors. The results are summarized in Table II.

Of the total of 12 palms tested in the above experiment, only one matured *P. thurstonii* exhibited typical LY symptoms. EM examination also revealed MLO's in the thin sections (Fig. 2).

V. ATTEMPTS TO TRANSMIT LETHAL YELLOWING MYCOPLASMA TO *CATHARANTHUS ROSEUS* BY *MYNDUS CRUDUS*

Another experiment was designed to test whether LY mycoplasma would be transmitted to *Catharanthus roseus* (Don) which is host to many other mycoplasmal diseases. The adult *Myndus* were collected from the LY epidemic areas, and were given various AAP's on LY-infected palms. They were then caged on

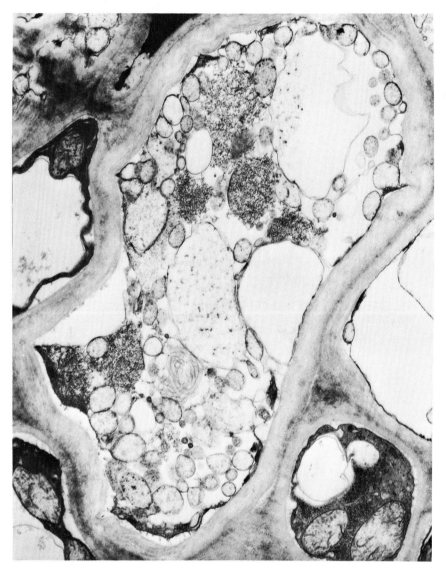

*Fig. 2. Transverse section through a sieve element of
Pritchardia thurstonii containing MLO.*

the potted periwinkles. Table III is a summary of these
tests.
   Two test plants of *C. roseus* from the above test showed
overall chlorosis of leaves and the reduction in size of leaf

TABLE III. Attempts to Transmit Lethal Yellowing Mycoplasma by *Myndus crudus* after Given Various AAP'S

AAP (day)	Number of insects transferred after AAP	Test plant	Longevity (days)
2	1053	*Catharanthus roseus*	7
2	600	*C. roseus*	9
2	363	*C. roseus*	7
2	528	*C. roseus*	8
2	480 (+940)[a]	*C. roseus*	8
0	1760	*C. roseus*	9
0	2989	*C. roseus*	7
0	2350	*C. roseus*	8
0	1000	*C. roseus*	7

[a]*These insects in the bracket were not given acquisition access feed.*

and flower, but samples taken from these two plants for electron microscopy examination revealed no MLO's.

VI. ATTEMPTS TO TRANSMIT LETHAL YELLOWING MYCOPLASMA TO *COCOS NUCIFERA* AND *VEITCHIA MERRILLII* BY OTHER INSECTS

Other species of homopterans were selected as test insects based on their ability to transmit other plant prokaryotes (Maramorosch, 1974; Maramorosch *et al.*, 1970; Whitcomb and Davis, 1970; Tsai, 1979), or they could feed on the palm foliages for various lengths of time. *Macrosteles fascifrons* (Stal.), *Dalbulus maidis* (DeLong and Wolcott), *Graminella* spp., *Spangbergiella vulnerata* (Wolcott), *Peregrinus maidis* (Ashmead), *Idioderma virescens* (Van Duzee) were tested in various transmission trials. After AAP or injection, the insects were transferred to their preferred hosts for an incubation period (IP) of 2-3 weeks. They were then transferred to a palm or spear leaf in a cage on a test palm. A series of transmission trials were conducted in the last two years, and are summarized in Tables IV, V, and VI.

All of those 19 potted palms including ten *V. merrillii*, eight *C. nucifera*, and one *P. eriostachya* in this series of tests, and only one *P. eriostachya* showed LY-like symptoms. However, no MLO's were found in the thin sections.

TABLE IV. Attempts to Transmit LY Agent by *Macrosteles fasci-frons*[a]

Test plant	Number of insects[b]		Longevity on palm (days)	
			Max.	Mean
Cocos nucifera	210/300	(70)	33	9.3
C. nucifera	2194/3800	(822)		
C. nucifera	197/1200	(142)	24	9.7
Veitchia merrillii	347/700	(208)	38	6.7
V. merrillii	42/280			
V. merrillii	222/400	(110)	43	

[a]*AAP = 5-9 days.*

[b]*No. of insects survived after AAP/No. of insects at start of experiment. Number in parenthesis, no. of insects transferred to palm.*

TABLE V. Attempts to Transmit LY Agent by *Macrosteles fasci-frons*

Injection inoculum	Number of insects injected[a]	Test plant
Lyophilized *Cocos* meristem	28/100	*Cocos nucifera*
Lyophilized *Cocos* meristem	41/265	*C. nucifera*
Phloem exudate	215/1186	*Veitchia merrillii*

[a]*No. of insects survived after 14 days/no. of insects injected.*

VII. ATTEMPTS TO TRANSMIT PERIWINKLE PHYLLODY MYCOPLASMA BY *MYNDUS CRUDUS* AND OTHER INSECTS

Periwinkles are perennial plants in southern Florida and Jamaica. They are found to grow abundantly in the undergrowth of plams and are infected with phyllody mycoplasm (Fig. 3). Searching for an alternate host of LY agent is one of the main research objectives (Tsai, 1980). Attempts were made in the last 2 years to identify the phyllody vector and its role in relation to LY of coconut palms. In one experiment, 3770

TABLE VI. Attempts to Transmit LY Agent by Various Insects

Test insect	AAP (day)	Number of insects[a]		Test plants	Longevity on palm (days)
Dalbulus maidis	1	2097/4200	(608)	Cocos nucifera	3
Dalbulus maidis	1	377/1280	(279)	V. merrillii	
Spangbergiella vulnerata	5	120/400	(59)	C. nucifera	
Spangbergiella vulnerata	3	110/300	(94)	V. merrillii	14
Peregrinus maidis	2-5	281/600	(163)	V. merrillii	
Peregrinus maidis	2-5	256/600	(159)	V. merrillii	
Peregrinus maidis	7	90/300	(10)	C. nucifera	
Graminella spp.	7	71/200	(14)	V. merrillii	
Graminella spp.	2-6	588/796		V. merrillii	
Idioderma virescens	0	32		Pritchardia eriostachya	5

[a]No. of insects survived after AAP/No. of insects at start of experiment. Number of parentheses, no. of insects transferred after 14-20 days incubation.

M. crudus adults were placed on a phyllody-infected periwinkle for 1-day AAP. At the end of AAP, 2116 survivals were caged on two 1-ft. tall P. eriostachya. Other experiments were conducted using D. maidis, M. fascifrons, and Graminella spp. as experimental vectors. Different groups of test insects were caged on the phyllody-infected periwinkles for various AAP's. At the end of AAP, the survivors were transferred to their preferred hosts (sweet corn for D. maidis; rye for M. fascifrons and Graminella spp.) for 2-3 weeks incubation period (IP). After IP, the remaining groups of insects were then caged on a potted periwinkle, respectively. Table VII summarizes the various trials.

Three periwinkle plants inoculated by D. maidis and one periwinkle plants inoculated by M. fascifrons had developed chlorosis of leaf and a reduction in size of leaf and flower. No MLO's were detected in the thin sections.

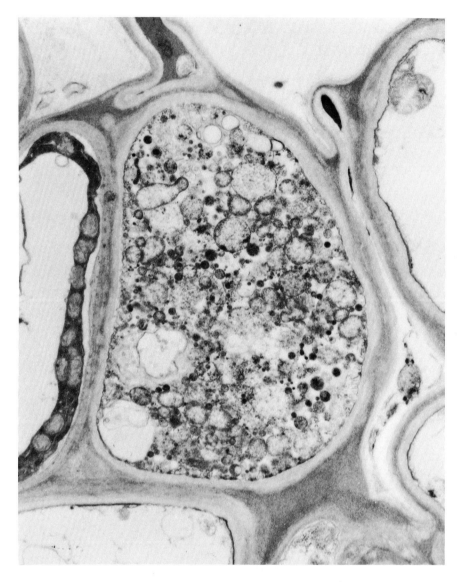

Fig. 3. Transverse section through a sieve element of Catharanthus roseus naturally infected with periwinkle phyllody mycoplasma.

TABLE VII. Attempts to Transmit Periwinkle Phyllody Mycoplasma by *Dalbulus maidis, Macrosteles fascifrons,* and *Graminella* spp.

Test insect	AAP (day)	No. of insects tested	No. of insects after AAP	No. of insects caged on periwinkle
D. maidis	1	1930	921	826
	1	1670	880	704
	1	1350	770	610
M. fascifrons	3	1200	944	822
	4	1160	886	710
	4	1180	904	726
	6-14	1960	710	476
Graminella spp.	2	860	432	205
	2	744	351	111

VIII.   DISCUSSION AND CONCLUSION

Although *M. crudus* do not breed on coconut and other palms, it is still the most common insect associated with palms in the LY-affected areas. The numbers of insects and test plants used in this study are shown in Tables I-VII. Of all the plants tested in the last two years, only 15 plants showed LY symptoms. The electron microscopic examinations of these symptomatic plants could only find one *P. thurstonii* (Table II), and one *P. remota* (Table I) containing mycoplasmas in the tissue. The failure to find MLO in thin sections did not prove that the results of transmission tests with *M. crudus* were negative. At best, the EM examinations can substantiate the positive results. However, more confirmatory experiments are being conducted by the author (JHT) using reared *M. crudus*. Hopefully, more information can be gained from the tests. The numerous failures of transmission of the LY agent from palm to palm by *M. crudus* reported both in Florida and Jamaica (Eden-Green and Schuiling, 1978; Eden-Green, 1979), and a few apparent transmissions in this study and in the study conducted by Howard (1980) indicate that this insect is an inefficient vector of LY. In most tests, a large number of test insects were used with no positive results. Nontransmission could be due to a number of factors, such as acquisition feeding site, time, titer of MLO in a given source plant, and susceptibility of test palms, as well as the involvement of other insect vectors. Generally, we found extremely low concentration of MLO in the mature leaves of the infected palms in EM

examination. The most likely tissue to find MLO is the pri-
mordial leaves near the meristem. Waters and Osborne (1978)
also reported that MLO did not occur in the mature leaves
other than flag leaf of diseased palms, but they were regularly
found in the immature leaves and in the inflorescence with
terminal necrosis. Samples taken from Togo were examined by
electron microscopists in five different laboratories over a
2-year period before the detection of MLO (Neinhaus and Steiner, 1976). Parthasarathy (1974) also reported his EM study of
coconut, *Pritchardia* and *Veitchia* palms affected by LY. The
MLO's were mainly confined to recently matured protophloem and
early metaphloem elements, and were apparently absent in the
matured late metaphloem sieve elements of expanded leaves and
inflorescences.

In view of these findings, they could account for the very
minimal success of acquisition by natural feeding on diseased
palm fronds. An insect vector is unlikely to be able to probe
and feed on the growing point or young inflorescence tissue
with a close spadix, where most MLO were found. In his study
of anatomical and physiological features of palms, Zimmermann
(1980) concluded that older leaves were unlikely to be acquisition sites for the LY vector.

It was observed that palms susceptible to LY disease were
usually at least 2 years old. However, in our tests, the
majority of test palms were young palms under 14 months of
age. Several of the inoculated palms had exhibited LY symptoms
and two of the symptomatic *Pritchardia* palms were confirmed to
contain MLO in the EM examination. It appears that *Pritchardia*
palms are better test plants than the other palms used in our
study.

Other possible vectors may be infrequent visitors of palms.
It might only occasionally feed on palms. The inflorescence
visiting insects may also involve the transmission of LY agent.
Clark (1977) suggested that spiroplasmas might be transmitted
via flowers. The removal of unopened inflorescences from the
matured *Veitchia* palms has resulted in reduction of LY incidence (Tsai, 1978). This experiment is being continued to
March, 1980. Of the total 75 *Veitchia* palms tested only 38
palms had succumbed to LY after 4 years. This could explain
why more flowering palms were affected by LY than immature
ones, as was also stated by Sherman and Maramorosch (1977).
Furthermore, there may be an alternate host which could be a
monocotyledon or dicotyledon for supporting the growth of both
LY mycoplasma and vector. Thus, it is especially significant
to transmit LY agent to a herbaceous host. Such a breakthrough in research could reduce space requirements and shor-
ten the experimental time.

The criteria used in selecting other leafhopper and plant-hopper species employed in this study were that they should: (1) be able to survive on palms for various AAP, (2) bear a broad phylogenetic relationship to *M. crudus,* (3) be a widely distributed species in south Florida, and (4) be a known vector of one or more plant disease agents. For laboratory tests, the reasons for selecting *M. fascifrons, D. maidis, Graminella* spp., *P. maidis, S. vulnerata,* and *I. virescens* were that, with the exception of latter, they could be bred in the laboratory so as to provide large numbers for transmission tests and could commonly be found and tested both in Florida and Jamaica (Eden-Green, 1979). Although these insects are fairly predominant in the undergrowth of palm groves, but are not known to breed on the foliages of coconut and other palms. Thus, they can only be considered occasional feeders and exclusively as adults. McCoy (1976) theorized that the spread of LY agent is from palm to palm based on the epidemiological evidence. Therefore, it would be unlikely for these species to be the natural vector. Nonetheless, it was our intention to find an experimental vector in hopes it would lead to the discovery of the natural vector. This search should be continued by (1) artificially feeding insects on diseased palms and/or on membrane feeding system, and (2) injecting the insects with MLO extracted or cultured from plants. However, this success will heavily depend on our ability to extract viable MLO from disease palms and the development of suitable culture media. Markham *et al.* (1974) reported using an unnatural vector to experimentally transmit the disease agent, even though the vector-disease agent relationship does not exist in the nature.

The finding of an LY natural vector must be unequivocally proved; knowledge of its biology and epidemiology are essential in determining the feasibility and possibility of controlling the LY vector. Furthermore, it would lead to a standardized and more reliable method of testing palm species resistant to LY, than is currently used, i.e., exposure to natural infection for an indefinite period.

ACKNOWLEDGMENTS

We thank Mr. Brian Houha and Ms. Ronelle Norris for their technical help. The research was supported in part by the Horticultural Research Institute's 1979 research grant (to J.H.T.).

REFERENCES

Beakbane, A. B., Slater, C. H., and Posnette, A. F. (1972).
   Mycoplasmas in the phloem of coconut, *Cocos nucifera* L.,
   with lethal yellowing disease. *J. Hort. Sci. 47*, 265.
Carter, W. (1966). Report to the Government of Jamaica on
   lethal yellowing disease of coconut. FAO. United Nations,
   Rome TA2158 24 pp.
Clark, T. B. (1977). *Spiroplasma* sp., a new pathogen in honey
   bees. *J. Invertebr. Pathol. 29*, 112-113.
De La Torre, C. (1906). La enfermedad de los cocoteros. *Rev.
   Letras Ciencias, Univ. Habana 2*, 269-281.
Dring, B. M. (1963). Rapport au gouvernment du Togo sur la
   maladie de Kaincope du cocotier, Rome, FAO. 29 pp.
Eden-Green, S. J. (1979). Attempts to transmit lethal yellow-
   ing disease of coconuts in Jamaica by leafhoppers (Homop-
   tera: Cicadelloidea). *Trop. Agr. (Trinidad) 56*, 185-192.
Eden-Green, S. J. and Schuiling, M. (1978). Root acquisition
   feeding transmission tests with *Haplaxius* spp. and *Proar-
   na hilaris,* suspected vectors of lethal yellowing of coconut palm in Jamaica. *Plant Dis. Rep. 62*, 625-627.
Fawcett, W. (1891). Report on the coconut disease at Montego
   Bay. *Bull. Bot. Dept. Jamaica 23*, 2.
Fisher, J. B. (1975). Environmental impact of lethal yellowing
   disease of coconut palms. *Environ. Conserv. 2*, 299-304.
Harries, H. C. (1971). 11th *Rep. Coconut Ind. Bd. Res. Dept.
   Jamaica,* pp. 13-35.
Harries, H. C. (1978). Susceptibility to lethal yellowing di-
   sease in global perspective. *Proc. 3rd Meeting Int. Coun-
   cil Lethal Yellowing, Univ. Fla.* FL-78-2, p. 29.
Heinze, K. G. (1972). Lethal yellowing disease of coconut.
   FAO Rep. No. TA3152 to the Government of Jamaica.
Howard, F. W. (1980). Lethal yellowing transmission experiments
   with Myndus crudus Van Duzee. *Proc. 4th Meeting Intl.
   Council Lethal Yellowing, Univ. Fla.* FL-80-1, p. 13.
Hunt, P., Dabek, A. J., and Schuiling, M. (1974). Remission of
   symptoms following tetracycline treatment of lethal yellow-
   infected coconut palms. *Phytopathology 64*, 307-312.
Johnson, G. G. and Eden-Green, S. J. (1978). The search for a
   vector of lethal yellowing of coconuts in Jamaica; A re-
   appraisal of the experiments from 1962 to 1971. *FAO Plant
   Prot. Bull. 26*, 137-149.
Kramer, J. P. (1979). Taxonomic study of the planthopper genus
   Myndus in the Americas (Homoptera: Fulgoroidea: Cixiidae).
   Trans. Amer. Entomol. Soc. 105, 301-389.
Leach, R. (1946). The unknown disease of the coconut palm in
   Jamaica. *Trop. Agr. (Trinidad) 23*, 50-60.

Leather, R. L. (1959). Further investigations into the Cape
    St. Paul Wilt of coconuts at Keta, Ghana. *Emp. J. Exp.
    Agr. 27*, 105.
McCoy, R. E. (1972). Remission of lethal yellowing in coconut
    palm treated with tetracycline antibiotics. *Plant Dis.
    Rep. 56*, 1019-1021.
McCoy, R. E. (1976). Comparative epidemiology of the lethal
    yellowing, Kaincope and cadang-cadang diseases of coconut
    palm. *Plant Dis. Rep. 60*, 498-502.
Maramorosch, K. (1964). "A Survey of Coconut Diseases of Un-
    certain Etiology," FAO, Rome. 39 pp.
Maramorosch, K. (1974). Mycoplasmas and Rickettsiae in rela-
    tion to plant diseases. *Annu. Rev. Microbiol. 282*, 301-
    324.
Maramorosch, K., Granados, R. R., and Hirumi, H. (1970). Myco-
    plasma diseases of plants and insects. *Advan. Virus Res.
    16*, 135-193.
Markham, P. G., Townsend, R., Bar-Joseh, M., Daniels, M. J.,
    Plaskitt, A., and Meddins, B. M. (1974). Spiroplasmas are
    the causal agents of citrus littleleaf disease. *Annu.
    Appl. Biol. 78*, 49-57.
Martinez, A. F. (1965). Lethal yellowing of coconut in Flori-
    da. *FAO Plant Prot. Bull. 13*, 25-29.
Mijailova, P. T. (1967). Informe sobre la investigacion de la
    enfermedad "Pudricion del cogollo del cocotero." *Rev.
    Agr. Habana 1*, 74-110.
Moore, H. E., Jr. (1973). Palms in the tropical forest eco-
    system of Africa and South America. *In* "Tropical Forest
    Ecosystems in Africa and South America: A Comparative
    Review" (B. J. Meggers, E. S. Ayensu, and W. D. Duckworth,
    eds.), pp. 63-68. Smithsonian Institution Press, Washing-
    ton, D. C. 350 pp.
Neinhaus, F. and Steiner, K. G. (1976). Mycoplasma-like organ-
    isms associated with Kaincope disease of coconut palms in
    Togo. *Plant Dis. Rep. 60*, 1000-1002.
Ollagnier, M. and Weststeijn, G. (1961). Coconut diseases in
    the islands of Caribbean: Comparison with Kaincope disease
    in Togo. *Oleagineaux 16*, 729-736.
Parthasarathy, M. V. (1973). Mycoplasma-like organisms in the
    pholoem of palms in Florida affected by lethal yellowing.
    Plant Dis. Rep. 57, 861-862.
Parthasarathy, M. V. (1974). Mycoplasma-like organisms asso-
    ciated with lethal yellowing disease of palms. *Phytopa-
    thology 64*, 667-674.
Parthasarathy, M. V. and Fisher, J. B. (1973). The menace of
    lethal yellowing to Florida palms. *Principes 17*, 39-45.

Plavsic-Banjac, B., Hunt, P., and Maramorosch, K. (1972). My-
    coplasma-like bodies associated with lethal yellowing di-
    sease of coconut palms. *Phytopathology 62,* 298-299.
Schieber, E. and Hichez-Frias, E. (1970). Lethal yellowing di-
    sease of coconut palms in the Dominican Republic. *Phyto-
    pathology 60,* 1542.
Seal, J. L. (1928). Coconut bud rot in Florida. Univ. of Flo-
    rida, *Fl. Agr. Exp. Stat. Bull. No. 109,* 22 pp.
Sherman, K. E. and Maramorosch, K. (1977). Present status of
    lethal yellowing disease of the coconut palm. *J. Planta-
    tion Crops 5,* 75-83.
Steiner, K. G. (1976). Studies on the Kaincope disease of co-
    conut palms in Togo. *Principes 20,* 68-69.
Thomas, D. L. (1974). Possible link between declining palms
    species and lethal yellowing of coconut palms. *Proc. Fla.
    State Hort. Soc. 87,* 502-504.
Thomas, D. L. (1976). Possible hosts of lethal yellowing.
    Principes 20, 59.
Thomas, D. L. (1979). Mycoplasma-like bodies associated with
    lethal declines of palms in Florida. *Phytopathology 69,*
    928-934.
Tsai, J. H. (1975). Transmission studies of three suspected
    insect vectors of lethal yellowing of coconut palm. *FAO
    Plant Prot. Bull. 23,* 140-145.
Tsai, J. H. (1977). Attempts to transmit lethal yellowing of
    coconut palms by the planthopper, *Haplaxius crudus, Plant
    Dis. Rep. 61,* 304-307.
Tsai, J. H. (1978). Vector studies in Florida. *Proc. 3rd Meet.
    Int. Council Lethal Yellowing, Univ. Fla.* FL-78-2, p. 24.
Tsai, J. H. (1979). Vector transmission of mycoplasmal plant
    diseases. *In* "The Mycoplasmas" (M. F. Barile, S. Razin,
    J. G. Tully, and R. F. Whitcomb, eds.), Vol. 3 pp. 265-307.
    Academic Press, New York.
Tsai, J. H. (1980). Lethal yellowing of coconut palms - Search
    for a vector. *In* "Vectors of Plant Pathogens" (K. F. Har-
    ris and K. Maramorosch, eds.), Vol. III, Chapt. 11, pp.
    177-200. Academic Press, New York.
Tsai, J. H. and Kirsch, O. H. (1978). Bionomics of *Haplaxius
    crudus* (Van Duzee) (Homoptera: Cixiidae). *Environ. Ento-
    mol. 7,* 305-308.
Tsai, J. H., Woodiel, N. L., and Kirsch, O. H. (1976). Rearing
    techniques for *Haplaxius crudus* (Homoptera: Cixiidae).
    Florida Entomol. 59, 41-43.
Waters, H. and Osborne, I. (1978). Preliminary studies upon
    lethal yellowing and the distribution of MLO in coconut
    palms. *Proc. 3rd Meet. Int. Council Lethal Yellowing,
    Univ. Fla.,* FL-78-2, p. 15.

Westwood, D. (1953). Coconut diseases, Kets. Ho. *Gold Coast, Dept. Agr.*, 25 pp.

Whitcomb, R. F. and Davis, R. E. (1970). Mycoplasma and phytarboviruses as plant pathogens persistently transmitted by insects. *Annu. Rev. Entomol. 15*, 405-464.

Woodroof, J. G. (1970). "Coconuts; Production, Processing, Products," 241 pp. AVI Publ., Westport, Conn.

Zimmermann, M. H. (1980). Lethal yellowing and translocation in palms. *Proc. 4th Meet. Int. Council Lethal Yellowing, Univ. Fla.* FL-80-1, p. 7.

ASSOCIATION OF MYCOPLASMA AND ALLIED PATHOGENS
WITH TREE DISEASES IN INDIA

S. K. Ghosh

Kerala Forest Research Institute
Kerala, India

I. INTRODUCTION

Mycoplasmas or rickettsia have been implicated in more
than twenty five tree diseases. Of these, two important diseases occur in India and have attracted worldwide attention.
One, sandal spike, is well known in the state of Karnataka,
the other, citrus greening, is widely distributed throughout
the citrus tracts of India. Both these diseases have a similar mode of transmission, sensitivity to tetracyclines, and
can be classified under yellows-type diseases. However, it
is now known definitely that the two organisms causing these
diseases differ very much in their ultrastructural morphology,
although both are intraphloemic and prokaryotic in nature.

A. *Sandal Spike*

Sandalwood *(Santalum album* L.*)*, is considered one of the
world's most precious woods. The wood is either used for
carving or extraction of the famous sandalwood oil or sandal
oil, used mainly in the soap, perfume, and cosmetic indus-
tries.
Although sandal trees have been found to grow in many
places, the best heartwood formation, which yields the essential oil, takes place only in the Indian sandal tract. In
India, natural forests of sandal are spread over 480 km from
Dharwar in the North to Nilgiris in the South and 400 km from
Coorg in the West to Kuppam in the East. Medium-size ever-

green trees (average 18 m height and 2.4 m girth), flourish
well from sea level up to 1350 m and the best heartwood formation takes place at between 600 to 900 m in areas with an
annual rainfall of 850 to 1200 mm.

The tree is normally resistant to diseases and pests.
However, its most destructive disease, spike, was first recorded by McCarthy in Coorg in 1899 (Barber, 1903). Inciden-
tally, this discovery was made just 1 year after mycoplasma
(Mycoplasma mycoides var. *mycoides)* was first cultured at the
Pasteur Institute in France by the associates of Louis Pasteur from bovine pleuropneumonia (Nocard and Roux, 1898).

1. *Symptoms*

The disease is characterized by severe reduction in
the size of leaves and internodes. As the disease advances,
the new leaves become smaller and smaller and show a tendency
to stand out stiffly from the branches; the whole shoot gives
an appearance of spike inflorescence. Spiked plants usually
do not bear any flower or fruits. Phylloid flowers are sometimes present. Heavy reduction of the total photosynthetic
area may lead to the death of the trees within 2-3 years after the appearance of visible symptoms. Sometimes the individual infected shoots show a pendulous or drooping type
of spike, which is characterized by the continuous apical
growth of the diseased shoot. In such a case, the dormant
buds usually do not grow.

2. *Diagnosis*

Although the spike symptom is very characteristic,
it is sometimes very difficult to diagnose the diseased
plants in the field. There is also frequent masking of the
symptom in nature, which may reappear with the new flush of
leaves. Mme. Dijkstra's group in the Netherlands (Dijkstra,
1968; Dijkstra and van der Want, 1970; Dijkstra and Hiruki,
1974) and the scientists at Forest Research Laboratory,
Bangalore (Parthasarathy *et al.*, 1966) tried to develop histopathological color reactions to be used to diagnose the disease. Unfortunately, however, none of these methods have
proved conclusive.

3. *Alternate/Collateral Hosts*

In the spiked areas, a number of plants with phyl-
loidy or little leaf symptoms have been recorded (Nayar and

Ananthapadmanabha, 1977). Some of these disease symptoms
could be produced when sandal spike was transmitted to such
hosts by grafting or doddar (Srimathi, personal communication).

4. Electron Microscopy

Three groups working independently with this disease,
almost simultaneously published electron micrographs of the
causal agent (Varma et al., 1969; Dijkstra and Ie, 1969;
Hull et al., 1969). Dijkstra and Lee (1972) could transmit
spike disease to Vinca rosea and vice versa, using the vegetative vector, dodder (Cuscuta sp.). Scientists from Indian
Agricultural Research Institute, Delhi and Forest Research
Laboratory, Bangalore, in collaboration with the group of
Prof. Maramorosch in the United States confirmed the presence
of mycoplasma-like bodies both in the naturally infected as
well as graft-transmitted spiked plants, thereby proving also
the graft transmissibility of these organisms (Plavsic-Banjac
et al., 1973). All electron micrographs of the spike-infected
tissues showed unicellular, prokaryotic inclusion bodies surrounded by a triple-layered typical unit membrane, approximately 10 nm thick.

5. Insect Transmission

Moonia albimaculata and Coelidea indica (=Jassus in-
dicus) were incriminated as the possible disease vectors
(Anonymous, 1933; Rangaswamy and Griffith, 1941). However
attempts to duplicate these experiments have proved unsuccessful. Nielson (1975) recently questioned the identity of the
Jassus indicus as the vector, since no type speciment was
available for study. Shivaramakrishnan and Sen-Sarma (1978)
obtained the transmission of spike disease by the green rice
leafhopper Nephotettix virescens Distant, under forced control feeding.

6. Antibiotic Therapy

Like other mycoplasmal diseases of plants, spike responds to temporary tetracycline therapy. The most effective
method of applying antibiotic is the girdling method, which
is a modification of tree injection (Raychaudhuri et al.,
1972; Raychaudhuri, 1977; Rao et al., 1975).

B. *Citrus Greening*

At present, data is unavailable as to the financial losses
due to greening disease of citrus in India, although the disease is prevalent throughout the country and is a threat to
the entire citrus industry.
Citrus greening disease was first recognized in South Africa (McClean and Oberholzer, 1965a). It is now known to be
widespread over India and other southeast Asian countries. Its
occurrence in India was first suspected in 1966. However, in
the later stages of the disease, diagnosis is complicated by
the attack of secondary pathogens which cause dieback of the
twigs (Fraser *et al.*, 1966; Raychaudhuri *et al.*, 1966; Raychaudhuri *et al.*, 1974).

1. *Symptoms*

Both the greening diseases from India and South Africa, produce the same type of symptoms, e.g., stunting, heavy
leaf and fruit drops, premature flowering, and formation of
secondary branches having young upright leaves with various
degrees of chlorosis or mottling, which resemble typical iron
or zinc deficiency. Chlorotic leaves usually abscise and the
twigs that bear them die (McClean and Schwarz, 1970).

2. *Epiphytological Studies*

Greening disease was found to be widespread throughout
India in all of the citrus-growing areas in almost all of the
budwood (Raychaudhuri *et al.*, 1969; Nariani *et al.*, 1971). By
grafting as well as through the vector, the disease could be
transmitted to budwood (Nariani *et al.*, 1973). The citrus
psylla, *Diaphorina citri,* is a very efficient vector; even a
single insect is capable of transmitting greening under controlled conditions. The incubation period in the vector after
acquisition is about 10-12 days (Nariani and Singh, 1971).
Symptoms of greening disease on *Citrus sinensis* (L.) Obs. in
India are maximized at 27o-31oC, whereas symptoms in the
African type are maximized at between 21o to 26oC (Bove *et al.*,
1974; Ghosh *et al.*, 1977). Both types of greening are transmitted by psyllid vectors, but by different species in the
two countries. In South Africa it is transmitted by *Trioza
erytreae* (McClean and Oberholzer, 1965b) and in India by *Diaphorina citri* (Capoor *et al.*, 1967).

3. Diagnosis

Anatomical studies show severe internal necrosis of phloem, which is usually accompanied by the formation of excessive phloem (Schneider, 1968).

Schwarz and his associates isolated a fluorescent marker compound, monoglucose ester of gentisic acid (gentisoyl-β-D-glucose), from greening-infected sweet orange trees. This fluorescent marker can be used effectively to detect greening-infected trees in the field (Schwarz, 1965; Schwarz et al., 1974).

4. Electron Microscopy

In France, Bové and his colleagues were the first to apply the technique of electron micrography to greening in infected trees. His electron micrographs showed mycoplasma-like bodies in the infected phloem (Lafleche and Bové, 1970; Saglio et al., 1971). Later, the same group reported that greening mycoplasmas differed from stubborn mycoplasmas (Saglio et al., 1972), based on their membrane thickness (10 nm for stubborn mycoplasmas and more than 20 nm for greening organisms). In 1974, during an International Congrees on mycoplasmas in Bordeaux, France, Moll and Martin from South Africa presented electron micrographs of greening organisms from the tissues of trees infected with South African greening disease. They concluded that these organisms structurally resembled the rickettsia-like organisms of clover club leaf disease, but differed from the mycoplasma-like organisms of plants and gram-negative bacteria, based on membrane morphology (Moll and Martin, 1974).

Recent studies on ultrastructural morphology of citrus greening confirm the observations of both Bové's and the South African group (Ghosh et al., 1977). We could transmit both African and Indian greening to the vegetative vector Cuscuta spp. The pleuromorphic prokaryotes in the infected phloem of dodder (Fig. 2) are sometimes elongated up to 5 to 6 μm with a diameter of 0.170 to 0.250 μm. The filaments are nonramifying and sometimes terminate in a bulblike structure, 0.5 to 0.9 μm in diameter. Cells are typical prokayotes with ribosomes and fibrillar nuclear material, which is surrounded by a double tripartite complex membrane, each measuring about 9 nm (Figs. 1 and 2). Utilizing various fixatives the two layers of the tripartite membrane could be well visualized. In addition, by using histochemical tests, polysaccharide could be detected on the outer membrane of the organisms. Hence, the outer membrane could be regarded as a rudimentary cell

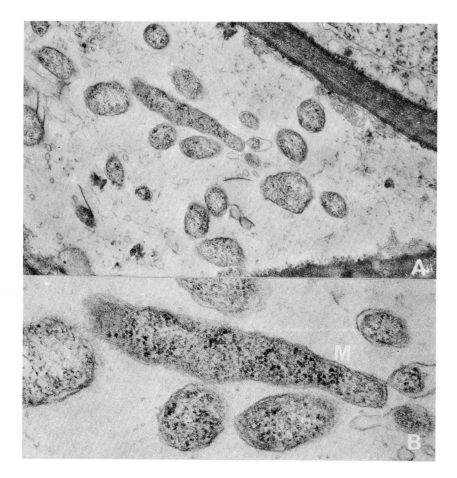

Fig. 1. Rickettsia-like organisms associated with citrus greening. (A) Organisms in the phloem cells of citrus; x 28, 800. (B) Higher magnification of the same organisms showing double envelope; x 66,000. R, rickettsia-like organisms; M, double unit membrane.

wall, allowing us tentatively to classify them with the rickettsia (Giannotti *et al.*, 1978).

        5. Chemo/Heat Therapy

        Like other yellows diseases, this disease also res-ponds to tetracycline (Nariani *et al.*, 1971) and heat thera-py. The South African group has developed an excellent meth-od of continuous tree injection (Schwarz *et al.*, 1974;

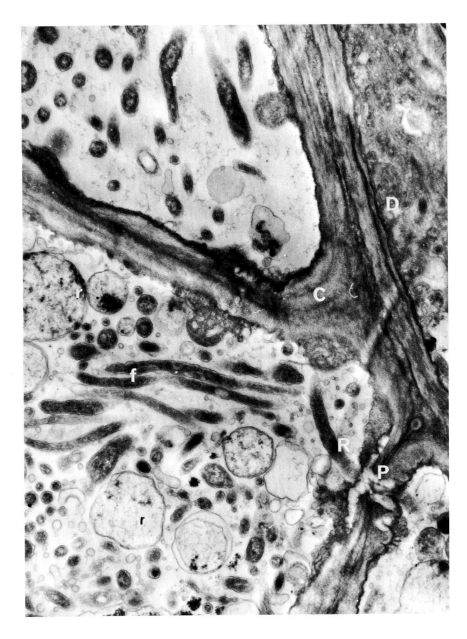

Fig. 2. Citrus greening organisms in the phloem cells of Cuscuta subinclusa. x45,000. R, rickettsia-like organisms; f, filamentous bodies; r, round bodies; D, degenerated bodies; C, cell wall; P, sieve pore.

van Vuuren *et al.*, 1977) for applying antibiotics or insecticides to infected trees. Nariani and his co-workers found
that heating infected budwood by moist hot air at 47ºC for
4 hours or 45ºC for 6 hours inactivates the greening pathogen.
In addition, potted plants kept in a heat-therapy chamber at
38º or 40ºC for 3 weeks could supress greening symptoms (Nariani *et al.*, 1975). Due to the presence of polysacharide in
the cell membrane of these pathogens, they are sensitive to
penicillin therapy, as in the case of European clover phyl-
loidy and clover club leaf disease (Windsor and Black, 1973)
where the rickettsial nature of the pathogen was confirmed
by the use of penicillin therapy (Giannotti *et al.*, 1978).

C. *Other Suspected Diseases of this Group*

Most of the diseases of this group show abnormal stunted
growth, reduced leaves and internodes, induction of axillary
buds, and sterility. A number of trees and under shrubs with
such disease symptoms have been occasionally reported, such
as, little leaf of *Eycalyptus, Toona ciliata, Dodonea, Stachy-
tarpheta, Dendrocalamus, Randia, Dichrostachys* and *Acacia.*
Some of these species are suspected to be the alternate hosts
for the sandal spike organism (Nayar and Ananthapadmanabha,
1977).

1. *Little Leaf of Eucalyptus*

Eucalyptus is a fast-growing species which was introduced in India, mostly for its soft wood, used for pulping,
and oil, which is extracted from leaves for limited medicinal
use.
Little leaf disease on *Eucalyptus* was first reported
from Karnataka by Sastry *et al.* (1971), and is now known to
occur in parts of Tamil Nadu and on the plains of Kerala.
Although, at present, the disease incidence in the field is
low, it may be a future menace.
The disease is characterized by stunted growth and
reduction of leaf areas. The new leaves become smaller in
size and sessile. The disease has been successfully transmitted by grafting from diseased to the young plants of
healthy *E. citriodora.*

2. *Little Leaf of Toona ciliata (Roxb.) Roem.*

Toona is an important timber-yielding plant in India,
distributed mainly in Sub-Himalayan tracts in Assam, West

Bengal, Bihar, Western Ghats, and other hills of the Deccan Peninsulas. It is sometimes grown in cities and towns as avenue trees.

This disease was observed in Kalyani township in West Bengal (Mukhopadhyay *et al.*, 1979). It is characterized by yellowing and reduction in the size of leaves. Leaves and branches in the new flush become extremely small, stiff, erect, and clustered, and eventually dry up. The whole tree may ultimately dry up as the disease advances.

III. SUMMARY

More than twenty tree diseases have been implicated to be caused by mycoplasma, rickettsia, or allied pathogens. In India, sandal spike and citrus greening are the two major tree diseases known to be caused by these prokaryotic agents.

Although in both these diseases the causal agents are intraphloemic prokaryotes, transmitted by insect vectors, and responding to tetracyclines, the electron microscopic studies showed that the two prokaryotes differ in their ultrastructural morphology. In sandal spike the organisms are bound by a unit trilaminar membrane which measures 10 nm, whereas, in citrus greening, the organism is bound by double unit membrane, each measuring about 9 nm. Cytochemical tests showed the presence of polysaccharides in the outer membrane of citrus greening agent, indicating its resemblance to rickettsia.

ACKNOWLEDGMENT

Part of the work was done at the Station de Recherches de Pathologie Comparee, 30380 Saint-Christol-les-Ales, France, with the collaboration of Prof. C. Vago, Monsieur J. Giannotti and Monsieur C. Louis.

REFERENCES

Anonymous (1933). Insect transmission of spike disease. *Nature, London 132*, 592-593.
Barber, C. A. (1903). Report on "spike" disease in sandalwood trees in Coorg. *Indian Forest. 29*, 21-31.
Bové, J. M., Calavan, E. C., Capoor, S. P., Cortez, R. E., and Schwarz, R. E. (1974). Influence of temperature on symptoms of Californian Stubborn, South African green-

ing, Indian citrus decline and Philippine leaf mottling diseases. *Proc. 6th Conf. Int. Organ. Citrus. Virol.*, pp. 12-15.

Capoor, S. P., Rao, D. G., and Viswanath, S. M. (1967). *Diaphorina citri* Kuway., a vector of the greening disease of *citrus* in India. *Indian J. Agr. Sci. 37*, 572-576.

Dijkstra, J. (1968). The occurrence of inclusion bodies in leaf epidermis cells of sandal affected with spike disease. *Neth. J. Plant Pathol. 74*, 101-105.

Dijkstra, J. and Ie, T. S. (1969). Presence of mycoplasma like bodies in the phloem of sandal affected with spike disease. *Neth. J. Plant Pathol. 75*, 374-378.

Dijkstra, J. and van der Want, J. P. H. (1970). Anatomical aspects of sandal plants affected with spike disease. *Neth. J. Plant Pathol. 76*, 174-178.

Dijkstra, J. and Lee, P. E. (1972). Transmission by dodder of sandal spike disease and the accompanying mycoplasma-like organisms via *Vipca rosea*. *Neth. J. Plant Pathol. 78*, 218-224.

Dijkstra, J. and Hiruki, C. (1974). A histochemical study on sandal *(Santalum album)* affected with spike disease and its diagnostic value. *Neth. J. Plant Pathol. 80*, 37-47.

Fraser, L. R., Singh, D., Capoor, S. P., and Nariani, T. K. (1966). Greening virus - the likely cause of citrus dieback in India. *FAO Plant Prot. Bull. 14*, 127-130.

Ghosh, S. K., Giannotti, J., and Louis, C. (1977). Multiplication intense des procaryotes associes aux maladie de type "Greening" des agrumes dans les cellules criblees de cuscutes. (Intense multiplication of citrus "greening" organism in the phloem cells of dodder). *Ann. Phytopathol. 9*, 520-530.

Giannotti, J., Benhamou, Niocole, and Ghosh, S. K. (1978). Comparative studies of different rickettsia-like organisms of plants in dodder. *Proc. 3rd Int. Congr. Plant Pathol., Munich, August, 1978*, p. 63. (Abstr.).

Hull, R., Horne, R. W., and Nayar, Radha (1969). Mycoplasma-like bodies associated with sandal spike disease. *Nature (London) 224*, 1121-1122.

Lafleche, Dominique and Bove, J. M. (1970). Structures de type mycoplasme dans les feuilles d'orangers atteints de la maladie du "Greening". *C. R. Acad. Sci. D270*, 1915-1917.

McClean, A. P. D. and Oberholzer, P. C. J. (1965a). Greening disease of the sweet orange: evidence that it is caused by a transmissible virus. *S. Afr. J. Agr. Sci. 8*, 253-276.

McClean, A. P. D. and Oberholzer, P. C. J. (1965b). Citrus psylla, a vector of the greening disease of sweet orange. *S. Afr. J. Agr. Sci. 8*, 297-298.

McClean, A. P. D. and Schwarz, R. E. (1970). Greening or
    blotchy-mottle disease of citrus. *Phytophylactica 2*,
    177-194.
Moll, J. N. and Martin, M. M. (1974). Comparison of the or-
    ganism causing greening disease with several plant patho-
    genic gram-negative bacteria, rickettsia-like organisms
    and mycoplasma-like organisms. *Colloq. Inserm, Mycoplas-
    mes, Inserm, Sept. 11-17, 1974*, Vol. 33, pp. 89-96.
Mukhopadhyay, S., Chowdhuri, A. K., and Tarafdar, P. (1979).
    A new nycoplasma-like disease of *Toona ciliata. Proc. All
    India Workshop Current Developments Res. Mycoplasma
    Allied Pathogens, Bangalore, Feb. 14-16.* (Abstr.)
Nariani, T. K. and Singh, G. R. (1971). Epidemiological stu-
    dies on the *Citrus* die-back complex in India. *Proc. In-
    dian Acad. Sci. B37*, 365-371.
Nariani, T. K., Raychaudhuri, S. P., and Viswanath, S. M.
    (1971). Response of greening pathogen of citrus to cer-
    tain tetracycline antibiotics. *Curr. Sci. 40*, 552.
Nariani, T. K., Raychaudhur, S. P., and Viswanath, S. M.
    (1973). Tolerance to greening disease in certain *Citrus*
    species *Curr. Sci. 42*, 513-514.
Nariani, T. K., Ghosh, S. K., Kumar, D., Raychaudhuri, S. P.,
    and Viswanath, S. M. (1975). Detection and possibilities
    of therapeutic control of the greening disease of citrus
    caused by mycoplasma. *Proc. Indian Nat. Sci. Acad. B41*,
    334-339.
Nayar, R. and Ananthapadmanabha, H. S. (1977). Little leaf
    disease in collateral hosts of sandal *(Santalum album
    L.). Eur. J. Forest Pathol. 7*, 152-158.
Nielson, M. W. (1975). The leafhopper vectors of phytopatho-
    genic viruses (Homoptera, Cicadellidae), taxonomy, bio-
    logy, and virus transmission. *U. S. Dept. Agr. Bull. No.
    1382*, pp. 386.
Nocard, E. and Roux, E. R. (1898). La microbe de la peripneu-
    monie. *Ann. Inst. Pasteur 12*, 240-262.
Parthasarathy, K., Gupta, S. K., and Rao, P. S. (1966). Stu-
    dies on sandal spike. Part VII. Some useful stain tests
    for the diagnosis of the spike disease. *Proc. Indian
    Acad. Sci. B64*, 152-156.
Plavsic-Banjac, Biljana, Maramorosch, K., Raychaudhuri, S. P.,
    Chenulu, V. V., Varma, A., and Ghosh, S. K. (1973). Elec-
    tron microscopy of graft transmitted sandal spike. *FAO
    Plant Prot. Bull. 21*, 25-26.
Rangaswamy, S. and Griffith, A. L. (1941). Demonstration of
    Jassus indicus (Walk.) as a vector of spike disease of
    sandal *(Santalum album L.). Indian Forest. 67*, 387-394.

Rao, P. S., Srimathi, R. A., Nag, K. C., Raychaudhuri, S. P.,
    Ghosh, S. K., Chenulu, V. V., and Varma, A. (1975). Res-
    ponse of spike disease of sandal to mixed treatment with
    antibiotics and fungicides. *Proc. Indian Nat. Sci. Acad.
    B41,* 340-342.
Raychaudhuri, S. P., Chenulu, V. V., Ghosh, S. K., Varma, A.,
    Rao, P. S., Srimathi, R. A., and Nag, K. C. (1972). Chemi-
    cal control of spike disease of sandal. *Curr. Sci. 41,*
    72-73.
Raychaudhuri, S. P., Nariani, T. K., Lele, V. C., and Singh,
    G. R. (1969). Greening and citrus decline problem in In-
    dia. *Proc. 5th Conf. Int. Organ. Citrus Virol., Tokyo.*
Raychaudhuri, S. P., Nariani, T. K., Ghosh, S. K., Viswanath,
    S. M., and Kumar, D. (1974). Recent studies on *Citrus*
    greening in India. *Proc. 6th Conf. Int. Organ. Citrus
    Virol.,* pp. 53-66.
Raychaudhuri, S. P. (1977). Sandal spike disease and its pos-
    sible control. *Euro. J. Forest Pathol. 7,* 1-5.
Saglio, P., Lafleche, Dominique, Bonissol, Christiane, and
    Bove, J. M. (1971). Isolement, culture et observation au
    microscope electronique des structures de type mycoplasma
    associaes a la maladie du stubborn des agrumes et leur
    comparison avec les structures observees dans le cas de la
    maladie du greening des agrumes. *Physiol. Veg. 9,* 569-582.
Saglio, P., Laflèche, Dominique, L'Hospital, Martine, Dupont,
    G., and Bové, J. M. (1972). Isolation and growth of citrus
    mycoplasmas. *In* "Pathogenic Mycoplasmas", (Elliot and
    Birch, ed.). pp. 187-203. Elsevier, Amsterdam.
Sastry, K. S. M., Thakur, R. N., Gupta, J. H., and Pandora,
    V. R. (1971). Three virus diseases of *Eucalyptus citrio-
    dora. Indian Phytopathol. 24,* 123-126.
Schwarz, R. E. (1965). A fluorescent substance present in
    tissues of greening-affected sweet orange. *S. Afr. J.
    Agr. Sci. 8,* 1177-1180.
Schwarz, R. E., Moll, J. N., and van Vuuren, S. P. (1974).
    Incidence of fruit greening on individual citrus trees in
    South Agrica. *Proc. 6th Conf. Int. Organ. Citrus Virol.*
Schneider, H. (1968). Anatomy of greening-diseased sweet
    orange shoots. *Phytopathology 58,* 1155-1160.
Shivaramakrishnan, V. R. and Sen-Sarma, P. K. (1978). Experi-
    mental transmission of spike disease of sandal, *Santalum
    album,* by the leaf-hopper *Nephotettix virescens* (Homopte-
    ra: Cicadellidae). *Indian Forest. 104,* 202-205.
Varma, A., Chenulu, V. V., Raychaudhuri, S. P. Prakash, N.,
    and Rao, P. S. (1969). Mycoplasma-like bodies in tissues
    infected with sandal spike and brinjal little leaf. *In-
    dian Phytopathol. 22,* 289-291.

van Vuuren, S. P., Moll. J. N., and da Graca, J. V. (1977).
    Preliminary report on extended treatment of citrus green-
    ing with tetracycline hydrochaloride by trunk injection.
    Plant Dis. Rep. 61, 358-359.
Windsor, I. M. and Black, L. M. (1973). Remission of symptoms
    of clover club leaf following treatment with penicillin.
    Phytopathology 63, 44-46.

BLUEBERRY STUNT

E. H. Varney

Cook College
Rutgers University
Department of Plant Pathology
New Brunswick, New Jersey

I. INTRODUCTION

The cultivation or at least partial cultivation of the
lowbush blueberry goes back to the American Indians who practiced burning as a pollarding technique on the blueberry bar-
rens of Canada and the northeastern United States (Eck and
Childers, 1966). No disease of the yellows type, however, was
recognized as a serious problem until the development of the
highbush blueberry industry based on the domestication and improvement of the Southeastern highbush blueberry *(Vaccinium
australe* Small), the northern highbush blueberry *(V. corymbosum* L.)*, and their hybrids. The new blueberry industry was
expanding rapidly by the 1940's in New Jersey and North Carolina, and at that time its future was seriously threatened by
the rapid spread of a new disease known as blueberry stunt. At
the time the causal agent was presumed to be a virus. Much of
what we know about stunt has been summarized in earlier reviews (Varney and Stretch, 1966; Stretch and Varney, 1970;
Varney, 1977).

The genus *Vaccinium* is a taxonomically difficult complex
of species and interspecific hybrids (Camp, 1945; Eck, 1966).
In general, diseases did not reach epidemic proportions in
wild habitats because of this great genetic diversity, including resistance to disease and potential vectors. The
new and initially disease-free blueberry crop, however, was
soon affected by a number of plant pathogens. The purpose of
this chapter is to reemphasize the dynamic nature of disease
development in cultivated crops and the high vulnerability of
a plant species to epidemics in the unnatural environment of

uniform plantings of individuals with the same genetic background. Blueberry stunt provides a unique opportunity to
study the epidemiology of a disease caused by a mycoplasmalike
organism introduced into a cultivated crop presumably from indigenous wild host species.

II. HISTORY AND GEOGRAPHICAL DISTRIBUTION

Stunt of highbush blueberry was described in 1942 by Wilcox who showed the causal agent was graft transmissible and
presumably a virus. The disease was first observed in 1928
(Tomlinson *et al.*, 1950), and a probable earlier record is a
report by Stevens in 1926 of virus-like symptoms in a wild
blueberry plant in New Jersey. Stunt is known from North
Carolina to eastern Canada and west to Michigan. In 1978,
Dale and Moore reported for the first time a stunt-like
disease and a mycoplasmalike organism associated with the
disease in Arkansas. Neither stunt nor the vector has been
reported from the blueberry-producing areas of the Pacific
Northwest. A witches'-broom of lowbush blueberry, *Vaccinium
myrtillus* L., has been known in Europe since 1925 (Blattny
and Blattny, 1970).

It has been suggested that the stunt causal agent was introduced at Whitesbog, New Jersey, in blueberry seedlings
grown by plant breeders in the USDA greenhouses at Beltsville,
Maryland. Survey data, however, support the view that stunt
is indigenous to the eastern United States and was not accidentally introduced through the distribution of contaminated
seedlings (Hutchinson *et al.*, 1960).

III. SYMPTOMS

Symptoms are variable, differing with the cultivar, time of
year, stage of growth, and age of infection (Varney and
Stretch, 1966). A characteristic symptom on cultivated and
wild highbush blueberry is a yellowing of the leaves along the
margins and between lateral veins. Normal green is generally
retained along the midrib and lateral veins, and affected
leaves are often cupped downward and reduced in size.
Chlorotic areas generally turn a brilliant red in late summer
and early fall, and this symptom is distinctive until the appearance of normal fall coloration. Diseased plants live for
many years but are much reduced in vigor. Branches may continue to set some fruit, but the berries are small and of
poor flavor. If an infected plant is cut back to ground level,
the new growth remains weak and severely stunted. Branches

are twiggy in appearance due to shortened internodes and the
growth of normally dormant lateral buds. Some cultivars are
more severely affected than others. Rancocas, for example,
seldom shows definite symptoms and usually produces apparently normal yields. MLO-infected plants in Arkansas do not
show severe stunting and loss of vigor (Dale and Moore, 1978).
The difference in symptom severity may be due either in part
to different environmental conditions and cultural practices
or to a different MLO.

   Symptoms on wild North American species of *Vaccinium* are,
in general, like those for cultivated highbush blueberry.
Symptoms have been observed on naturally infected *V. vacillans*
Torr., *V. atrococcum* Heller, *V. stamineum* L., and *V. myrtilloides* Michx. and graft-inoculated *V. amoenum* Ait., *V. alto-
montanum* Ashe, and *V. elliottii* Chap. In Europe, new growth
of infected *V. myrtilloides (V. myrtillus* L.) is erect rather
than of the plagiotropic-umbrella type characteristic of
healthy plants and frequent branching results in a dense,
broomy growth (Blattny and Blattny, 1970). *Vaccinium vitis-
idaea* L., *V. uliginosum* L., and *V. oxycoccus* L. show similar
symptoms--upright growth of shoots, brooming, stunting, reduced
leaf size, leaf reddening, and sterility.

IV. CAUSAL AGENT

   Although originally described as a viral disease, stunt is
now known to be caused by a mycoplasmalike organism (MLO).
Pleomorphic bodies resembling mycoplasma microorganisms are
easily detected in phloem tissues of stunt-infected blueberry
plants, but are never found in healthy blueberry. These
bodies bounded by a unit membrane, contain ribosomes and deoxyribonucleic acid (DNA)-like strands and lack nuclear enve-
lopes. The typical appearance of MLOs in sieve elements is
illustrated in Fig. 1. Chen (1971) examined with the electron
microscope the leaf tissues of Jersey blueberry plants infected with stunt and cranberry plants *(V. macrocarpon* Ait.) with
false blossom. He found MLOs only in the sieve tube elements
of diseased plants. The blueberry MLO ranged in diameter from
160-700 nm in contrast to 80-300 nm for the MLO observed in
cranberry. These results suggest that stunt and false blossom
are caused by two distinct MLOs as indicated by earlier host
range and vector studies. Hartmann *et al.* (1972) sectioned
leaves of the Coville and Concord blueberry cultivars and they
found pleomorphic bodies ranging from 135 to 530 nm for spherical and 100-225 nm by 400-850 nm for elongate forms. They,
like Chen, did not find MLOs in healthy control plants and did
not find any virus-like particles in either infected or healthy

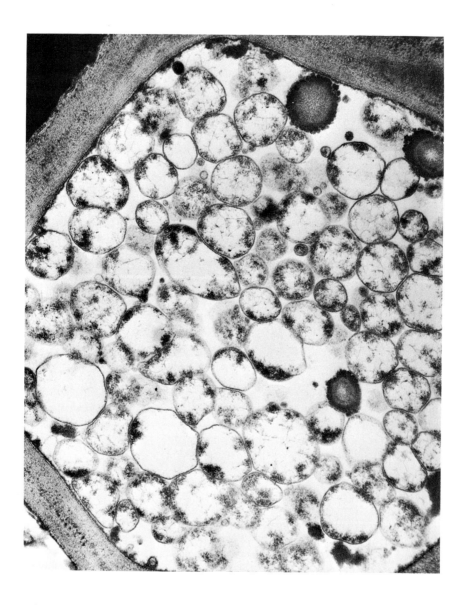

Figure 1

tissue. Hartmann *et al.*, however, did find large, electron-dense inclusions, angular in outline, in companion cells adjacent to degenerate sieve tubes containing small, electron-dense forms of MLO. Inclusions were found in approximately 50% of the diseased samples examined and never in healthy phloem tissue. So far as known, this is the first report of inclusions associated with any yellows-type disease. Dale and Moore (1978) found typical pleomorphic bodies in a blueberry in Arkansas with symptoms suggestive of stunt. All efforts to cultivate the causal organism have failed so far.

MLOs have also been associated with witches'-broom of *V. myrtilloides (V. myrtillus)* in Europe (Blattny and Vana, 1974; Kegler *et al.*, 1973; de Leeuw, 1975). Blattny and Vana reported diameters of 150-700 nm in Czechoslovakia, and de Leeuw reported diameters of 60-800 nm in the Netherlands. Since it is unwise to exchange infected plant material for comparative studies, any relationship between the European and North American MLOs will remain a mystery until the *Vaccinium* MLOs are successfully cultured and serological comparisons can be made through the exchange of antisera.

V. TRANSMISSION

Stunt is readily transmitted by grafting. Plants grafted in the fall will show symptoms the following spring. Plants grafted at or prior to budbreak may show symptoms the same year, often within two to three months. There is no evidence of transmission by pruning tools or through seed and pollen. L. O. Kunkel was the first to use dodder (*Cuscuta* sp.) to transmit the disease to periwinkle (*Vinca rosea* L.) in which the symptoms were distinct from those caused by other yellows-type disease agents (Tomlinson *et al.*, 1950). Leaves were pale green and marked with patches or transverse bands of dark green, and the flowers were smaller but virescence did not occur. Kunkel concluded stunt was caused by a yellows-type virus and that a leafhopper was the probable vector.

Fig. 1. Typical appearance of mycoplasma-like organisms (MLO) in sieve elements of stunt-infected highbush blueberry, Vaccinium corymbosum. Normal pleomorphic spherical bodies are bounded by a unit membrane, the bodies varying in electron density from small compact ones with high electron-dense cytoplasm, approximately 100 nm in diameter, to large bodies of 800 nm in diameter. Ribosomes and nuclear strands can be distinguished in several bodies. Magnification x 28,800. Original electron micrograph by Dr. F. Kondo.

C. A. Doehlert in 1944 used a mixed colony of leafhoppers to transmit stunt from one diseased to four healthy plants (Marucci *et al.,* 1947). Tomlinson *et al.* (1950) reported that a complex of the sharp-nosed leafhopper (*Scaphytopius magdalensis* Prov.) and blunt-nosed leafhopper (*S. verecundus* Van Duz.) transmitted stunt. Hutchinson (1955) and Maramorosch (1955) later found that *S. magdalensis* was a vector and that *S. verecundus* was not. Many other insects have been tested and so far *S. magdalensis,* the sharp-nosed leafhopper, is the only known vector. It occurs along the Atlantic seaboard from Florida to Canada. *Scaphytopius magdalensis* has not been identified in Arkansas, but other species of *Scaphytopius* have been collected in that state where there is natural spread of a stuntlike disease of highbush blueberry (Dale and Moore, 1978). In North Carolina, stunt continues to spread rapidly even though the sharp-nosed leafhopper population is usually very low, and Milholland (Varney, 1977) questions whether it is the only vector involved. Neither stunt nor the only known vector is known to occur in the blueberry-growing areas of the Pacific Northwest. Additional vector studies are needed.

In Europe, the leafhopper *Idiodonus cruentatus* Panz is a vector of witches'-broom. The disease, however, also occurs abundantly in areas where the known vector has not been found. It has been observed that the excessive felling of trees in the forest supports the occurrence of leafhoppers and causes an increase in the number of diseased plants (Blattny and Blattny, 1970).

VI. CONTROL

Certification programs, roguing infected plants, and vector control have greatly reduced but have not eliminated losses caused by the stunt MLO. In New Jersey, a certification program was initiated when stunt threatened the blueberry industry in the 1940's. This has prevented the introduction of stunt to disease-free areas. Because of the relatively slow spread of stunt, roguing continues to be very effective when combined with vector control. Growers need to be reminded every growing season of the importance of removing diseased bushes. When this control measure is neglected, stunt begins to spread again. A few leafhoppers survive the insecticides and there is a constant migration of vectors into cultivated bushes from the wild. It is essential to spray diseased bushes with an appropriate insecticide prior to removing them from a planting. Unless the vector is killed, the act of removing the stunted bushes will distribute the immature wingless leafhoppers to nearby healthy plants. If diseased bush-

es are removed early, there is no source of inoculum for winged adults to distribute through the fields later in the
summer. In areas where stunt is indigenous, it is advisable
to destroy or spray wild hosts in the vicinity of commercial
plantings.

VII. CONCLUSION

    The control of the MLO-induced stunt disease of highbush
blueberry is dependent on an effective certification program,
the roguing of infected plants, and control of the vector.
Certification programs have helped prevent the spread of stunt
as well as virus diseases into new areas. If we are to learn
from our experiences with the northern highbush blueberry, a
search should be started now for mycoplasma, virus, and other
diseases in the wild populations of *V. ashei* and other southern
species which are forming the basis for a highbush blueberry
industry in the southern United States. It may then be possible
to initiate control programs before potential epidemics occur
and before MLOs, viruses, and other disease agents have been
accidentally spread by plant breeders and propagators to new
areas.

REFERENCES

Blattny, C., Sr., and Blattny, C., Jr. (1970). Blueberry
    witches'-broom. *In* "Virus Diseases of Small Fruits and
    Grapevines" (N. W. Frazier, ed.), pp. 177-179. Univ.
    California Div. Agr. Sciences, Berkeley.
Blattny, C. and Vana, V. (1974). Mycoplasmalike organsims in
    Vaccinium myrtillus L. infected with blueberry witches'-
    broom. *Biol. Plant. 16*, 476-478.
Camp, W. H. (1945). The North American blueberries with notes
    on the other groups of Vacciniaceae. *Brittonia 5*, 203-
    275.
Chen, T. A. (1971). Mycoplasmalike organisms in sieve tube
    elements of plants infected with blueberry stunt and cran-
    berry false blossom. *Phytopathology 61*, 233-236.
Dale, J. L. and Moore, J. N. (1978). Blueberry Disorder Pos-
    sibly Caused by a Mycoplasma. Arkansas Agricultural Ex-
    periment Station, Fayetteville, Arkansas.
de Leeuw, G.T.N. (1975). Presence of mycoplasmalike organisms
    in the phloem of witches'-broom diseased *Vaccinium myrtil-
    lus* plants in the Netherlands. *Phytopathol. Z. 83*, 91-94.
Eck, P. (1966). Botany. *In* "Blueberry Culture" (P. Eck and
    N. F. Childers, eds.), pp. 14-44. Rutgers University
    Press, New Brunswick, NJ.

Eck, P. and Childers, N. F. (1966). The blueberry industry. *In* "Blueberry Culture" (P. Eck and N. F. Childers, eds.), pp. 3-13. Rutgers University Press, New Brunswick, NJ.

Hartmann, J. X., Hopper, G. R., and Bath, J. E. (1972). Occurrence and nature of mycoplasmalike organisms in stunt disease of Michigan highbush blueberry. *Mich. Acad. 4*, 461-467.

Hutchinson, M. T. (1955). An ecological study of the leafhopper vectors of blueberry stunt. *J. Econ. Entomol. 48*, 1-8.

Hutchinson, M. T., Goheen, A. C., and Varney, E. H. (1960). Wild sources of blueberry stunt virus in New Jersey. *Phytopathology 50*, 308-312.

Kegler, H. Muller, H., Kleinhempel, H., and Verderevskaja, T.D. (1973). Investigations on cherry decline and bilberry witches' broom. *Nach. Pflanz. DDR 27*,5-8 (*Rev. Plant Pathol. 53*, 37, 1974).

Maramorosch, K. (1955). Transmission of blueberry stunt virus by *Scaphytopius magdalensis*. *J. Econ. Entomol. 48*, 106.

Marucci, P. E., Tomlinson, W. E., Jr., and Doehlert, C. A. (1947). Cage tests for possible carriers of blueberry stunt disease. *Proc. 16th Ann. Blueberry Open House 16*, 9-11.

Moore, J. N. (1966). Breeding. *In* "Blueberry Culture" (P. Eck and N. F. Childers, eds.), pp. 45-74. Rutgers University Press, New Brunswick, NJ.

Stevens, N. E. (1926). The false blossom situation. *Proc. Annu. Conv. Amer. Cranberry Growers' Assoc. 57*, 20-26.

Stretch, A. W. and Varney, E. H. (1970). Blueberry stunt. *In* "Virus Diseases of Small Fruits and Grapevines" (N. W. Frazier, ed.), pp. 175-176. Univ. California Div. Agr. Sciences, Berkeley.

Tomlinson, W. E., Jr., Marucci, P. E., and Doehlert, C. A. (1950). Leafhopper transmission of blueberry stunt disease. *Econ. Entomol. 43*, 658-662.

Varney, E. H. (1977). Viruses and mycoplasmalike diseases of blueberry. *Hort. Sci. 12*, 476-478.

Varney, E. H. and Stretch, A. W. (1966). Diseases and their control. *In* "Blueberry Culture" (P. Eck and N. F. Childers, eds.), pp. 236-279.

Wilcox, R. B. (1942). Blueberry Stunt, a virus disease. *Plant Dis. Rep. 26*, 211-213.

SANDAL *(SANTALUM ALBUM)* SPIKE DISEASE

J. C. Varmah

Forest Research Institute
and
Colleges, Dehra Dun

I. INTRODUCTION

In India, natural stands of sandal occur primarily in the
southern states of Karnataka and Tamil Nadu and, to a limited
extent, in Andhra Pradesh. Sandalwood is valued for its fra-
grant oil obtained from its heartwood. Although sandal trees
have also been planted in various parts of India outside its
natural belt, heartwood obtained from these trees is poorly
scented and, therefore, uneconomical. India derives considerable revenue from the exportation of sandalwood and its products. In 1976-1977, this amounted to $53,200,000. However,
the supply of sandal trees from the natural forests is progressively diminishing primarily due to spike, which is a
"yellows-type" disease. In Indian forestry, we are particularly concerned with its devastating effects. Consequently,
the Government of India has recently opened a separate
"Sandal Research Centre" at Bangalore with a view to tackling
the problem more effectively. The Research team is comprised
of pathologists, entomologists, silviculturists, geneticists,
plant physiologists, and plant biochemists. Thus, a multi-
disciplinary approach is envisaged in solving this long-stand-
ing problem. In this chapter, an endeavor is made to highlight
the current status of research on this disease. A brief
review of the past work on various aspects of disease investigation has been given for a better understanding of the
present status.

MYCOPLASMA DISEASES
OF TREES AND SHRUBS
253

II. DISEASE SYMPTOMS

Like most yellows diseases, spike is characterized by excessive proliferation of buds leading to a "witches-broom"
effect, shortening of internodes, reduction in the size of the
leaf and the chromosomes, phyllody of flowers, and inhibition
of fruiting. Initially, only a part of the affected tree
shows these symptoms. However, these symptoms slowly, but
invariably, spread to the remainder of the tree. Death occurs
within 2-3 years after the manifestation of external symptoms.

III. SPIKE PATHOGEN

Butler (1903) was the first to point out the similarities
of spike disease and yellows diseases of other plants. The
epoch-making discovery of Doi *et al.* (1967) that yellows diseases are caused by mycoplasma-like organisms (MLO's) also
proved true in the case of spike disease of sandal. Electron
microscopic examination of diseased leaves and twigs of sandal
by three groups of investigators working independently in
India, The Netherlands, and the United Kingdom revealed the
presence of MLO's in the sieve tubes (Dijkstra and Ie, 1969;
Hull *et al.*, 1969; Varma *et al.*, 1969). The MLO etiology of
the disease was further confirmed by the temporary remission
of the disease symptoms by tetracycline treatment (Hull *et al.*,
1969; Raychaudhuriy *et al.*, 1972; Nayar *et al.*, 1973). Further confirmation was obtained by electron microscopy of graft-
transmitted diseased leaves (Plavsic-Banjac *et al.*, 1973).
Penicillin treatment does not bring about remission of symptoms and this excludes the possibility of association of
rickettsia-like organisms (RLO's) with the disease. A recent
discovery of helical forms of corn stunt MLO's and citrus
little leaf MLO's in thick sections under electron microscopy
(Markham *et al.*, 1974) calls for reexamination of spiked leaves
and twigs in order to determine if *Spiroplasma* is the causal
agent.

IV. ALTERNATE HOSTS OF SPIKE PATHOGENS

Plant host ranges of yellows agents are often broad
(Kunkel, 1931; Freitag and Severin, 1933). Several plants
growing in the sandal ecosystem exhibit symptoms of little
leaf with "witches'-broom" effect. Electron microscopic examination of leaves of some of these plants showed the presence of MLO's (Hull *et al.*, 1970). Several agricultural and
horticultural crops also show mycoplasma disease symptoms

(Ghosh and Raychaudhury, 1974). Mycoplasmal disease of *Ziz-phus oenoplia* and *Catharanthus roseus* (= *Vinca rosea)* was transmitted to healthy sandal and vice versa through dodder (Srimathi, unpublished; Dijkstra and Lee, 1972). Thus, these plants seem to act as the alternate hosts of the spike patho-gen (Sen-Sarma, 1978). Since spike disease may be transmitted by haustorial contact (6-7% according to Coleman, 1923) and insect vectors (Rangaswami and Griffith, 1940; Shivaramakrish-nan and Sen-Sarma, 1978), presence of these alternate sources of infection is important in the ecology of the sandal spike particularly, with reference to their role in vector efficien-cy.

V. DISEASE INCIDENCE AND RANK VEGETATION

It has long been suspected that rank vegetation plays an important role in the spread of the disease. In a classical study, Venkata Rao (1935) clearly demonstrated that incidence and speed of spread of the disease are directly proportional to the density of undergrowth. The lesser the density, the lower the disease incidence. He explained this by surmising that underbushes might furnish harborage for insect vectors (leafhoppers). An intensive entomological survey has conclu-sively proved that no leafhopper breeds on sandal but they breed and thrive on the rank vegetation. This aspect is currently being reinvestigated.

VI. INSECT VECTOR/VECTORS

In nature, the majority of yellows agents are transmitted to healthy plants by insect vectors (Whitcomb and Davis, 1970). Earlier investigations conclusively proved that sandal spike is primarily vector transmitted. Insect vectors incriminated so far are *Moonia albimaculata, Coelidia indica (=Jassus in-dicus)*, and *Nephotettix virescens* (Anon. 1933; Rangaswami and Griffith, 1940, Shivaramakrishnan and Sen-Sarma, 1978). Trans-mission by *Moonia* and *Coelidia* could not be subsequently con-firmed. At least 60 adults of *Nephotettix* are required for experimental transmission of the disease in potted healthy plants after an acquisition feeding for 24 hours and symptoms of the disease appear after 69 days. Successful transmission by *N. virescens* was confirmed by biometrical studies on ratio of length and breadth of leaves, Mann's stain (violet stain of the phloem, see Parthasarathi *et al.,* 1966), and leaf and bud grafting. However, no electron microscopical confirmation of the presence of MLO's in the vector body could be carried out.

The discovery of *Nephotettix virescens* as vector is important, as this species also causes MLO disease in rice (rice yellow dwarf), which is extensively cultivated in Karnataka and Tamil Nadu. Earlier, Sreenivasaya and Rangaswami (1931) reported constant association of agricultural operations with the primary site of the first disease attack. It is possible that some other species of leafhoppers might also be involved in transmitting the disease in nature and leafhoppers occurring in the sandal ecosystem are being screened for their ability to transmit the disease.

VII. PATHOLOGICAL PHYSIOLOGY

Over the last few decades, a lot of work has been carried out on the biochemistry and physiology of diseased sandal. The disease results in higher starch, total nitrogen, and phosphorus content, higher levels of phenolic bodies including tannins, etc., and lower ash, potash, fat, and lime contents (Rao, 1965). The presence of a toxinlike substance in the diseased leaves, which induced a certain disease syndrome in bioassay plants, including healthy sandal, was demonstrated by Gurumurti *et al.* (1979). Auxin-like activity was higher in diseased stem tissues than in healthy tissues. In addition, diseased leaf tissues showed a higher level of cytokinin-like substances. Similar results were also obtained in diseased *Catharanthus roseus* where disease symptoms could be obtained by spraying healthy plants with a mixture of auxins and cytokinins (K. Gurumurti, unpublished).

VIII. CHEMOTHERAPY

Temporary remission of the disease symptoms could be brought about by the antibiotics belonging to tetracycline group, namely Ledermycin, Terramycin, Achromycin, and Aureomycin. Among the fungicides, diseased trees respond to treatment by benylate (Rao *et al.*, 1972). However, a mixture of 300 mg of tetracycline and 1 gm of benylate applied by the girdling method gave the best result (Raychaudhuriy *et al.*, 1972). However, the remission was only temporary and repeated treatment was required. Chemotherapy thus has limited scope until a new chemical which will cure the disease with a single application is discovered.

IX. BREEDING SANDAL FOR DISEASE RESISTANCE

Breeding of resistant varieties is perhaps the best method
to combat plant diseases. In heavily spiked areas, a few
trees often remain healthy. It is, however, not clear whether
these trees have merely escaped infection or have developed
inherent resistance to the disease. This has stimulated renewed interest in developing spike-resistant strains either
through selection or resistant breeding through intra- and
interspecific hybridization, induced polyploidy (Kapoor and
Kedharnath, 1974), or mutation breeding. Colchicine-induced
autotetraploids have altered leaf morphology, such as thicker
leaves, which might prevent leafhopper transmission (Kapoor
and Kedharnath, 1976). For mutation breeding, radiosensitivity
studies have been carried out (Kedharnath, 1970). Further
research on the above lines is being intensified.

X. CONCLUSION

The motto "prevention is better than cure" is applicable
for all plant diseases, irrespective of their causal organisms.
Pending development of spike-resistant strains, an integrated
approach, like elimination of diseased trees (sanitation),
collateral hosts of pathogens, vector host plants, vector
control through chemicals and ecological interference, and
chemotherapy, is recommended for combating the disease in the
natural stands.

REFERENCES

Anonymous (1933). *Nature (London) 132*, 592-593.
Butler, E. J. (1903). *Indian Forest. (Appendix Ser.), 29*,
    1-11.
Coleman, L. C. (1923). *Indian Forest. 49*, 6.
Dijkstra, J. and Ie, T. S. (1969). *Neth. J. Plant Pathol.*
    75, 374-378.
Dijkstra, J. and Lee, P. E. (1972). *Neth. J. Plant Pathol.*
    78, 213-224.
Doi, Y., Taranaka, M., Yora, K., and Asuyama, M. (1967). *Ann.*
    Phytopathol. Soc. Jap. 33, 259-266.
Freitag, J. H. and Severin, H. M. P. (1936). *Hilgardia 10*,
    263-302.
Ghosh, S. K. and Raychaudhuri, S. P. (1974). In "Current
    Trends in Plant Pathology" (S. P. Raychaudhuri and J.
    Verma, eds.), pp. 112-121. Lucknow Univ., India.

Gurumurti, K., Raju, P. S., and Sen-Sarma, P. K. (1979).
    Indian J. Forest., in press.
Hull, R., Horne, R. W., and Nayar, R. (1969). *Nature (London)*
    224, 1121-1122.
Hull, R., Plaskitt, A., Nayar, R., and Ananthapadmanabha, H.
    S. (1970). *J. Indian Acad. Wood Sci. 1* (1), 62.
Kedharnath, S. (1970). *Final Techn. Rep., PL 480 Project,*
    129 pp.
Kapoor, M. L., and Kedharnath, S. (1974). *Indian J. Genet.*
    34A, 408-412.
Kapoor, M. L. and Kedharnath, S. (1976). *Indian Forest. 102,*
    495-497.
Kunkel, L. O. (1931). *Contrib. Boyce Thompson Inst. 3*, 85-123.
Markham, P. G., Townsend, R., Bar-Joseph, M., Daniels, M. J.,
    Plaskit, and A. Meddins, B. M. (1974). *Ann. Appl. Biol.*
    78, 49-57.
Nayar, R., Shyamsunder, S., and Prasad, I. V. (1973). *J.*
    Indian Acad. Wood Sci. 4, 33-37.
Parthasarathi, K., Gupta, S. K., and Rao, P. S. (1966). *Proc.*
    Indian Acad. Sci. B64, 152-156.
Plavsic-Banjac, B., Maramorosch, K., John, V. T., Raychaud-
    huri, S. P., Chenulu, V. V., Varma, A., and Ghosh, S. K.
    (1973). *FAO Plant Prot. Bull. 21*, 289.
Rangaswami, S. and Griffith, A. L. (1940). *India Forest. Rec.*
    Entomol. (N.S.) 6 (4), 85-196.
Rao, P. S.(1976). IN "Advancing Frontiers in the Chemistry of
    Natural Products." pp. 237-250. Hindusthan Publ. Corp.
    Delhi-7.
Rao, P. S. Srimathi, R. A., and Nag, K. C. (1972). *Curr. Sci.*
    41, 221-222.
Raychaudhuri, S. P., Chenulu, V. V., Ghosh, S. K., Varma, A.,
    Rao, P. S., Srimathi, R. A., and Nag, K. C. (1972).
    Curr. Sci. 41, 72-73.
Sen-Sarma, P. K. (1978). *Indian J. Forest. 1*(2), 100-104.
Shivaramakrishnan, V. R. and Sen-Sarma, P. K. (1978). *Indian*
    Forest. 104, 202-205.
Sreenivasaya, M. and Rangaswami, S. (1931). *J. Indian Inst.*
    Sci. A14, 59-65.
Varma, A., Chenulu, V. V., Raychaudhuri, S. P., Prakash, N.,
    and Rao, P. S. (1969). *Indian Phytopathol. 22*, 289-291.
Venkata Rao, M. G. (1935). *Indian Forest. 61*, 169-188.
Whitcomb, R. F. and Davis, R. E. (1970). *Annu. Rev. Entomol.*
    15, 405-464.

THE ENIGMA OF THE FLOWER SPIROPLASMAS

Robert E. Davis

Plant Virology Laboratory,
Plant Protection Institute,
Agricultural Research,
Science and Education Administration,
U. S. Department of Agriculture
Beltsville, Maryland

I.   INTRODUCTION

The widely varied microbial flora present on surfaces of
aerial plant parts includes a newly discovered component --
the spiroplasmas -- helical, motile cell wall-less prokaryotes
that were first recognized in and cultivated *in vitro* from
internal tissues of diseased plants *(4, 23, 29-31, 35, 36, 42,
67-69)*. Investigations of epiphytic microflora have focused
principally on nonpathogenic algae and on phytopathogenic and
saprophytic bacteria and fungal species and have generally
included studies of leaf, stem (trunk), and bud surface inhabitants *(1, 2)*. In contrast, the epiphytic spiroplasmas
have no known plant pathogenic potential and, in their association with plants, are known only from surfaces of flowers
(14). Among the intriguing questions provoked by the discovery of spiroplasmas in this specialized ecological niche
are those related to the origins of these spiroplasmas, their
relationships with spiroplasmas from other habitats, the nature of their vectors and/or hosts, their possible specifici-
ties for flowers and arthropods, possible pathogenicity in
vector and nonvector hosts including vertebrates as well as
arthropods, and the nature of their associations with nectar
and plant tissue, and with other microorganisms in the flower
site. Most of these questions cannot yet be answered.

TABLE I. Spiroplasmas and Their Natural Habitats, Some Examples[a]

Strain(s)	Habitat(s)	Reference(s)[b]
"Rio Grande" corn stunt spiroplasma	Zea spp. and certain leafhoppers	6, 23, 24, 29-31, 35, 36, 79
Spiroplasma citri	Citrus, horseradish, other plants, and certain leafhoppers	4, 11, 42, 67-69
Sex ratio organism and non-male-killing spiroplasma[c]	Drosophila	75-78
Suckling mouse cataract agent, closely related strains, and the unrelated strain 277F	Rabbit tick	10, 26, 47, 73, 77
Strains AS576, G1, BW, BC3, TT-15, MG-10, and other strains	Honeybee and on surface of flowers of several plant species	8, 9, 16, 33, 65, 77
Strain 23-6 (27-31, OBMG, BNRl), and the unrelated strain brevi	Surface of flowers of tulip tree	8, 15, 18, 19, 26, 27, 32, 73, 77
Strains SR3 (SR9), PPS1, L89, B13, B-B-1, F1, F2, Bee 5 and others	Several insect species (including honeybee) and on surfaces of flowers of several plant species	15, 18, 19, 26, 32, 47, 58, 59, 72
Strains LB10 and LB12	Green leaf bug, Trigonotylus ruficornis Geoffroy	53
Oncometopia spiroplasma	Oncometopia nigricans, a xylem-feeding leafhopper from plants	60

^aNatural habitat(s) signifies source(s) in which the indicated strains have been found and does not imply that a strain is necessarily limited to the habitat(s) listed.

^bOnly references cited in the text are listed. For additional literature, the reader is referred to the several reviews included in the literature cited.

^cReference to non-male-killing spiroplasma in D. hydei; Ota, T., M. Kawabe, K. Oishi, and D. F. Poulson. 1979. J. Heredity 70: 211-213.

II. SPIROPLASMAS AND THEIR HABITATS

The original discovery of spiroplasmas occurred during research on a plant disease known as corn stunt (35, 36). Early
investigations on this disease had led to a presumption of
viral etiology (49, 50), but through years of research no
causal virus could be isolated from, or visualized in, either
diseased plants or insect vectors (for review, see 28). However, soon after Japanese workers reported that the phloem
cells of certain plants affected by yellows diseases contained
bodies resembling mycoplasmas (38, 46), mycoplasmalike organisms (MLO) were recognized in phloem cells of plants with
corn stunt disease and in corn stunt inoculative leafhopper
vectors (43-45, 56, 57). Like mycoplasmas previously known
to occur in animals, the MLO in plants are prokaryotic microorganisms that are bounded only by a single trilaminar membrane. Unlike ordinary bacteria, mycoplasmas and MLO have no
rigid cell wall.
Although the unusual prokaryotes associated with corn
stunt disease did resemble mycoplasmas in their lack of cell
wall, it was reported in 1971 (36) that the corn stunt microorganisms were unlike any previously reported mycoplasmas in
that they were helical in cell shape. During the same period,
cultivation *in vitro* of a MLO from citrus plants with stubborn
disease was achieved, and the cultivated organism was reported
to be a mycoplasma (4, 42, 68, 69). Later it was found that
the MLO from citrus was helical in cell shape (11), resembling
the helical organism in corn stunt disease, and exhibited
motility similar to that previously observed in the corn stunt
microorganism (11, 22, 24). Motility has since been the ob-
ject of several studies (12, 18, 19, 34). The term spiroplasma, originally proposed (30) for the corn stunt disease organism and for similar helical, motile cell wall-free prokaryotes, was subsequently adopted as the genus name of *Spiroplasma citri* — the prokaryote that had been isolated from
citrus (4, 68, 69).
Following these pioneering studies of plant pathogenic
spiroplasmas, spiroplasmas have been found in a wide array of
ecological niches (see Table I). These include the bodies of
ticks from mammalian hosts, as well as several species of insects (9, 10, 21, 47, 53, 60, 75, 76, 78). In some cases,
spiroplasmas are known to induce disease in their insect hosts
(8, 9, 75, 76). More recently, in 1976, an unusual minute,
helical microorganism was discovered in diseased individuals
of the honeybee *(Apis mellifera ligustica)* in Maryland (9).
Morphological, ultrastructural, cultural, serological, and
biochemical characterization showed it to be a spiroplasma
(20, 26, 27, 33). Examination of naturally infected bees

revealed high concentrations of the spiroplasma in hemolymph.
Cultivation *in vitro* permitted tests proving the pathogenicity
of the new spiroplasma in both worker bees and queen bees by
intrahemocoelic injection or by acquisition of the cultured
spiroplasma through feeding (9). Extensive studies have
shown that this spiroplasma is related to, but distinct from,
both *S. citri* and the corn stunt spiroplasma (26, 27, 33, 47,
51, 77). The presence of this new spiroplasma in honeybees
encouraged speculation on its possible association with plants
(9, 33) and led to attempts to detect it in nectar-bearing
flowers (8, 15-19).

Even though the mode of natural transmission of the Maryland honeybee spiroplasma has not yet been conclusively demonstrated, circumstantial evidence now strongly favors an
insect-flower-insect transmission cycle (14). Initial attempts
to detect the honeybee spiroplasma in flowers, however, yielded some surprising results. The first spiroplasmas found on
flowers were not related to the honeybee spiroplasma, nor to
any other known spiroplasma.

III. DISCOVERY OF SPIROPLASMAS (AND MYCOPLASMAS) IN FLOWERS

Spiroplasmas were first discovered in flowers of the tulip tree *(Liriodendron tulipifera* L.*)* growing in Maryland and
Connecticut (18, 32). In that work, not only spiroplasmas,
but also (nonhelical) mycoplasmas were obtained by rinsing
nectar from surfaces of petals and bracts. This procedure
targeted the location where it was thought spiroplasmas might
occur. It also avoided extensive release from plant tissues
of inhibitors that are often encountered in plant extracts
(54, 55) and that might prevent isolations of spiroplasmas
from flowers by homogenizing tissues in culture medium. The
isolations of spiroplasmas from Maryland tulip tree flowers
were soon confirmed and extended to flowers of *Magnolia grand-
iflora* L. in Maryland by adopting the method involving rinsing
of flowers (8). Not surprisingly, the method has since also
been successfully applied by other researchers in the isolation of spiroplasmas from flowers of several additional plant
species (16, 26, 42a, 47, 58, 59, 65). Indeed, possible problems associated with maceration techniques in attempts to isolate epiphytic bacterial and fungal microflora in culture have
been discussed by Dennis (37), and the importance of avoiding
excessive injury to flower tissues has recently been reempha-
sized by Raju *et al.* (65), who have isolated and cultivated
spiroplasmas from tulip tree and magnolia flowers in California.

Following the early recoveries of spiroplasmas from flowers, it was assumed (8) that the spiroplasmas isolated

from tulip tree and magnolia flowers in Maryland were isolates
of the Maryland honeybee spiroplasma, and it was postulated
that flowers of the Magnoliaceae were important sources from
which bees acquired the honeybee spiroplasma in Maryland (8).
Other work showed, however, that a Maryland tulip tree flower
spiroplasma, strain 23-6, was serologically unrelated (by
tests of membrane antigens) to the honeybee spiroplasma (18,
32). Indeed, among the spiroplasmas isolated originally from
flowers of tulip tree, none has been found to be serologically related to the Maryland honeybee spiroplasma (18, 25-27, 47,
72). This finding led to an early appreciation of the flower
site as the repository of a wide variety of spiroplasma
strains.

IV. THE VARIETY OF WALL-FREE PROKARYOTES FOUND ON FLOWERS

     In the first successful isolation and cultivation *in vitro*
of cell wall-free prokaryotes from surfaces of flowers, 21
isolates of spiroplasmas and 30 nonhelical mycoplasma isolates
were obtained from individually assayed flowers of tulip tree
(15, 17-19, 32). Cultural studies indicated that several distinct strains might be among the spiroplasma isolates (15, 18),
and serological studies showed that at least one strain,
strain 23-6 (from tulip tree flower in Maryland), was a new
spiroplasma (18). Work confirming the presence of spiroplasmas on tulip tree flowers, and recovering spiroplasmas also
from flowers of magnolia (8), yielded isolates that were later
reported as BNRI and OBMG (77) and identified as strain 23-6
type (47). Additional early studies on the 21 spiroplasma
isolates originally obtained from flowers of tulip tree revealed a second new spiroplasma represented by two strains
from Connecticut, termed SR9 (15) and SR3 (26, 27), that are
indistinguishable from one another by serology and DNA-DNA
hybridizations (26, 27, 51, 52). More recently, a third new
spiroplasma, strain brevi, was reported from among the spiroplasmas originally isolated from Maryland tulip tree flowers
(25).
     Although the honeybee spiroplasma was not identified among
any isolates from flowers of tulip tree or magnolia in Maryland, strains closely related to the Maryland honeybee spiroplasma were found on flowers of the basswood tree *(Tilia
americana* L.*)* in Maryland and of *Bidens pilosa* L. in Florida
(15, 16, 26). Flowers of *B. pilosa* yielded several strains
(including G1 and G2) that proved to be closely related to the
Maryland honeybee spiroplasma (AS576 and related strains)
(15, 16, 26, 51), and flowers of basswood tree yielded the

closely related strain BW (26). Interestingly, flowers of
tulip tree and magnolia in California have been found to harbor strains closely related to the Maryland honeybee spiroplasma (65) whereas, strains related to the SR3 tulip tree
spiroplasma group were found in honeybees in California (65)
and in honeybees in Morocco (47, 73a).

Flowers of other plant species have yielded additional
spiroplasma strains. Flowers of *Calliandra hematocephala* in
Florida, for example, have been found to harbor a spiroplasma
(59) that was termed PPS1 (58) and found to be related to the
tulip tree flower spiroplasma strain SR3 (and, therefore, also to SR9) (72). Spiroplasma strains recently isolated from
surfaces of flowers in France and Corsica have also been found
to be serologically related to the spiroplasma strain SR3 from
tulip tree flowers in Connecticut (47, 72), and several
strains from tulip tree flower in Maryland have been found to
be similar to strain SR3 (62).

During our work with spiroplasmas isolated from surfaces
of flowers, we became convinced that serological criteria
could prove applicable for distinguishing *Spiroplasma* species.
In suggesting serology as an approach to spiroplasma systematics at the 1978 meetings of International Organization for
Mycoplasmology and American Phytopathological Society, we proposed that species could be diagnosed on the basis of serological reactions and proposed that each major group defined
in our work represented at least one *Spiroplasma* species (26,
27). These and additional serological groupings of spiroplasma strains by ourselves and others (72, 73, 77), and polyacrylamide gel electrophoretic (PAGE) analyses of cellular
proteins (3, 25, 61, 64) are in mutual agreement, and DNA-DNA
homology studies have indicated that that the classifications
based on serological properties reflect phylogenetic relationships (7, 47, 51, 52). Taxonomic ranking of separate
major groups and of distinct subgroups as separate *Spiroplasma* species has been proposed in a classification scheme
(Table II) based initially on serological properties and
supported principally by DNA-DNA homology studies and PAGE
analyses of proteins (25).

V. THE FLOWER SITE

It has long been known that leaf surfaces are populated
by bacteria and fungi. This epiphytic microflora includes
potentially plant pathogenic organisms (some of which are in
a state of saprophytic existence on the leaf surface) and
saprophytes of no known potential plant pathogenicity (2).
The plant pathogenic microflora and their interactions with

saprophytes on the leaf surface have received major emphasis
in investigations on the microbiology of aerial plant parts.
The flower, consisting of modified leaf structures, also
bears a varied microflora. Indeed, the microflora and micro-
fauna present on surfaces of flowers might encompass a greater
variety and concentration of microorganisms than that found
on surfaces of leaf and bud. As discussed, recent investigations have revealed that a part of this epiphytic microflora
on flowers consists of spiroplasmas of no known potential
plant pathogenicity. Evidence indicates that these microorganisms are simply in a casual, transient association with
flowers (14, 18). What attribute(s) of the flower account(s)
for the presence of spiroplasmas in this specialized niche?
The answer seems to be mainly nectar, but pollen or edible
fleshy flower tissues may also account for visitation of
flowers by various insects and other fauna carrying spiroplasmas. It is presumed that many species of nectar-feeding insects deposit spiroplasmas in, or acquire them from nectar
(14). If flowers serve as important foci for dissemination
of some spiroplasmas, nectar may serve as a major vehicle for
their spread (15). Flowers of the tulip tree (L. tulipifera
L.), in which spiroplasmas have been discovered (8, 18, 32, 65),
produce copious amounts of nectar and are visited by a wide
variety of arthropods. Factors such as these seem likely to
favor frequent occurrence of cell wall-less prokaryotes — as
well as other microflora — on flower surfaces.

   If nectar is a factor determining presence of cell wall-
less procaryotes on flowers, it seems likely that these microorganisms can also occur on surfaces of extrafloral nectaries
found in some plant species. Assuming that sugars in nectar
are a major factor influencing presence (or survival) of spiroplasmas in nectar, it may be further speculated that nectar-
bearing surfaces may not be the only sites casually associated
with spiroplasmas on plants. It is easily hypothesized, for
example, that spiroplasmas may be found in plant exudates,
other than nectar, that are rich in sugars. Indeed, if one
supposes that these prokaryotes are commonly carried by insects that visit exudates rich in sugars, various types of
sugar-rich exudates (e.g., inflorescence scars in certain
tropical fruits, and other wound sites) might be potential
sites for chance occurrence of spiroplasmas. To my knowledge,
epiphytic wall-less prokaryotes (spiroplasmas and (nonhelical)
mycoplasmas) have been searched for experimentally only in
flowers and on leaves, but the finding of mycoplasmas in rotting tissues of coconut (Cocos nucifera L.) (39, 40) indicates
a new site where nonphytopathogenic wall-less prokaryotes may
be found in association with plants.

VI. NATURE OF THE SPIROPLASMA-FLOWER ASSOCIATION

In all cases studied thus far, plant pathogenic spiroplasmas occur in association with plants only as internal parasites and specifically inhabit phloem tissue in diseased
plants (14). No saprophytic period of existence outside of
living plant or insect host tissues has been revealed for any
of the known plant pathogenic spiroplasmas. Although certain
non-plant pathogenic mycoplasmas are known to exist as saprophytes, the finding of spiroplasmas on the surfaces of flowers
gave the first evidence that certain spiroplasmas can exist
outside of living host tissues in nature. Adaptation to this
environment may select for properties uncommon in spiroplasmas
that have no period of existence outside their hosts.

The occurrence of spiroplasmas on surfaces of flowers seems
to represent a casual, transient association of spiroplasmas
with flowers, presumably with the nectar. Because previous
work failed to detect spiroplasmas within tissues of flowers
bearing these microorganisms on their surfaces (18), it appears most likely that these spiroplasmas arrive at the flower
site from outside the plant. There is no evidence to indicate
that any spiroplasma strain moves from the interior of a plant
through the nectary to the surface of flowers. Nevertheless,
the possibility that this might occur in some as yet undis-
covered circumstances cannot be discarded entirely. It should
be noted, however, that none of the strains thus far isolated
from surfaces of flowers has been shown to be capable of inhabiting the interior of plants.

Even though it seems unlikely that flower spiroplasma
strains reached the flower surface from the plant interior, it
is possible that some strains normally found on flowers may be
capable of at least limited invasion of plant tissues (19).
The vigorous translational motility of which spiroplasmas are
capable (12, 18, 19, 34) could conceivably aid such a process.
It may be, for example, that spiroplasmas on flowers can occasionally act as secondary invaders of damaged floral tissues
in situations analogous to the process presumed to account for
the presence (39, 40) of *Acholeplasma axanthum* in rotting coconut tissue and in sound tissues short distances from the
rotting portions. Because phloem cells are generally in relatively close proximity to the outer surface of nectary tissue
(41), one wonders whether spiroplasmas on flower surfaces can
sometimes invade plant tissues and eventually reach the phloem.

Nectar contains a wide array of components (48, 63, 71)
that might provide a nutritionally adequate milieu for growth
of certain spiroplasmas, but it is still not known whether
flower spiroplasmas are capable of multiplying in nectar.
Multiplication of the spiroplasmas in the flower site may not

TABLE II. Proposed Classification of Several Cultivated Spiroplasmas[a]

Group	Subgroup	Spiroplasma strain(s)	G + C (mol %)	Proposed taxonomic rank
I	A(1-1)	*Spiroplasma citri*	26	Species
	B(1-2)	Honeybee strains; AS576, G1, BC3, B63	26	Species
	C(1-3)	Corn stunt strains; I747, PU8-17, E275, Miss.	26	Species
	(1-4)	Tick strain; 277F	26	Species
II (III)		Flower strains; 23-6, 27-31, OBMG, BNR1	25	Species
III(IV)	A	Flower strains; SR3, SR9	29	Species
	B[b]	Flower strain; PPS1	29	Possible species(?)[b]
	Unclassified	Flower and insect strains; L89, F1, F2, B13	29-31	
IV (V)		Tick strain; SMCA (suckling mouse cataract agent)	29-31	Species
V		Flower strain; brevi	28	Species

268

^{a}From Davis and Lee (ref. 25). Placement of strains in serogroups and subgroups, and base compositions [mol % guanine + cytosine (G + C)], are based upon data from several laboratories (3, 7, 16, 18, 25–27, 47, 67, 72, 73, 77). Only strains that can be cultivated in vitro and are capable of adequate description are included, and uncultivated strains are excluded even though tentative grouping of such strains may be possible. Group and subgroup designations are according to R. E. Davis et al. (25–27). Designations after P. Junca et al. (47) are in parentheses.

^{b}Preliminary data (G. Basham et al., personal communication) have suggested that strain PPS1 is related to, but may be distinct from, strain SR3 and may therefore represent a separate and distinct subgroup in serogroup III. Additional research will be necessary to clarify the relatedness between PPS1 and SR3, but if it proves correct that PPS1 is distinct from SR3, strain PPS1 could represent a separate species on the basis of recently proposed criteria for diagnosis of species (see ref. 25).

269

be required for their acquisition by insects or other hosts/
vectors, however. Sufficient concentration of viable spiroplasma cells in liquid suspension in nectar would presumably
be the major requirement for acquisition by nectar-feeding
fauna. Whether this concentration is achieved during seeding
of nectar by flower-visiting fauna or by multiplication of the
spiroplasmas in nectar is not known. Since the flower surfaces from which spiroplasmas have been isolated have often
been dry (18), spiroplasmas evidently can remain viable in a
dried state between periods of nectar flow.

VII. TRANSMISSION OF FLOWER SPIROPLASMAS

    Although other possible vectors may be considered, insects
are presumed to play the major role in transmission of the
flower spiroplasmas to flowers. Presumably, some flower-visiting insects also acquire spiroplasmas at the flower site.
However, the best evidence for an insect-flower-insect transmission cycle is only indirect. This evidence consists mainly
of the occurrence of the honeybee spiroplasma (strain AS576
subgroup, see Table II) both in bees and on surfaces of certain nectar-bearing flowers (9, 16, 18) and, concomitantly,
the ability of bees to acquire the spiroplasma by feeding in
the laboratory (9). It is not known whether modes of transmission may exist other than those suspected to involve feeding on nectar. As more research is directed toward understanding these spiroplasmas, we may learn how they are disseminated: whether by transovarial passage in insects or other
arthropods, by deposition in feces, by consumption of infected
insects by predatory arthropods, by transport of insects or
wind of pollen that is surface-contaminated with nectar containing spiroplasmas, by rain splash in opened flowers, by
consumption of flower tissues by phytophagous fauna, by mechanical transmission on the bodies of flower-visiting fauna,
or by other phenomena.
    Because many species of insects (and possibly other fauna)
may deposit spiroplasmas in, or acquire them from, flowers, it
would be of interest to inquire whether or not any specificity
exists between plant or insect species and spiroplasma strain.
For example, since the Maryland honeybee spiroplasma is found
on surfaces of some flowers as well as in bees (16, 18), it
may be asked whether insects other than bees carry this spiroplasma. Moreover, several distinct groups of spiroplasma
strains have been isolated from surfaces of flowers in Maryland during the period that honeybees carry spiroplasmas.
Thus far, however, spiroplasmas isolated from bees in Maryland
have been reported to belong to only one serological subgroup

(26, 27, 77). What may explain this apparent specificity is
not known.

The honeybee spiroplasma provokes some additional intriguing questions not yet adequately addressed experimentally.
Thus far, for example, no clear explanation has appeared for
the reported presence of spiroplasmas in Maryland bees for only
a brief period during the year (May— mid-July) (9). If infected bees in Maryland eventually die without transmitting spiroplasmas directly to other bees (i.e., excluding passage of the
spiroplasmas through intervening flower), a partial explanation for seasonal disappearance of the spiroplasma from Maryland bees could be at hand. It may be worthwhile to investigate whether this apparent disappearance is due to inability
to detect the spiroplasma. However, if bees in Maryland do
not carry the bee spiroplasma year round (9), there must be a
primary source of the spiroplasma outside the Maryland bee
population. Since no evidence indicates that flowering plants
alone act as this source, (i.e., spiroplasmas have not been
found within tissues of plants bearing flowers that harbor
them), one may assume that flower-visiting fauna, presumably
insects, are the primary source that deposits spiroplasmas
(in flowers) where they are acquired by Maryland bees. It
thus may well be that, in a given season, local honeybees are
not the earliest carriers of the "honeybee spiroplasma". Bees
instead may initially acquire this spiroplasma in spring from
flowers into which it is deposited by other insect species.

Alternatively, is it possible that in eastern United
States the Maryland honeybee spiroplasma (AS576 subgroup) is
mainly carried by bees and that it gradually "moves" northward
each year, passed from infected bee to flower to healthy bee
population as the blooming period of appropriate spring-flow-
ering plants advances northward? It is known that strains of
this spiroplasma are found in Florida in November (16) and in
southern California in March (26).

VIII. VERTEBRATES AS POSSIBLE CARRIERS OF FLOWER SPIROPLASMAS

As noted, nectar-feeding insects are presumed to be primarily responsible for deposition of spiroplasmas on surfaces
of flowers. It must be seriously considered, however, that
other fauna -- including vertebrates -- could be active in
depositing spiroplasmas in, or acquiring them from, flowers
in certain circumstances.

A number of flower-visiting vertebrates, including mammals,
are known, including lizards, birds, bats, rodents, and pri-
mates (66, 70, 74). Indeed, it appears that some plants may
have evolved floral characteristics adapted to pollination by

certain vertebrates (66, 70). The flowers visited by some
vertebrates are partially or totally eaten, but in other cases
the animals feed on nectar (66, 70, 74). These associations
would seem to offer numerous possibilities for vertebrates to
deposit spiroplasmas (or mycoplasmas) in or acquire them from
flowers. It would be of interest to learn whether a flower -
vertebrate (oral or other) niche exists for any spiroplasmas
or whether any spiroplasmas may be transferred between flower
and vertebrate by arthropods such as mosquitoes and flies.
Conceivably, flowers could serve as foci for dissemination of
pathogens among both arthropods and vertebrates, thus providing
an important common ground for transfer of widely different
agents between divergent groups of fauna.

IX. CONCLUDING REMARKS

    Discoveries of spiroplasmas in increasingly diverse habi-
tats have begun to shape a new area in the science of microbial ecology; the discovery of spiroplasmas on flowers has
opened an intriguing new topic in the microbiology of aerial
plant surfaces. A vast array of challenging questions is
provoked by a new appreciation of the diversity of micro-
flora--and one surmises, of microfauna--present on surfaces of
flowers. As discussed, the flower may be viewed as a potentially important focus for dissemination of spiroplasmas and
mycoplasmas and other agents among arthropods and possibly
among other fauna also. Indeed, the flower and other sites of
sugar-rich exudates have been long known to carry, and to act
as infection courts for, certain insect-transmitted plant
pathogenic bacteria (5, 48, 71). It may prove worthwhile to
expand investigations of the flower niche as a harbor not only
of cell wall-less prokaryotes, but also of plant or animal
pathogenic agents including bacteria, fungi, and protozoa.

REFERENCES

1. Billing, E. (1976). The taxonomy of bacteria on the aerial
   parts of plants. In "Microbiology of Aerial Plant Surfaces" (C. H. Dickinson and T. F. Preece, eds.), pp. 223-
   273. Academic Press, New York.
2. Blakeman, J. P. and Brodie, I. D. S. (1976). Inhibition
   of pathogens by epiphytic bacteria on aerial plant surfaces. In "Microbiology of Aerial Plant Surfaces" (C. H.
   Dickenson and T. F. Preece, eds.), pp.529-557. Academic
   Press, New York.

3. Bové, J. M. and Saillard, C. (1979). Cell biology of spiroplasmas. *In* "The Mycoplasmas, Vol. III: Plant and insect mycoplasmas" (R. F. Whitcomb and J. G. Tully, eds.), pp. 83-153. Academic Press, New York.

4. Bové, J. M., Saglio, P., Tully, J. G., Freundt, A. E., Lund, Z., Pillot, J., and Taylor-Robinson, D. (1973). Characterization of the mycoplasma-like organism associated with stubborn disease of citrus. *Ann. N. Y. Acad. Sci. 225,* 462-470.

5. Buddenhagen, I. W., and Elsasser, T. A. (1962). An insect-spread bacterial wilt epiphytotic of bluggoe banana. *Nature (London) 194,* 164-165.

6. Chen, T. A., and Liao, C. (1975). Corn stunt spiroplasma: Isolation, cultivation, and proof of pathogenicity. *Science 188,* 1015-1016.

7. Christiansen, C., Askaa, G., Freundt, E. A., and Whitcomb, R. F. (1979). Nucleic acid hybridization experiments with *Spiroplasma citri* and the corn stunt and suckling mouse cataract spiroplasmas. *Current Microbiol. 2,* 323-326.

8. Clark, T. B. (1978). Honeybee spiroplasmosis: A new problem for beekeepers. *Amer. Bee J. 118,* 18-23.

9. Clark, T. B. (1977). Spiroplasma sp., a new pathogen in honey bees. *J. Invert. Pathol. 29,* 112-113.

10. Clark, H. F., and Rorke, L. B. (1979). Spiroplasmas of tick origin and their pathogenicity. Pages 155-174. *In* "The Mycoplasmas, Vol. III: Plant and Insect Mycoplasmas" (R. F. Whitcomb and J. G. Tully, eds.), Academic Press, New York (351 pages).

11. Cole, R. M., Tully, J. G., Popkin, T. J., and Bové, J. M. (1973). Morphology, ultrastructure, and bacteriophage infection of the helical mycoplasma-like organism (*Spiroplasma citri* gen. nov., sp. nov.) cultured from "stubborn" disease of citrus. *J. Bacteriol. 115,* 367-383.

12. Daniels, M. J., Longland, J. M., and Gilbart, J. (1980). Aspects of motility and chemotaxis in spiroplasmas. *J. Gen. Microbiol. 118,* 429-436.

13. Davis, R. E. (1981). Antibiotic sensitivities *in vitro* of diverse spiroplasma strains associated with plants and insects. *Appl. Environ. Microbiol. 41,* in press.

14. Davis, R. E. (1979). Spiroplasmas: Newly recognized arthropod-borne pathogens. *In* "Leafhopper Vectors and Plant Disease Agents" (K. Maramorosch and K. F. Harris, eds.), Chapter 13. Academic Press, New York.

15. Davis, R. E. (1979). Spiroplasmas: Helical cell wall-free prokaryotes in diverse habitats. Pages 59-64. *Proc. Republic of China-United States Cooperative Sci. Program, Joint Sem. on Mycoplasma Diseases of Plants, Taipei, 27-31, March, 1978.* (NSC Symposium Series No. 1.) National Science Council, Taipei, Taiwan.

16. Davis, R. E. (1978). Spiroplasmas from flowers of *Bidens pilosa* L. and honeybees in Florida: Relationship to honey-bee spiroplasma AS576 from Maryland. *Phytopathol. News* *12*(7), PO-7.

17. Davis, R. E. (1978). New spiroplasma strains from nectar-bearing flowers. *Zentralbl. Bakteriol., Parasitol. Infektskrank. Hyg. Abt. 1, Orig., A241*(2), 192.

18. Davis, R. E. (1978). Spiroplasma associated with flowers of the tulip tree (*Liriodendron tulipfera* L.). *Can. J. Microbiol. 24,* 954-959.

19. Davis, R. E. (1978). Cell wall-free prokaryotes harbored in flowers. *Proc. 3rd Meeting Intern. Council Lethal Yellowing, Oct. 30-Nov. 3, 1977.* Univ. Florida Publ. Fl. 78-2, p.19.

20. Davis, R. E. (1977). Spiroplasma: Role in the diagnosis of corn stunt disease. Pages 92-98. *Proc. Maize Virus Disease Colloq. Workshop, 16-19 August 1976.* pp. 92-98. Ohio Agricultural Research and Development Center, Wooster, Ohio.

21. Davis, R. E. (1974). Spiroplasma in corn stunt-infected individuals of the vector leafhopper *Dalbulus maidis. Plant Dis. Rep. 58,* 1109-1112.

22. Davis, R. E. (1974). New approaches to diagnosis and control of plant yellows diseases. Pages 289-302. *3rd. Intern. Virus Diseases Ornamental Plants 11-15 Sept. 1972,* pp. 289-302. College Park, Md. Tech. Commun. Intern. Soc. Hort. Sci., Acta Horticulturae No. 36.

23. Davis, R. E. (1973). Occurrence of a spiroplasma in corn stunt-infected plants in Mexico. *Plant Dis. Reptr. 57,* 333-337.

24. Davis, R. E., Chen, T. A., and Worley, J. F. (1981). Corn stunt spiroplasma. *In* "Virus and Viruslike Diseases of Maize and Sorghum in the United States" (D. T. Gordon, J. K. Knoke, and G. E. Scott, eds.), in press. Ohio Agricultural Research and Development Center, Wooster, Ohio.

25. Davis, R. E., and Lee, I.-M. (1981). Spiroplasmas: Comparative properties and emerging taxonomic concepts--a proposal. *Proc. 3rd Conf. Intern. Organ. Mycoplasmology, Sept. 3-9, 1980, Rev. Inf. Diseases,* in press.

26. Davis, R. E., Lee, I.-M., and Basciano, L. K. (1979). Spiroplasmas: Serological grouping of strains associated with plants and insects. *Can. J. Microbiol. 25,* 861-866.

27. Davis, R. E., Lee, I.-M., and Basciano, L. K. (1978). Spiroplasmas: Identification of serological groups and their analysis by polyacrylamide gel electrophoresis of cell proteins. *Phytopathol. News 12*(9), 512.

28. Davis, R. E., and Whitcomb, R. F. (1971). Mycoplasmas,
    rickettsiae, and chlamydiae: Possible relation to yellows
    diseases and other disorders of plants and insects. *Ann.
    Rev. Phytopathol. 9,* 119-154.
29. Davis, R. E., Whitcomb, R. F., Chen, T. A., and Granados,
    R. R. (1972). Current status of the etiology of corn
    stunt disease. *In* "Pathogenic Mycoplasmas" CIBA Found.
    Symp. London, England, 25-27 January 1972. (J. Birch,
    ed.), pp. 205-225. Elsevier and North-Holland, Amsterdam.
30. Davis, R. E., and Worley, J. F. (1973). Spiroplasma:
    Motile, helical microorganism associated with corn stunt
    disease. *Phytopathology 63,* 403-408.
31. Davis, R. E. and Worley, J. F. (1972). Motility of helical
    filaments produced by a mycoplasmalike organism associated
    with corn stunt disease. *Phytopathology 62,* 752-753.
32. Davis, R. E., Worley, J. F., and Basciano, L. K. (1977).
    Association of spiroplasma and mycoplasmalike organisms
    with flowers of tulip tree, *Liriodendron tulipifera* L.
    Proc. Amer. Phytopathol. Soc. 4, 185-186.
33. Davis, R. E. Worley, J. F., Clark, T. B., and Moseley, M.
    (1976). New spiroplasma in diseased honeybee (*Apis Melli-
    fera* L.): Isolation, pure culture, and partial characteri-
    zation *in vitro. Proc. Amer. Phytopathol. Soc. 3,* 304.
34. Davis, R. E., Worley, J. F., and Moseley, M. (1975).
    Spiroplasmas: Primary isolation and cultivation in cystine-
    tryptone media, and translational locomotion in semi-solid
    versions. *Proc. Amer. Phytopathol. Soc. 2,* 153.
35. Davis, R. E., Worley, J. F., Whitcomb, R. F., Ishijima,
    T., and Steere, R. L. (1972). Helical filaments produced
    by a mycoplasmalike organism associated with corn stunt
    disease. *Science 176*(4034), 521-523.
36. Davis, R. E., Worley, J. F., Whitcomb, R. F., Ishizima,
    T., and Steere, R. L. (1972). Helical filaments associ-
    ated with a mycoplasmalike organism in corn stunt-infected
    plants. *Phytopathology 62,* 494.
37. Dennis, C. (1976). The microflora on the surface of soft
    fruits. *In* "Microbiology of Aerial Plant Surfaces"
    (C. H. Dickinson and T. F. Preece, eds.), pp. 419-432.
    Academic Press, New York.
38. Doi, Y., Teranaka, M., Yora, K., and Asuyama, H. (1967).
    Mycoplasma- or PLT group-like microorganisms found in the
    phloem elements of plants infected with mulberry dwarf,
    potato witches' broom, aster yellows, or Paulownia witches'
    broom. *Ann. Phytopathol. Soc. Jap. 33,* 259-266.
39. Eden-Green, S. J. (1978). Isolation of acholeplasmas from
    coconut palms affected by lethal yellowing disease in
    Jamaica. Proc. 2nd Conf. Internatl. Organiz. Mycoplasmo-
    logy. Aug. 28-Sept. 1, 1978. *Zentralbl. Bakteriol. Para-
    sitol. Infekskrank. Hyg. Abt. 1, Orig. A241*(2), 226.

40. Eden-Green, S. J. (1980). Acholeplasmas and lethan yellow-
    ing disease. I. Present status. *Proc. 4th Meet. Intern.
    Council on Lethal Yellowing, 13-17 August 1979, Fort
    Lauderdale, Florida*, p. 11. University of Florida, Fort
    Lauderdale. Publication FL 80-1.
41. Frey-Wyssling, A. (1955). The phloem supply to the nec-
    tarines. *Acta Botan. Neerl. 4*(3), 358-369.
42. Fudl-Allah, A. E.-S. A., Calavan, E. C., and Igwegbe,
    E. C. K. (1972). Culture of a mycoplasmalike organism
    associated with stubborn disease of citrus. *Phytopathol-
    ogy 62*, 729-731.
42a. Giannotti, J., Giannotti, D., and Vago, C. (1980). Multi-
    plication intracellulaire et action pathogene de mollicutes
    isolés de plantes chez des insectes non vecteurs. *C. R.
    Acad. Sci. Ser. (Paris) D290*, 417-419.
43. Granados, R. R. (1969). Electron microscopy of plants and
    insect vectors infected with the corn stunt disease agent.
    Contrib. Boyce Thompson Inst. 24, 173-187.
44. Granados, R. R. (1969). Chemotherapy of corn stunt
    disease. *Phytopathology 59*, 1556.
45. Granados, R. R., Maramorosch, K., and Shikata, E. (1968).
    Mycoplasma: Suspected etiologic agent of corn stunt.
    Proc. Nat. Acad. Sci. U.S.A. 60, 841-844.
46. Ishiie, T., Doi, Y., Yora, K., and Asuyama, H. (1967).
    Suppressive effects of antibiotics of tetracycline group
    on sympton development of mulberry dwarf disease. *Ann.
    Phytopathol. Soc. Jap. 33*, 267-275.
47. Junca, P., Saillard, C., Tully, J., Garcia-Jurado, O.,
    Degorce-Dumas, J. R., Mouches, C., Vignault, J.-C., Vogel,
    R., McCoy, R., Whitcomb, R., Williamson, D., Latrille, J.,
    and Bové, J. M. (1980). Caracterisation de spiroplasmes
    isolées d'insectes et de fleurs de France continentale,
    de Corse et du Maroc. Proposition pour une classification
    des spiroplasmes. *C. R. Acad. Sci. (Paris) Ser. D290*,
    1209-1212.
48. Keitt, G. W. (1941). Transmission of fire blight by bees
    and its relations to nectar concentration of apple and
    pear blossoms. *J. Agr. Res. 62*, 745-753.
49. Kunkel, L. O. (1948). Studies on a new corn virus disease.
    Arch. Virusfor. 4, 24-46.
50. Kunkel, L. O. (1946). Leafhopper transmission of corn
    stunt. *Proc. Nat. Acad. Sci. U.S.A. 32*, 246-247.
51. Lee, I.-M., and Davis, R. E. (1980). DNA homology among
    diverse spiroplasma strains representing several serolog-
    ical groups. *Can. J. Microbiol.*, in press.
52. Lee, I.-M., and Davis, R. E. (1980). DNA homologies among
    spiroplasma strains representing different serological
    groups. *Phytopathology 70*, 464.

53. Lei, J. D., Su, H. J., and Chen, T.-A. (1979). Spiroplas-
    mas isolated from green leaf bug, *Trigonotylus ruficornis*
    Geoffroy. *Proc. Republic of China - U. S. Cooperative
    Sci. Program, Joint Sem. Mycoplasma Dis. Plants, Taipei,
    27-31 March, 1978* (NSC Symposium Series No. 1), pp.90-98.
    National Science Council, Taipei, Taiwan.
54. Liao, C. H., Chang, C. J., and Chen, T.-A. (1979). Spiro-
    plasmastatic action of plant tissue extracts. Pages 99-
    103. *Republic of China - U. S. Cooperative Sci. Program,
    Joint Sem. on Mycoplasma Dis. Plants, Taipei, 27-31 March
    1978,* NSC Symposium Series No. 1, pp.99-103. National
    Science Council, Taipei, Taiwan.
55. Liao, C. H., and Chen, T.-A. (1975). Inhibitory effect of
    corn stem extract on the growth of corn stunt spiroplasma.
    Proc. Amer. Phytopathol. Soc. 2, 53.
56. Maramorosch, K., Granados, R. R., and Hirumi, H. (1970).
    Mycoplasma diseases of plants and insects. *Advan. Virus
    Res. 16,* 135-193.
57. Maramorosch, K., Shikata, E., and Granados, R. R. (1968).
    Structures resembling mycoplasma in diseased plants and
    in insect vectors. *Trans. N. Y. Acad. Sci. 30,* 841-855.
58. McCoy, R. E. and Basham, H. G. (1980). Isolation and
    characterization of mycoplasmas from floral surfaces.
    *Proc. 4th Meet. Intern. Council on Lethal Yellowing,
    13-17 August, 1979, Fort Lauderdale, Florida,* p. 11.
    University of Florida, Fort Lauderdale. Publication
    FL 80-1.
59. McCoy, R. E., Williams, D. S., and Thomas, D. L. (1979).
    Isolation of mycoplasmas from flowers. *Proc. Republic of
    China - U. S. Cooperative Sci. Program, Joint Sem. Myco-
    plasma Dis. Plants, Taipei, 27-31 March 1978,* NSC Symposium
    Series No. 1, pp.75-81. National Science Council, Taipei,
    Taiwan.
60. McCoy, R. E., Tsai, J. H., and Thomas, D. L. (1978).
    Occurrence of a spiroplasma in natural populations of the
    sharpshooter *Oncometopia nigricans. Phytopathol. News
    12*(9), 217.
61. Mouches, C., Vignault, J. C., Tully, J. G., Whitcomb, R. F.,
    and Bové, J. M. (1979). Characterization of spiroplasmas
    by one- and two-dimensional protein analysis on polyacry-
    lamide slab gels. *Current Microbiol. 2,* 69-74.
62. Muniyappa, V. and Davis, R. E. (1980). Occurrence of
    spiroplasmas of two serogroups on flowers of the tulip
    tree (*Liriodendron tulipifera* L.) in Maryland. *Current
    Sci. 49,* 58-60.
63. Nunez, J. (1977). Nectar flow by melliferous flora and
    gathering flow by *Apis mellifera ligustica. J. Insect.
    Physiol. 23,* 265-275.

64. Padhi, S. B., McIntosh, A. H., and Maramorosch, K. (1977). Characterization and identification of spiroplasmas by polyacrylamide gel electrophoresis. *Phytopathol. Z. 90,* 268-272.

65. Raju, B. C., Nyland, G., Meikle, T., and Purcell, A. H. (1981). Helical, motile mycoplasmas associated with flowers and honey bees in California. *Can. J. Microbiol.,* in press.

66. Rourke, J. and Wiens, D. (1977). Convergent floral evolution in South African and Australian Proteaceae and its possible bearing on pollination by nonflying mammals. *Ann. M. Botan. Gard. 64,* 1-17.

67. Saglio, P., L'Hospital, M., Lafleche, D., Dupont, G., Bové, J. M., Tully, J. G., and Freundt, E. A. (1973). *Spiroplasma citri* gen. and sp. n: a mycoplasma-like organism associated with "stubborn" disease of citrus. *Int. J. Syst. Bacteriol. 23,* 191-204.

68. Saglio, M. P., Lafleche, D., Bonissol, C., and Bové, J. M. (1971). Isolement et culture in vitro des mycoplasmes associés au stubborn des agrumes et leur observation au microscope électronique. *C. R. Acad. Sci. (Paris) Ser. D272,* 1387-90.

69. Saglio, P., Lafleche, D., L'Hospital, M., Dupont, G., and Bové, J.-M. (1972). Isolation and growth of citrus mycoplasmas. Pages 187-198. *In* "Pathogenic Mycoplasmas," CIBA Foundation Symposium, London, England: 25-27 January 1972 (J. Birch, ed.), pp.187-198. Elsevier and North-Holland, Amsterdam.

70. Sussmann, R. W. and Raven, P. H. (1978). Pollination by lemurs and marsupials: An archaic coevolutionary system. *Science 200,* 731-736.

71. Thomas, H. E. and Ark, P. A. (1934). Nectar and rain in relation to fire blight. *Phytopathology 24,* 682-685.

72. Tully, J. G., Rose, D. L., Garcia-Jurado, O., Vignault, J.-C., Saillard, C., Bové, J. M., McCoy, R. E., and Williamson, D. L. (1970). Serological analysis of a new group of spiroplasmas. *Current Microbiol. 3,* 369-376.

73. Tully, J. G., Williamson, D. L., Rose, D. L., and Whitcomb, R. F. (1979). Serological analysis of spiroplasmas from various hosts. *Abstr. Annu. Meet. Amer. Soc. Microbiol.,* p. 83.

73a. Vignault, J. C., Bové, J. M., Saillard, C., Vogel, R., Farro, A., Venegas, L., Stemmer, W., Aoki, S., McCoy, R., Al-Beldawi, A. S., Larue, M., Tuzco, O., Ozsan, M., Nhami, A., Abassi, M., Bonfils, J., Moutous, G., Fo., A., Poutiers, F., and Viennot-Bourgin, G. (1980). Mise en culture de spiroplasmes à partir de matériel végètal et d'insectes provenant de pays circummediterranéens et du Proche-Orient. *C. R. Acad. Sci. (Paris) Ser. D290,* 775-778.

74. Wiens, D. and Rourke, J. P. (1978). Rodent pollination in southern African *Protea* spp. *Nature (London) 276* (5683), 71-73.
75. Williamson, D. L. (1969). The sex ratio spirochete in *Drosophila robusta*. *Jap. J. Genet. 44* (Suppl. 1), 36-41.
76. Williamson, D. L. and Poulson, D. F. (1979). Sex ratio organisms (spiroplasmas) of *Drosophila*. Pages 175-208. *In* "The Mycoplasmas, Vol. III: Plant and Insect Myco-plasmas" (R. F. Whitcomb and J. G. Tully, eds.), pp. 175-208. Academic Press, New York.
77. Williamson, D. L., Tully, J. G., and Whitcomb, R. F. (1979). Serological relationships of spiroplasmas as shown by combined deformation and metabolism inhibition tests. *Intern. J. Syst. Bacteriol. 29*, 345-351.
78. Williamson, D. L. and Whitcomb, R. F. (1974). Helical, wall-free prokaryotes in *Drosphila,* leafhoppers, and plants. *Colloq. Inst. Nat. Sante Rech. Med., Les Myco-plasmas, Bordeaux, France, 11-17 Sept. 1974.* I.N.S.E.R.M. *33*, 282-290.
79. Williamson, D. L. and Whitcomb, R. F. (1975). Plant myco-plasmas: A cultivable spiroplasma causes corn stunt disease. *Science 188*, 1018-1020.

TREES AS VECTORS OF SPIROPLASMAS AND MYCOPLASMA
AND RICKETTSIA-LIKE ORGANISMS

Robert P. Kahn

Plant Protection and Quarantine Programs,
Animal and Plant Health Inspection Service,
U. S. Department of Agriculture

I.   INTRODUCTION

A. *Plant Disease Control Methods*

Plant disease control methods are placed traditionally in
broad categories, one of which usually pertains to regulatory
aspects. For example, Walker (1969) uses: (1) exclusion,
(2) eradication, (3) protection, and (4) resistence. Agrios
(1978) prefers: (1) regulatory, (2) cultural, (3) biological,
(4) physical, and (5) chemical methods. The "exclusion" of
Walker and the "regulatory" of Agrios are comparable terms.

B. *Objectives*

The objective of the chapter is to review the extent to
which regulatory control such as exclusion or prohibition of
either the host or the agent serves as a control or safeguard
against mycoplasma-like organisms (MLO's) that are associated
with diseases of woody plants. Also included are rickettsia-
like organisms (RLO's) and spiroplasmas since some agents of
these types are also associated with disease in trees. For
example, both MLO's and RLO's have been associated with the
apple proliferation and citrus greening diseases (Hopkins,
1977; Borges, 1975; Hull, 1971). All 3 types of agents are
often associated with diseases known as "yellows" and sometimes are referred to as "virus-like agents" (McCoy, 1979;
Nienhaus and Sikora, 1979; Whitcomb and Davis, 1970, Davis
and Whitcomb, 1971).

To meet the objective, the author surveyed the scientific
literature and the quarantine regulations of 125 countries to
determine to what extent exclusion or prohibition of the disease agent or its tree host is practiced. At the onset of
the survey, the author did not expect that MLO's. which were
first reported in plants in 1967 (Doi *et al.*, 1967), would be
a prominent feature of plant quarantine regulations on an international basis, since many of the regulations or Acts were
promulgated prior to 1967 (USDA, 1933-1980). However, some of
the older regulations do name diseases or agents thought at
the time to be virus diseases or viruses but which are now
known to be associated with RLO's or MLO's. Examples include
the agents of elm phloem necrosis, phony peach, pear decline,
and apple proliferation. In addition, older regulations are
often amended to reflect biological updating (USDA, 1933-1980).

II. LITERATURE REVIEW OF CONTROL METHODS FOR DISEASES ASSO-
     CIATED WITH MYCOPLASMA-LIKE ORGANISMS

     A survey of the scientific literature for MLO's, RLO's and
spiroplasmas showed that control measures fall primarily in
thecultural, chemical, and, to a limited extent, physical categories, but not in the regulatory category.
     Cultural control, according to Whitcomb and Davis (1970),
may be the most effective method since known chemical treatments are neither eradicative nor protective. Cultural methods would include management of wild hosts of vectors and of
perennial plants which serve as reservoirs of the agents
(Tsai, 1979; Whitcomb and Davis, 1970), rotation of crops,
roguing, insertion of barriers to migrating insect vectors,
and other methods as reviewed by Broadbent (1969).
     Chemical methods of control would include the environmen-
tally sound use of insecticides to reduce populations of MLO
vectors and the use of antibiotics against the agents (Davis,
1979; Maramorosch, 1974; Sinha, 1979; Davis and Whitcomb, 1971;
Whitcomb and Davis, 1970.) Reduction of vectors is recommended under pest management programs (Tsai, 1979). Antibiotics
such as tetracycline, chlortetracycline, and oxytetracycline,
have therapeutic value (Sinha, 1979). Such compounds serve,
in effect, to suppress symptoms rather than to eradicate the
agents although the concentration of the agent in the host is
often reduced.
     Physical methods of control include the use of hot water
or hot air treatments to eradicate agents from propagative
material (Nienhaus and Sikora, 1979). The practical application of these control measures for vegetatively propagated

crops is the development of agent-free mother plants to serve as the foundation stock for the production of planting stock to be used by growers (Nyland and Goheen, 1969).

In recent review articles (Davis, 1979; Davis and Whitcomb, 1971; Hopkins, 1977; Nienhaus and Sikora, 1979, Whitcomb and Davis, 1970), concerning diseases incited by MLO's, RLO's, and spiroplasmas, regulatory methods of control are not discussed. Yet, the author's survey (Kahn, 1980) of the regulations of 125 countries showed that regulatory control methods (exclusion or prohibition) were in effect in some countries.

III. REVIEW OF EXCLUSION OR PROHIBITION AS A REGULATORY CONTROL MEASURE

The author reviewed (Kahn, 1980) eleven methods by which rules, regulations and policy, as well as decisions by quarantine officers at the time of entry of plant material, may result in the prohibition or exclusion of hosts or agents. These methods range from outright prohibition by regulation of plants because they are potential hosts of hazardous exotic organisms to denying entry at ports of arrival because these plants have been found to carry or harbor hazardous pests. In addition, some countries allow the entry of plants provided the exporting country certifies that a specified organism is not known to occur in a designated area. If the exporting country cannot so certify, the plants become prohibited *ipso facto*.

Many countries, as a matter of policy or regulation, exclude from prohibition plants which are imported in small quantities for scientific purposes provided the safeguards are adequate (Kahn, 1977, 1980). However, these exceptions are not taken into account in this chapter when prohibitions are tabulated — as long as the crop or genera are prohibited to the general or commercial public.

Two of the eleven methods of exclusion which this chapter deals with are: (1) excluding or prohibiting *the host* because it is a potential vector; and (2) excluding *the pathogen* either as a culture for scientific purposes or when found at ports of entry on imported articles, such as plants or commodities.

Before discussing these two methods of exclusion as applied to MLO's, RLO's and spiroplasmas associated with woody plants, it would be useful to have a yardstick by which to measure the frequency of such regulatory practices. For the purpose of comparison, the regulatory action by 125 countries against all pests and pathogens including insects, mites, fungi, bacteria, viroids, viruses, RLO's, spiroplasmas, and MLO's is presented in this section.

A.   *Exclusion as a Host*

The author reported (Kahn, 1980) the results of a survey
of the quarantine regulations of 125 countries in which he
tabulated the frequency that a crop or genus was listed by
name because the plants and/or seeds were prohibited from one
or more countries. The basis of the prohibition was that the
plants or seeds could harbor or carry pests and/or pathogens
that were considered as hazardous by the importing country.

The survey showed that 246 genera or crops were on the
lists of prohibited articles published by one or more of the
125 countries. A total of 1802 such prohibitions was tabulated. The 40 most frequently prohibited genera or crops were
18 genera or crops listed in Table I plus the following: *Cocos, Fragaria, Prunus, Ribes, Ipomoea, Acer, Crategus, Salix,
Sorbus, Helianthus, Hevea, Nicotiana,* oil palm, rice, *Rosa,
Theobroma,* tea, *Saccharum,* and conifers (including *Picea,
Larix, Pinus,* and *Abies*).

TABLE I. Ranking of Herbaceous and Woody Crops Based on the
Extent to which Importation of Plants and/or Parts is Prohibited in the Quarantine Regulations of 125 Countries

Rank	Percentage basis[a]		Frequency[b]	
	Crop	*Percentage*	*Crop*	*Percentage*
1	*Ulmus*	64	*Citrus*	62
2	*Coffea*	62	*Gossypium*	52
3	*Citrus*	60	*Coffea*	49
4	*Quercus*	54	*Solanum*	48
5	*Populus*	54	*Theobroma*	43
6	*Castanea*	49	*Vitis*	41
7	*Gossypium*	48	*Saccharum*	40
8	*Juglans*	47	*Musa*	39
9	Pome fruit	47	Pome fruit	37
10	Tea	47	*Castanea*	34

[a] $100 \times \dfrac{\textit{number of countries which prohibit importation}}{\textit{number of countries which can grow crop.}}$

[b] *Percentage of countries which prohibit importation.*

B. Exclusion of Pests and Pathogens

The author's survey (Kahn, 1980) also included a tabulation of the number of times pests and pathogens were listed
by name in the quarantine regulations of 125 countries. The
total number of listings was 5303 with duplications because
of synonymy estimated not to exceed 1%. There were 1588 different species of pests and pathogens as follows: Insects,
614 species cited 2475 times; nematodes, 46 species cited
255 times; fungi, 537 species with 1444 listings; and bacteria,
96 species with 514 listings. For viruses, MLO's, RLO's, viroids, and spiroplasma as a group 292 names were listed 645
times.

Some general statistics about pest and pathogen listings
are as follows: The numbers of different species excluded from
importation ranges from 0 to 275 with an average of 35 per
country on a worldwide basis. By geographic regions, the
average per country in a given region was as follows: North
America, including Central America, 12; Europe, 54; Southeast
Pacific, 25; South America, 21; Africa, 46; and Asia, 28.

The 10 most frequently mentioned viruses or virus-like
agents (Kahn, 1980), from the most frequently cited agent to
the least were as follows: (1) plum pox virus; (2) potato
spindle tuber viroid; (3) apple proliferation agent; (4) peach
yellows agent; (5) citrus tristeza virus; (6) rose wilt agent;
(7) strawberry yellow-edge agent; (8) strawberry witchs'-broom
agent; (9) strawberry green petal agent; and (10) potato
witchs'-broom agent. Of these 10 agents, 2 agents of the MLO
category (i.e., those of peach yellows and apple proliferation)
were associated with tree diseases.

C. Designation of Specific Pathogens for Tree Species

The author's survey (Kahn, 1980) showed that pests and
pathogens were named along with prohibited entry of a specified crop or genus in about one-third of the listings for
tree species. Eighteen tree genera or crops were listed among
the 40 most frequently cited genera. These 18 genera and the
percentage of countries which both prohibit a genus plus name
an organism in the same rule or regulation are listed in the
tabulation below:

Plant host	Percentage (%)
Acer	43
Castanea	34
Citrus	62
Cocos	28
Coffea	49
Conifers	27
Crategus	14
Juglans	21
Oil palm	
Pome fruits	37
Populus	27
Prunus	32
Quercus	25
Salix	22
Sorbus	24
Theobroma	43
Ulmus	32

The range for tree genera was 16-62%, while the average was about 32%. Thus, only about one-third of the countries which prohibit tree genera also name the organism upon which the prohibition is based.

IV. MLO'S, RLO's, AND SPIROPLASMAS ASSOCIATED WITH TREE DISEASES

A. Hosts

A survey of the literature showed that there were at least 40 genera of trees reported as hosts for RLO's, MLO's, and spiroplasmas. Nineteen of these were palm genera which were reported (Thomas, 1974, 1979) as hosts of coconut palm lethal yellowing (MLO). Four of these palm genera plus 29 other non-palm species were reported as hosts for 33 MLO's, RLO's, or spiroplasmas (Table II).

B. Agents

Havens (1979) compiled a list of pests and pathogens mentioned in domestic (Federal and State) and foreign regulations

in which 9 out of the 33 agents shown in Table II are listed.
These 9 include 3 listings as viruses (apple proliferation,
apple rubbery wood, and elm phloem necrosis), 1 as a RLO
(peach phony peach), and 7 listed as MLO's (citrus greening,
citrus decline, coconut palm lethal yellowing, elm phloem necrosis, peach western X, apple proliferation, and pear decline).

The author's survey of regulations (Kahn, 1979) showed
that these organisms plus two others (*Ziziphus* spike and *Zizi-*
phus witchs'-broom agents) were mentioned a total of 40 times.
However, some countries prohibit all viruses or "virus-like"
agents of a plant genus. When this broad prohibition is considered as including MLO's, RLO's, and spiroplasmas, the total
listings may be increased to 92.

The 9 agents listed on 92 occasions constitute a relatively small proportion of total listings. Earlier in this chapter, it was pointed out that 292 names of viruses or "virus-
like" agents were listed 645 times. Thus, the 9 agents represent only 3% of the total virus or "virus-like" listing;
the 92 citations represent 14% of the total. In terms of the
listing of 1588 species of all pests and pathogens listed
5303 times, the figures of 9 and 40 represent 0.05 and 1.6%,
respectively.

C. *Prohibition or Exclusion of Tree Hosts of MLO's, RLO's,*
   and Spiroplasmas

An analysis of the quarantine regulations of 125 countries
revealed that of the 40 most frequently prohibited genera or
crops (see Section III, A), regardless of the biological reason for their prohibition, 10 genera are tree genera affected
by diseases with which MLO's, RLO's or spiroplasmas are associated (Table III). It should be noted, however, that except
in the case of *Ulmus*, RLO, MLO, or spiroplasmas usually are
not the agents specified for such prohibitions.

V. CONCEPTS RELATING TO EXCLUSION OF THE HOST AS A REGULATORY
   ACTION

A consensus of 12 textbooks of plant pathology or entomology (author's unpublished data) showed that quarantine actions were considered as justified when (1) man rather than
nature is the prime mover of the agent; (2) the action is
based on biology or pest risk analysis; (3) the action taken
is the least drastic action which provides the necessary safeguard.

TABLE II. List of Mycoplasma-Like Organisms (MLO), Rickettsia-Like Organisms (RLO), and Spiro-plasma Associated with Tree Diseases[a]

Host				
Scientific name	Common name	Disease	Agent	Reference[b]
Aracia catechu (L.F.) Willd.	Catechu	Sandal spike	MLO	7, 17
Areca catechu L.	Betel nut	Areca catechu yellow leaf	MLO	8
Breynia cernua L.	-	Breynia bunch	MLO	2
Carica papaya L.	Papaya	Papaya bunchy top	MLO	1, 17
Carya illinoensis (Wangenh.) K. Koch	Pecan	Pecan bunch	MLO	17
Castanea spp.	Chestnut	Chestnut yellows	MLO	10
Citrus spp.	Citrus	Citrus greening (citrus likubin)	MLO	1, 4, 17
		Citrus stubborn (little leaf)	RLO	3
			Spiroplasma	1, 4, 17
Cocos nucifera L.	Coconut	Coconut palm lethal yellowing[c]	MLO	1, 17
Cornus amomum, C. stolon-ifera	Dogwood	Dogwood witches'-broom	MLO	12
Corylus avellana	Filbert	Hazel line pattern	MLO	13
Dichrostachys cinerea W.&A.	-	Sandal spike	MLO	7
Dodonea viscosa	-	Sandal spike (Dodonea viscosa spike)	MLO	7, 17, 18

288

Host		Disease	Agent	Reference[b]
Scientific name	Common name			
Eucalyptus grandis	Eucalyptus	Sandal spike	MLO	7
Fraxinus americana L.	Ash	Ash witches'-broom	MLO	4, 17
Juglans spp.	Walnut	Walnut bunch	MLO	17
Larix spp.	Larch	Larch witches'-broom	MLO	9
Malus sylvestris Mill	Apple	Apple chat fruit	MLO	4, 17
		Apple proliferation	MLO	1, 4, 17, 18
			RLO	3, 9
		Apple rubbery wood	MLO	17
Morus spp.	Mulberry	Mulberry dwarf	MLO	1, 4, 17, 18
Paulownia tomentosa (Thumb.) Steud	Paulownia	Paulownia witches'-broom	MLO	1, 17, 18
Phoenix canariensis Hort ex Chaub	Date palm	Coconut palm lethal yellowing[c]	MLO	14, 15
Phoenix dactylifera L.	Date palm	Coconut palm lethal yellowing[c]	MLO	14, 15
Phoenix reclinata Jacq.	Date palm	Coconut palm lethal yellowing[c]	MLO	14, 15
Pritchardia pacifica L.	—	Coconut palm lethal yellowing[c]	MLO	14, 15
Prunus spp.	(Stone fruit)	Apricot chlorotic leaf roll	MLO	6
		Peach rosette	MLO	5
		Peach yellows (little peach)	MLO	18

289

TABLE II (continued)

Host				
Scientific name	Common name	Disease	Agent	Reference[b]
Prunus domestica L.		Phony peach	RLO	3
		X disease (Western X)	MLO	1, 4, 18
Prunus dulcis (Mill.) D. A. Webb	Plum	Plum leaf scald	RLO	3
	Almond	Almond leaf scorch	RLO	3
Pyrus communis L.	Pear	Apple rubbery wood	MLO	17
		Pear decline	MLO	1, 4
Robinia pseudoacacia L.	Black locust	Black locust witches'-broom	MLO	1
Santalum album L.	Sandalwood	Sandal spike	MLO	1, 17, 18
Salix rigida Muhl.	Willow	Willow witches'-broom	RLO	1, 17
Ulmus americana L.	Elm	Elm phloem necrosis	MLO	1, 17
Zizyphus jujuba Mill.	Jujube	Jujube witches'-broom	MLO	8
Zizyphus mauritiana Lam.	Jujube	Jujube witches'-broom	MLO	8
Zizyphus oenoplea Lam.	Jujube	Sandal spike	MLO	7

[a] In addition to the palms listed, 19 other members of the Palmaceae are known to be hosts of the agent of coconut palm lethal yellowing.

[b] Key to References: (1) Borges, 1975; (2) Dabek and Jackson, 1977; (3) Hopkins, 1977; (4) Hull, 1971; (5) Kirkpatrick et al., 1975; (6) Morvan, 1977; (7) Nayar and Ananthapadmanabha 1977; (8) Nayar and Seliskar, 1978; (9) Nienhaus and Sikora, 1979; (10) Okuda et al., 1974; (11) Pandey et al., 1976; (12) Ragozzino et al., 1973; (13) Raju et al., 1976; (14) Thomas, 1974; (15) Thomas, 1979; (16) Tsai, 1979; (17) Wilson and Seliskar, 1976; (18) Whitcomb and Davis, 1970.

[c] Coconut palm lethal yellowing is also known as Kaincopa disease, Cape St. Paul wilt, and bribe disease.

TABLE III. Results of a Survey of Plant Quarantine Regulations of 125 Countries Showing, for the Most Frequently Cited Crops or Genera, the Extent to which They are Prohibited

Genera[a]	Estimated number in which genera will grow	Percentages of countries in which genera are prohibited[b]	% of countries in six geographic areas prohibited					
			North America	Europe	Southwest Pacific	South America	Africa	Asia
Castanea	69	49	50	59	29	12	82	25
Citrus	103	60	62	67	60	54	66	52
Cocos	83	34	24	0	60	9	40	37
Juglans	64	47	20	65	0	12	83	0
Larix	64	42	60	28	80	12	75	20
Malus	81	47	40	52	50	40	47	33
Prunus	81	40	40	45	50	20	47	33
Pyrus	81	47	40	52	50	40	47	33
Salix	64	34	40	34	40	25	42	20
Ulmus	50	64	40	90	33	20	75	20

[a] All other hosts (listed in Table II) are prohibited from entry into none to ten countries.

[b] $100 \times \dfrac{\text{number of countries which prohibit importation}}{\text{number of countries which can grow the crop.}}$

*A. Man versus Nature*as the Prime Mover of Agents*

Exclusion or prohibition as a plant quarantine action is most effective when man rather than nature acts as the prime mover in relocating pests and pathogens. Obviously, if pests and pathogens move over long distances unaided by man, there is not much to be gained by regulating man and his activities. However, in some crop/pest interactions, a delaying action through regulation may delay spread and, thus, buy time for research on control.

Nature can be considered as the prime mover of pests and pathogens if they or their vectors (usually arthropods) can be transmitted over long distances either by air currents and wind or by their own means of locomotion. Migratory birds, parasitic plants, and animals, including nematodes, also may serve as vectors of disease agents. Some of these vectors may carry seeds which, in turn, carry disease agents. Disease agents are also disseminated when floating plant materials (e.g., coconuts) harboring pests and pathogens are washed ashore. Drainage of surface water may account for the spread of soil-borne pests and pathogens. Although natural grafting of roots accounts for some plant-to-plant spread, this means dissemination is not a factor in long distance spread.

Man may serve as a vector when he moves articles from one area to another. If these articles are plants or plant parts, man provides the means by which the pest or pathogen is kept in association with its host. This is significant for the spread of obligate parasites such as viruses and rust fungi. If the articles are plants or plant parts, agricultural or nonagricultural cargo, baggage, containers, dunnage, mail, or a carrier serving as a means of transport, man provides a means of transportation by which some pests and pathogens become hitchhikers. Man also transmits pathogens through nursery practices such as grafting if the nursery stock is subsequently moved over long distances.

*
"Nature," as used herein, is a term of convenience that refers to the summation of all the natural factors that influence or relate to the movement of pests or pathogens to new areas and their subsequent establishment including climate, inoculum potential or population density, presence of susceptible hosts, life cycles, etc.

Both nature and man can move pests and pathogens over long
distances in a short time frame, but both can also move these
organisms over long distances by the cumulative effect of a
series of short-distance movements.

The known insect vectors of MLO's, RLO's, and spiroplasmas
are planthoppers (Fulgoroidea), leafhoppers (Cicadellidae),
and psyllids (Psyllidae) (Tsai, 1979). Although aphids and
scale insects have been implicated, their role as vectors
requires confirmation (Tsai, 1979). Tsai (1979) suggests that
leafhoppers may migrate over long distances or are moved pas-
sively by the wind. However, of the seven examples given,
only one, *Circulifer tenellus*, the vector of the citrus stubborn spiroplasma, infests tree hosts. He also points out that,
with the exception of sandal spike and elm phloem necrosis
agents, the biology of these disease agents has been neglected
in contrast to the prokaryote disease agent of cereals which
has been intensely studied.

Assuming that insect vectors of RLO's, MLO's, and spiroplasmas that infect trees are leafhoppers, planthoppers, or
psyllids (whatever their specific identity might be) and that
these vectors are not efficient long-distance movers of the
agents, then nature does not appear on the surface to have a
significant role in long-distance spread of these agents; except perhaps as noted previously where agents may be moved by
vectors over about distances in the form of a relay. The
other means of natural transmission described above including
seed transmission do not seem to be involved in long-distance
spread (McCoy, 1979; Nienhaus and Sikora, 1979).

Man, on the other hand, can be an efficient "vector" in
moving infected plants or plant parts imported for propagation from a region where these agents occur to another region
or continent where they do not occur. Thus, it is not likely
that the agent of apple proliferation could be moved by nature
from Europe where it occurs to North America or Australia
where it is not known to occur. However, man could import
infected scions, cuttings, or plants. These could be established in a nursery where further spread by man through vegetative propagation or by nature if a domestic vector were present could take place.

B. *Pest Risk Analysis*

Pest risk analysis is an important prerequisite to the
promulgation of rules and regulations. The author (Kahn, 1979)
reviewed a concept of pest risk analysis in which the entry
status of plants is "matched" with the pest risk associated
with the importation. Thus, if the importation of a plant

species from an area where a pest or pathogen of quarantine
significance occurs is considered to be a high risk, the entry
status (e.g., rules, regulations, policy) should be corres-
pondingly conservative. Conversely, if the risk is low, the
entry status should be liberal. However, safeguards are often
instituted which can alter the entry status. Plants which may
be prohibited to the general or commercial public because of
risk may often be imported for scientific purposes provided
there are adequate safeguards such as inspection, treatment,
quarantine, and isolation (Kahn, 1977; USDA, 1933-1980).

C. Least Drastic Action

The concept of least drastic action has been reviewed
(McCubbin, 1956). In general, when several quarantine actions
are available, each of which provides an equivalent safeguard,
the one which represents the least drastic action and has the
least economic or social impact should be utilized. For example, we could assume that a pest or pathogen of quarantine
significance occurs in the country of origin on the plant
material to be imported and that an effective eradicant treatment, which can be administered at a port of entry, is known.
Treatment of the imported plant material rather than prohibition would be "a least drastic action."

VI. REGULATORY CONTROL MEASURES

Prohibition or exclusion of the host from countries where
MLO's, RLO's or spiroplasmas of quarantine significance are
known to occur appears to be a meaningful control measure assuming that nature is an inefficient mover of these agents and
their vectors over long distances.
Exclusion assumes an important role as a first line of
defense in view of the lack of effective backup control
measures should the agent enter and become established in a
new area. As pointed out earlier, chemical control measures
employed against the agents may only result in symptom remission rather than eradication of the agent. Against known vectors, integrated pest management may serve only to reduce populations of vectors. Cultural control measures also serve to
reduce populations of vectors or reservoirs of the agent.
Roguing of diseased plants has its limitations once the agent
has become established, especially if it has a long incubation
period or tendency toward latentcy. Furthermore, even though

growers are furnished pathogen-free planting stock, such plants are still susceptible should the pathogen be present.

As pointed out previously, movement of many of the important tree crops which serve as hosts for MLO's, RLO's and spiroplasmas are already regulated by prohibition or exclusion. However, these trees are not so regulated because of the threat of these agents per se but rather because of the potential risk of other pathogens such as bacteria, fungi, or viruses. Thus, regulatory control measures are in force in a number of countries although such regulation is, for the most part, fortuitous in its effect on the spread of MLO's, RLO's and spiroplasmas.

Regulatory control measures often allow for the importation of prohibited plant materials in small quantities for scientific purposes provided the safeguards are adequate. Safeguards at quarantine stations have been reviewed (Kahn, 1977). Hot water or hot air treatments administered to vegetative propagations at the time of entry may eradicate MLO's, RLO's or spiroplasmas. Heat therapy plus meristem tip culture may produce agent-free plantlets. In the absence of eradicant treatments, safeguards may consist of indexing or other agent detection methods such as electron microscopy, serology, and susceptibility to antibiotic therapy.

REFERENCES

Agrios, G. W. (1978). "Plant Pathology," 2nd ed. Academic Press, New York.
Borges, Maria de Lourdes V. (1975). Mycoplasmas, Rickettsias e Doencas das Plantas Coleccao "Natura," Nova Serie, Vol. 2.
Broadbent, L. (1969). Disease control through vector control. In "Viruses, Vectors, and Vegetation" (K. Maramorosch, ed,). pp. 593-630. Wiley (Interscience), New York.
Dabek, A. J., and Jackson, G. V. H. (1977). Association of mycoplasma-like organisms with Breynia bunch, a newly described disease on Malaita, Solomon Islands. *Phytopathol. Z. 90,* 132-138.
Davis, R. E. (1979). Spiroplasmas: newly recognized arthropod-borne pathogens. In "Leafhopper Vectors and Plant Disease Agents" (K. Maramorosch and K. F. Harris, eds.). Academic Press, New York.
Davis, R. E. and Whitcomb, R. F. (1971). Mycoplasmas, rickettsiae, and chlamydiae: possible relation to yellows disease and other disorders in plants and insects. *Annu. Rev. Phytopathol. 9,* 119-154.

Doi, Y., Teranaka, M., Yora, K., and Asuyama, H. (1967). Myco-
    plasma-PLT group-like organisms found in the phloem ele-
    ments of plants infected with mulberry dwarf, potato
    witches'-broom, aster yellows or paulownia witches'-broom.
    Ann. Phytopathol. Soc. Jap. 33, 259-266.
Havens, O. (1979). "Plant Pests of Phytosanitary Significance
    to Importing Countries and States," First revision (loose-
    leaf) May 11, 1979. State of California, Department of
    Food and Agriculture, Sacramento, California.
Hopkins, D. L. (1977). Diseases caused by leafhopper-borne,
    rickettsia-like bacteria. *Annu. Rev. Phytopathol. 17,*
    277-94.
Hull, R. (1971). Mycoplasma-like organisms. *Rev. Plant Pathol.
    50,* 121-130.
Kahn, R. P. (1977). Plant quarantine: principles, methodology,
    and suggested approaches. *In* "Plant Health and Quarantine
    in International Transfer of Genetic Resources" (W. B.
    Hewitt and L. Chiarappa, eds.). CRC Press, Cleveland,
    Ohio.
Kahn, R. P. (1979). A concept of pest risk analysis. *Bull.
    Eur. Medit. Plant Prot. Organ. 9,* 119-130.
Kahn, R. P. (1980). The host as a vector: exclusion as a con-
    trol. *In* "Ecology and Control of Vector-Borne Pathogens."
    Academic Press, New York, in press.
Kirkpatrick, H. C., Lowe, S., and Nyland, G. (1975). Peach
    rosette: the morphology of an associated mycoplasma-like
    organism and the chemotherapy of the disease. *Phytopatho-
    logy 65,* 864-870.
McCoy, R. E. (1979). Mycoplasmas and Yellows Diseases. *In*
    "The Mycoplasmas," Vol. III. Plant and Insect Mycoplasmas.
    (R. F. Whitcomb and J. H. Tully, eds.). Academic Press,
    New York.
McCubbin, W. A. (1956). "The Plant Quarantine Problem." Ejnar
    Munksgaard, Copenhagen.
Maramorosch, K. (1974). Mycoplasma and rickettsia in relation
    to plant diseases. *Annu. Rev. Microbiol. 28,* 301-324.
Morvan, G. (1977). Apricot chlorotic leafroll. *Bull. Eur.
    Medit. Plant Prot. Organ. 7,* 37-55.
Nayar, R. and Ananthapadmanabha, H. (1977). Little leaf disease
    in collateral hosts of sandal *(Santalum album). Eur. J.
    Forest. Res. 7,* 152-164.
Nayar, R. and Seliskar, C. E. (1978). Mycoplasma-like organisms
    associated with yellow leaf disease of *Areca catechu* L.
    Eur. J. Forest. Pathol. 8, 125-128.
Nienhaus, F. and Sikora, R. A. (1979). Mycoplasmas, spiroplas-
    mas, and rickettsia-like organisms as plant pathogens.
    Annu. Rev. Phytopathol. 17, 37-58.

Nyland, G. and Goheen, A. C. (1969). Heat therapy of virus
    diseases of perennial plants. *Annu. Rev. Phytopathol. 7,*
    331-354.
Okuda, S., Doi, Y., and Yora, K. (1974). Mycoplasma-like
    bodies associated with chestnut yellows. *Ann. Phytopathol.
    Soc. Jap. 40,* 464-468.
Pandey, P. K., Singh, A. B., Nimbalkar, M. R., and Marathe,
    T. S. (1976). A witches'-broom disease of jujube from
    India. *Plant Dis. Rep. 60,* 301-303.
Ragozzino, A., Iaccarino, F. M., and Viggiana, G. (1973).
    *Proc. 3rd Conf. Phytiatry Phytopharmcol. Medit. Reg.
    "Maculatura lineare" Hazel 385-386 RAM 54,* p. 2624.
Raju, B. C., Chen, T. A., and Varney, E. H. (1976). Mycoplas-
    ma-like organisms associated with witches'-broom diseases
    of Cornus amomum. *Plant Dis. Rep. 60,* 642-464.
Sinha, R. C. (1979). Chemotherapy of mycoplasmal plant diseases.
    In "The Mycoplasmas", Vol. III. Plant and Insect Mycoplas-
    mas, (R. F. Whitcomb and J. G. Tully, eds.). Academic
    Press, New York.
Thomas, D. L. (1974). Possible link between declining palm
    species and lethal yellowing in coconut palms. *Proc.
    Fla. St. Hort. Soc. 87,* 502.
Thomas, D. L. (1979). Mycoplasma-like bodies associated with
    lethal decline in Florida. *Phytopathology 69,* 928-934.
Tsai, J. H. (1979). Vector transmission of mycoplasmal agents
    of plant diseases. *In* "The Mycoplasmas," Vol. III. Plant
    and Insect Mycoplasmas (R. F. Whitcomb and J. G. Tully,
    eds.). Academic Press, New York.
U. S. Department of Agriculture (USDA) (1933-1980) "Export
    Certification Manual," Vols. 1 and 2.
Walker, J. C. (1969). "Plant Pathology," 3rd ed. McGraw-Hill,
    New York.
Whitcomb, R. F. and Davis, R. E. (1970). Mycoplasma and phyta-
    arboviruses as plant pathogens persistently transmitted
    by insects. *Annu. Rev. Entomol. 15,* 405-464.
Wilson, C. L. and Seliskar, C. E. (1976). Mycoplasma asso-
    ciated diseases of tress. *J. Arboicult. 2,* 6-12.

STRAINS AND STRAIN INTERACTION OF THE CITRUS
GREENING PATHOGEN IN INDIA*

R. *Naidu and H. C. Govindu*

Citrus Experiment Station,
Gonicoppal, Kodagu District,
Karnataka State, India

and

College of Agriculture
University of Agricultural Sciences
Bangalore, India

I. INTRODUCTION

Citrus greening disease has been recognized as a major
threat to citrus culture, particularly, in South Africa and
Asia. The disease has been known since 1929 in South Africa
as "Geeltak." The first evidence on its transmission by grafting was recorded by Oberholzer and Hofmeyr (1955). "Blotchy
mottle" or "Greening" disease in South Africa, "Yellow shoot"
in China, "likubin" disease in Taiwan, "leaf mottling" or
"leaf mottle yellows" in the Philippines, "dieback" in Australia, citrus "greening" in India, Pakistan, and Thailand are
known to be caused by single pathogen or closely related
strains (Oberholzer *et al.*, 1965; Fraser *et al.*, 1966; Knorr
et al., 1970; Chen *et al.*, 1973; Broadbent and Fraser, 1976).
In India, greening disease is considered to be the principal
component of citrus decline and has been shown to be transmitted by the citrus psyllid *Diaphorina citri* Kuway (Fraser

*This paper formed the part of the Doctoral thesis sub-
mitted by the Senior author to the University of Agricultural
Sciences, Bangalore, India.*

and Daljit Singh, 1968; Capoor *et al.*, 1967; Nariani *et al.*, 1967; Nariani *et al.*, 1967). As there was no information on the existence of strains of the greening pathogen and their interaction, investigations were conducted and the results are discussed briefly below.

II. MATERIALS AND METHODS

Budwood and bark samples were collected from mandarin and sweet orange trees suspected of having greening disease in Andhra Pradesh, Karnataka, and Tamil Nadu States. All the isolates were established by means of bark or leaf patch grafts on sweet orange CV Sathgudi seedlings in the glass house. Disease-free colonies of psylla were maintained on curryleaf plants, *Murraya koenigii* L. (Chakravarthy *et al.*, 1976). The method used in the insect transmission study was that of Capoor *et al.* (1967). In the absence of insect transmission all the isolates were indexed on various indicator plants for the presence of tristeza, psorosis, vein enation and woody gall, crinkly leaf mosaic, yellow corky vein, and citrus exocortis viruses. A reliable and rapid indexing method developed by Schwarz (1968) was also used for diagnosing greening infection. Symptomatology of different isolates was studied on Sathgudi seedlings by inoculating through patch grafting. Various growth parameters such as plant height, stem girth, leaf size, root and shoot weights, and the degree of leaf chlorosis were used in grouping of the isolates. Seedlings of different citrus varieties/clones were tested against three groups to develop a set of differential hosts to identify the strains of the greening pathogen.

III. EXPERIMENTAL RESULTS

A. *Survey, Collection, Indexing, and Establishment of Isolates*

In all, sixty-eight budwood samples showing symptoms of greening disease were collected from three States. Of these, 48 isolates were established on Sathgudi seedlings. Indexing results clearly showed the widespread occurrence of the greening disease in many citrus orchards in South India. Mandarin and sweet orange varieties showed mild to severe symptoms in the field. Most of the isolates, when inoculated to Sathgudi seedlings in the glasshouse showed the same symptoms that are observed in the field. Tristeza, psorosis, and citrus mosaic

viruses were also found to be present, either alone or associated with the greening pathogen and such samples were eliminated from further studies. A majority of the isolates showed
violet fluorescent spots with an R_f value of 0.8-0.9 when the
chromatograms were examined under an ultraviolet lamp. Thus
the chromatographic technique employed in the present study
provides additional evidence for identification of greening
pathogen.

Out of 48 greening pathogen isolates, 11 showed mild, 16
severe, 8 very severe, and remaining 13 showed mild to moderate symptoms. It was, therefore, evident that the greening
pathogen exhibited different degrees of virulence on the common host, namely, Sathgudi.

1. Transmission

Mild and severe isolates were successfully transmitted
from diseased to healthy plants through leaf and bark-patch
grafts. However, repeated attempts to transmit 35 isolates
through citrus psylla gave negative results.

2. Symptomatology

Symptoms were usually noticeable 60-90 days after inoculations under glass house conditions. Mild chlorosis on
the young leaves was the first symptom. As the leaves expanded and matured, they showed a range of chlorotic patterns
varying from interveinal chlorosis to complete chlorosis with
green islands (Fig. 1). The leaf symptoms were similar to
those of zinc deficiency. The leaves were reduced in size,
distorted, leathery, and thickened. Infected seedlings showed
shorter internodes with upright growth. In very severe isolates, corky eruptions on veins and midrib were noticed (Fig.
2). Such leaves were found to be brittle and curled downward.
Abnormal bud sprouting and profuse branching were noticed in
addition to stunting. In general, mild isolates produced
slight interveinal chlorosis without affecting the growth,
severe isolates caused severe chlorosis with complete cessation of the growth, and trees infected with severe isolates in
the field exhibited poor growth with only a few lopsided and
uneven colored fruits. Seeds in such fruits were discolored
and aborted.

B. Grouping of Isolates

The 35 isolates that showed greening symptoms were inoculated onto 60-day-old sweet orange CV Sathgudi seedlings by

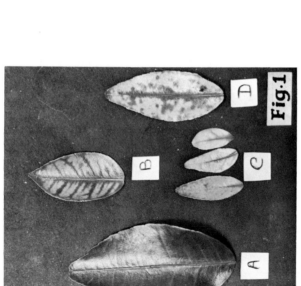

Fig. 2. Corky growth on the veins and leaf curling symptoms on Sathgudi seedlings infected by corky vein isolate of greening pathogen.

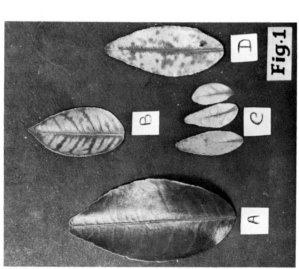

Fig. 1. Leaf symptoms of greening pathogen on sweet orange CV Sathgudi. A, Healthy; B-C, infected leaves showing different degrees of chlorosis and deformation; D, leaf showing irregular distribution of green islands.

bark-patch grafting. Each isolate was inoculated on five uniform seedlings and the data on various growth characters were
recorded after 18 months. The data were analyzed statistically. In order to group all the isolates into a few categories,
a grand mean and a standard deviation were calculated for each
character. The grand mean was considered as a middle point.
Grand mean ($\bar{x}$) + standard deviation ($\bar{x}$SD) form one group and similarly $\bar{x}$ -SD form a second group. In between these two groups
a third group was formed. The range of height, stem girth,
leaf chlorosis and size, root and shoot weights of each group,
and number of isolates in each group are presented in Table I.

1. Height and Girth

Great variations were produced in Sathgudi seedlings
inoculated by different isolates and these variations were
found to be statistically significant. The mean height of
mild isolate inoculated plants was 63.9 as against 72.2 cm in
uninoculated healthy plants. Plants inoculated with severe
isolates recorded mean height of 18.0 cm. Similarly, the mean
girth was 37 mm with mild isolates as against 37 mm in healthy,
while severe isolates showed 18 mm girth only.

2. Leaf Size and Chlorosis

Severe isolates caused greater reduction in leaf size
than found in mild isolates or in healthy plants. Mild isolates produced mild interveinal chlorosis while severe isolates caused severe yellowing of the leaves.

3. Root and Shoot Weight

Seedlings inoculated with mild isolate showed 17 and
25 gm mean root and shoot weights, respectively, as against
5.6 and 9.7 gm in severe isolate inoculated plants. Three
distinct groups, namely, mild (G-M), severe (G-S), and very
severe (G-C) were classified based on the symptoms and effect
on the growth. They are as follows:
  Group 1-G-M: Mild interveinal chlorosis; no significant
    effect on reduction of growth
  Group 2-G-S: Moderate to severe stunting; leaves chlorotic;
    leathery, thickened, reduced in size, abnormal branching
    with bushy growth

TABLE I. Grouping of Greening Pathogen Isolates On Sweet Orange CV Sathgudi Seedlings Based On Different Characters

Character	Group	Group range	No. of isolates	Isolate numbers
Height (cm)	I	More than 53.95	7	GH, G2, G6, G9, G14, G21, G24
	II	53.94-22.96	23	G1, G3, G4, G5, G7, G11, G12, G15, G16, G17, G18, G19, G20, G22, G23, G25, G26, G28, G29, G31, G33, G34
	III	Less than 22.95	6	G8, G10, G13, G27, G32, G35
Stem girth (mm)	I	More than 33.37	8	GH, G2, G6, G9, G14, G19, G21, G24
	II	33.56-21.72	20	G1, G3, G4, G5, G10, G11, G12, G15, G16, G17, G20, G22, G23, G25, G26, G28, G29, G30, G31, G33
	III	Less than 21.71	8	G7, G8, G13, G18, G27, G32, G34, G35
Root weight (gm)	I	More than 13.58	6	GH, G9, G14, G17, G21, G24
	II	13.57-26.52	26	G1, G2, G4, G5, G6, G7, G8, G11, G12, G13, G15, G16, G18, G19, G20, G22, G23, G25, G26, G27, G28, G29, G30, G31, G32, G33, G34
	III	Less than 7.51	4	G3, G8, G10, G35
Shoot weight (gm)	I	More than 20.64	6	GH, G2, G6, G9, G14, G24
	II	20.63-10.96	24	G1, G3, G5, G7, G11, G12, G13, G15, G16, G17, G18, G19, G20, G21, G23, G25, G26, G27, G28, G29, G30, G31, G33, G34
	III	Less than 10.95	6	G4, G8, G10, G22, G32, G35

(Continued Table I)

Character	Group	Group range	No. of isolates	Isolate numbers
Leaf chlorosis rating (Nos.)	I	Less than 1.87	6	GH, G2, G6, G9, G14, G21
	II	1.88–3.08	26	G1, G4, G5, G7, G8, G11, G12, G13, G15, G16, G17, G18, G19, G20, G22, G23, G24, G25, G26, G28, G29, G30, G31, G32, G33, G34
	III	More than 3.09	4	G3, G10, G27, G35
Leaf size (cm²)	I	More than 21.51	9	GH, G2, G6, G9, G14, G17, G19, G21, G28
	II	21.50–16.60	5	G1, G12, G15, G20, G30
	III	Less than 16.15	22	G3, G4, G5, G7, G8, G10, G11, G13, G16, G18, G22, G23, G24, G25, G26, G27, G29, G31, G32, G33, G34, G35

^{a}GH, healthy uninoculated.

Group 3-G-C: Severe stunting; cessation of growth, abnormal
   bud proliferation; corky growth on midrib and veins;
   leaves leathery and curled downward

4. *Correlations between Different Characters*

        Correlation between different characters used in the
grouping of the isolates was studied. Height of the plant with
stem girth, shoot and root weights, and leaf size was found to
be positively correlated showing concurrent increase or decrease. A negative correlation was found with leaf chlorosis
and other characters used.

C. *Differential Behavior of Three Groups of the Greening Pathogen on Various Citrus Species and Varieties*

The reaction pattern of 32 citrus varieties/clones was
studied against three different isolates of greening pathogen
to identify the strains. Based on the severity of symptoms
and effect on the growth, five infection types ranging from
immune to highly susceptible were used in this study.
The screening results clearly showed that the same variety
showed various degrees of severity of disease with different
isolates. This suggests that several strains may be present
in nature and a study on host-parasite reactions will be useful for differentiating the strains. The clones showing similar reaction pattern were grouped together. In all, 13 infection type groups were observed (Table II). Clones in Group
I showed highly resistant reaction to three isolates, while
clones in Group XIII were highly susceptible to all the isolates. Intermediate reaction patterns were observed in the
remaining groups. A representative clone from each group was
chosen and a set of 7 differential hosts for identification of
three strains in greening pathogen was proposed. The clones
suggested as a set of differential host and their reaction to
each strain are presented in Table III.
Troyer citrange and Baduvapuli showed highly resistant and
moderately resistant reactions to all the strains, respectively, Sathgudi, Rangpur lime, Narangi, and Meyer lemon showed
different reactions to different strains. Thus these seven
differential hosts could be used for testing new isolates.
The identification is complete if the infection type agrees
with those indicated for a particular strain. If not, the
strain is considered as new.

TABLE II. Grouping of Different Citrus Varieties Based on their Reaction to Isolates of Greening Pathogen

Group	Isolates and infection types[a]			Clones
	GM	GS	GC	
I	1	1	1	Sweet lime, Trifoliate, Troyer citrange
II	1	1	2	Jullunder Khatti, Lime Karna, Citrumon
III	1	1	4	Jambheri-Kodur, Citrus valkameriana
IV	1	2	2	Rough lemon, Assam lemon
V	1	3	3	Sour orange, Naichakotha, Hazara, citron, Tangerine-dancy, Adajamir, Limocana Brazil
VI	1	3	4	Nakoor lemon, Acid lime, Jatti Khatti, Coorg lemon
VII	1	3	5	Belladakithuli, Rangpur lime
VIII	1	4	4	Cleopatra mandarin, Kinnow mandarin, Hill lemon
IX	1	4	5	Sathgudi, Mosambi, Coorg orange
X	2	2	2	Baduvapuli
XI	4	1	4	Narangi
XII	4	2	4	Desert orange
XIII	5	4	5	Meyer lemon

[a] 1, Highly resistant; 2, moderately resistant; 3, moderately susceptible; 4, susceptible; 5, highly susceptible.

TABLE III. Reaction Pattern of Proposed Set of Differential Hosts for Identifying Greening Strains

	Varieties proposed and their infection types[b]						
Strains[a]	Troyer citrange	Baduvapuli	Sour orange	Rangpur lime	Narangi	Mayer lemon	Sathgudi
G-M	1	2	1	1	4	5	1
G-S	1	2	3	3	1	4	4
G-C	1	2	3	5	4	5	5

[a] G-M, Greening mild strain; G-S, greening severe strain; G-C, greening corky vein strain.

[b] Based on 1-5 scale where 1, highly resistant; 2, moderately resistant; 3, moderately susceptible; 4, susceptible; 5, highly susceptible.

*D. Cross-Protection Test with Mild Strain Against Severe
Strain*

A cross-protection test was conducted in the greenhouse
with mild strain (G-M) against severe strains (G-S). Sathgudi
seedlings, 6 months old, were first inoculated with mild
strains and these seedling were challenge-inoculated 12 weeks
later with severe strain. The extent of protection offered by
the mild strain was measured by using various growth parameters.

The results obtained in the pot culture experiment on
cross-protection suggested that the greening mild strain had
offered only partial protection against severe strain.

IV. DISCUSSION

Greening disease was observed to be widespread in most of
the mandarin and sweet orange orchards in South India. The
general symptoms of greening disease includes yellowing of
leaves similar to that of zinc deficiency. Few isolates
showed very severe symptoms consisting of cessation of grwoth,
corky vein, and leaves curling downward. McClean and Schwarz
(1970) also observed occasional corky vein symptoms in green-
house-inoculated plants. These observations suggest that the
greening disease exists in nature in the form of numerous
strains. Severe strains, besides causing leaf symptoms also
cause dieback in inoculated seedlings (Fraser and Daljit Singh,
1968; Nariani *et al.,* 1967).

The citrus psylla *(Diaphorina citri)* was reported earlier
as a vector of citrus greening (Capoor *et al.,* 1967; Martinez
and Wallace, 1967; Chen *et al.,* 1973). In the present investigation, it was not possible to transmit any isolates of
greening disease by this vector. The reason for failure of
citrus psylla to transmit the disease may be due to the low
temperature which prevailed throughout the year around Bangalore. In India, earlier transmission studies were conducted
at Punjab, Delhi, and Poona where the temperature is very
high. However, in the absence of vector transmission, all the
isolates were indexed on the indicator plants for the presence
of other citrus viruses reported in India. The presence of
fluorescent marker substance and response to tetracyline treatment confirms the association of greening pathogen in all
these isolates.

Experimental evidence on the existence of strains with
rapid and slow systemic spread and latent strains were reported from South Africa (Schwarz, 1972). The results of the

present studies showed the existence of strains in greening
pathogen, some of which caused severe yellowing and stunted
growth while others induced mild interveinal chlorosis with
little or no effect on growth. All the 35 isolates were
grouped on the basis of severity of symptoms and their effect
on growth of Sathgudi seedlings. In all, three groups, namely
mild (G-M), severe (G-S), and very severe with corky vein
(G-C) were proposed. Statistical interpretation also confirmed
the possible existence of 3 strains of greening pathogen.
This is the first attempt to classify the greening pathogen
strains, beyond the earlier concept of identification as severe and mild, based on symptoms only.

Host range and differential host reaction are normally
used to differentiate and also identify the strains in the
plant viruses (Pierce, 1935; Zitter, 1972; Anjaneyulu and
John, 1972). Balaraman and Ramakrishnan (1977) developed an
analytical key consisting of nine differential citrus clones
for identification of strains in tristeza virus. Similarly,
in the present studies attempts were made to explore the possibilities of developing a set of differential hosts for identification of strains in greening pathogen. McClean and
Schwarz (1970) observed variation in symptom pattern of various
citrus varieties to greening pathogen in South Africa. Studies were undertaken with 32 varieties/species of citrus a-
gainst three greening pathogen isolates for identification of
strains. It was interesting to note, that the so-called greening mild strain (G-M) of Sathgudi [*Citrus sinensis* (L.) Osbeck],
induced severe symptoms on Narangi (*C. madurensis* Lour.) and
on desert orange (*C. paradisi* Macf.). However these two
hosts showed highly resistant reaction to the severe strain.
This shows that the mildness of the pathogen is a host-parasite
interaction, but is not due to either host or pathogen alone.
Based on differential reaction of various citrus clones, 13
different infection groups were noticed. Further, studies on
type of infection suggest that seven differential hosts namely
Sathgudi *(Citrus sinensis)*, Meyer lemon [*C. limon* (L.) Burnof.],
Narangi *(C. madurensis)*, Rangpur lime *(C. limonia* Osbeck)*, sour
orange *(C. aurantium* L.)*, Baduvapuli *(C. pennivesiculata* Tan.)*,
and Troyer citrange *(Poncirus trifoliata x C. sinensis)* were
identified to differentiate three strains in greening pathogen.

A pot culture experiment on cross-protection was undertaken with a view to exploit the possibility of utilizing the
greening mild strain (G-M) for preimmunization against severe
strain. Growth parameters recorded clearly indicate that the
mild strain (G-M) failed to offer a high degree or complete
protection against severe strain when challenge-inoculated
through grafting.

Similar results were also noticed in citrus stubborn disease in which mild strain failed to offer complete protection when challenge-inoculated with severe strain (Calavan, 1969). The partial protection observed in the present investigation may be due to incomplete invasion of the mild strain in the host.

V.   SUMMARY

Greening disease is widespread in mandarin and sweet orange orchards in South India. Different isolates of the greening pathogen exhibited varying degrees of virulence when grafted on the common host, sweet orange, C. V. Sathgudi under identical greenhouse conditions. Based on the use of growth parameters, thrity five greening isolates were grouped as mild, severe, and very severe (corky vein). There was high correlation between different growth characters. Statistical analysis suggested the possible existence of three strains of the greening pathogen.

Based on the differential reaction of citrus species and varieties a set of 7 differential hosts, namely, Sathgudi (*Citrus sinensis*) Meyer lemon (*C. limon*), Narangi (*C. madurensis*), Rangpur lime (*C. limonia*), Baduvapuli (*C. pennivesiculata*), sour orange (*C. aurantium*), and Troyer citrange were suggested for identification of the three strains of the greening pathogen.

Possibilities of utilizing the mild strain of the greening pathogen for preimmunization against the severe strain were studied in a pot culture experiment. Growth parameters suggested that the mild strain did not offer complete or even protection against the severe strain when challenge inoculated through grafting.

ACKNOWLEDGMENT

The authors are grateful to the University of Agricultural Sciences, Hebbal, Bangalore for providing facilities for conducting the research.

REFERENCES

Anjaneyulu, A. and John V. T. (1972). Strains of rice tungro virus. *Phytopathology 62,* 1116-1119.
Balaraman, K. and Ramakrishnan, K. (1977). Studies on strains

and strain interaction in citrus tristeza virus. *U.A.S.*
Tech. Series No. 19, 62.

Broadbent, P. and Fraser, L. R. (1976). Citrus die-back.
Agr. Gaz. N.S.W. 87, 10-17.

Calavan, E. C. (1969). Investigations of stubborn disease in
California, indexing, effect on growth and production and
evidence for virus strains. *Proc. Int. Citrus Symp. Univ.*
Calif., pp. 1403-12.

Capoor, S. P., Rao, D. G., and Viswanath, S. M. (1967). *Dia-*
phorina citri Kuway, a vector of the greening disease of
citrus in India. *Indian J. Agr. Sci. 37,* 572-576.

Chakravarthy, N. K., Pondey, P. K., Chatterjee, S. N., and
Singh, A. B. (1976). Host preference in *Diaphorina citri*
Kuway. Vector of greening disease in citrus in India.
Indian J. Entomol. 38, 196-197.

Chen, M. H., Miyakawa, T., and Matsui,C. (1973).Citrus likubin
pathogen in salivary glands of *Diaphorina citri*. *Phyto-*
pathology 63, 1944-1954.

Fraser, L. R. and Daljit Singh (1968). Greening virus, a new
threat to citrus industry. *Indian J. Hort. 10,* 21-22.

Fraser, L. R., Daljit Singh, Capoor, S. P., and Nariani, T. K.
(1966). Greening virus, the likely cause of citrus dieback in India. *FAO Plant Prot. Bull. 14,* 127-130.

Knorr, L. C., Gupta, O. P., and Ahmed, S. (1970). Occurrence
of greening virus disease in West Pakistan. *Plant Dis.*
Rep. 54, 630-631.

McClean, A. P. D. and Schwarz, R. E. (1970). Greening or blotchy mottle disease of citrus. *Phytophylactica 2,* 177-194.

Martinez, A. L. and Wallace, J. M. (1967). Citrus leaf mottle
yellows disease in the Philippines and transmission of the
causal virus by a psyllid, *Diaphorina citri*. *Plant Dis.*
Rep. 51, 692-695.

Nariani, T. K., Raychaudhuri, S. P., and Bhalla, R. B. (1967).
Greening virus of citrus in India. *Indian Phytopathol.*
20, 146-150.

Nariani, T. K. and Raychaudhuri, S. P. (1968). Occurrence of
tristeza and greening viruses in Bihar, West Bengal and
Sikkim. *Indian Phytopathol. 21,* 343-344.

Oberholzer, P.C.J. and Hofmeyr, J. D. J. (1955). The nature
and control of clonal senility in commercial varieties of
citrus in South Africa. *Univ. Pretoria Publ. Ser. 1,*
p. 46.

Oberholzer, P. C. J., Von Staden, D. F. A., and Basson, W. J.
(1965). Greening disease of sweet orange in South Africa,
Proc. 3rd Conf. Int. Organ. Citrus Virol. Univ. Fla.,
pp. 213-219.

Pierce, W. H. (1935). Identification of certain viruses affecting leguminous plants. *J. Agr. Res. 51,* 1017-1039.

Schwarz, R. E. (1968). Thin-layer chromatographical studies on phenolic marker substances of the greening virus in various citrus species. *S. Afr. J. Agr. Sci. 11,* 797-801.
Schwarz, R. E. (1970). Seasonal graft transmissibility and identification of gentisoyl glucoside marker of citrus greening in the bark of infected trees. *Phytophylactica 2*(2), 115.
Schwarz, R. E. (1972). Strains of greening pathogen. *Proc. 5th Conf. Int. Organ. Citrus Virol. Univ. Fl.* pp. 40-44.
Zitter, T. A. (1972). Naturally occurring pepper virus in South Florida. *Plant Dis. Rep. 56,* 586-590.

CHEMOTHERAPY OF PLANT MYCOPLASMA DISEASES

S. P. Raychaudhuri and Narayan Rishi

International Union of Forestry Research Organization
W. P. Plant Mycoplasma Diseases, India
and
Department of Plant Pathology
Haryana Agricultural University, India

I. INTRODUCTION

The literature is replete with the effect of chemicals
upon diseases caused by mycoplasma and rickettsia in plants.
These chemicals range from antibiotics, sulfa drugs, fungicides, and insecticides. In most of the reports these chemicals had static effect. In a few reports, very encouraging
results have been obtained of the total cure of some mycoplasma diseases.

II. ANTIBIOTICS

In a majority of cases, antibiotics of only the tetracycline group (Fig. 1) have been used against mycoplasma diseases. Tetracyclines are of value in the diagnosis of mycoplasma diseases because they are mycoplasma-static not -cidal.
The role of mycoplasma in inciting plant diseases was shown by
Prof. Doi and co-workers in 1967. Ishiie et al. (1967) was
the first to use chlortetracycline hydrochloride, tetracycline
hydrochloride, and kanamycin on mulberry dwarf disease. Only
antibiotics of the tetracycline group could suppress the symptoms of mulberry dwarf, but not kanamycin. After this, ela-
borate work on the chemotherapeutic effect of tetracyclines on
yellows diseases was carried out in Japan by a group of scientists. An account of this work was given by Asuyama and Iida
(1973). This group used mulberry dwarf, rice yellow dwarf, potato witches'-broom, aster yellows and *Cryptotaenia* witches'-

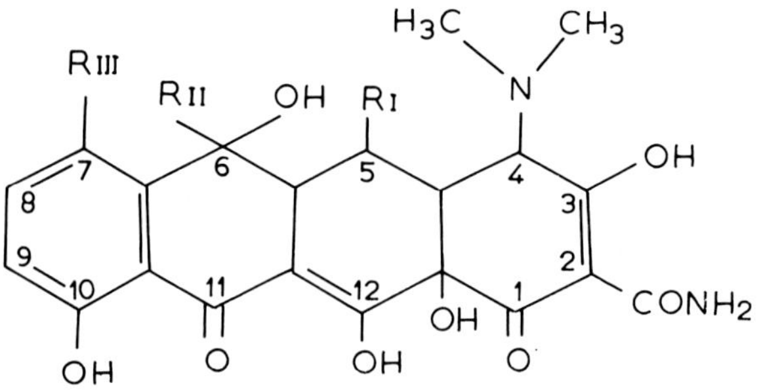

R I	R II	R III	
H	CH₃	H	TETRACYCLINE
OH	CH₃	H	OXYTETRACYCLINE
H	CH₃	Cl	CHLORTETRACYCLINE
H	H	Cl	DIMETHYLCHLORTETRACYCLINE

Figure 1

broom diseases. For comparison rice dwarf, rice stripe, and yellow dwarf diseases of broad bean and pea, which are caused by viruses, were also tested. Tetracycline, chlortetracycline, dimethylchlortetracycline, and oxytetracycline, along with their salts and chelates, were used. The following methods of application of tetracycline were attempted: (1) application shortly before and after inoculation (pre- and postinoculation application); (3) application in the form of lanolin paste on created wounds; (4) foliar application in the field; (5) application on leafhopper vector (pre- and postacquisition application, and application by injection); (6) root application.

There was no effect of these treatments on virus diseases. The following conclusions were drawn by Asuyama and Iida (1973) on the use of antibiotics, which was in general agreement with the studies performed in other countries: (1) Tetracyclines caused temporary remission but not permanent cure of plant diseases suspected of being caused by mycoplasma-like agents. When applied to plants immediately before or after inoculation with the disease agents, however, infection appeared to be prevented to some extent; (2) antibiotics other than tetracyclines, including several macrolides, were found to be ineffective; (3) in plants showing full symptoms, particularly older plants, the effect of tetracyclines was much slighter; (4) tetracyclines were absorbed more readily from roots than from foliage. Repeated root immersion of hydroponically cultured plants was most effective; (5) symptoms once suppressed reappeared occasionally even when repeated tetracycline applications were being continued. Whether this was due to an induced tolerance on the part of the disease agent is not clear; (6) the effectiveness of various tetracyclines differed only slightly and according to different diseases, but dimethylchlortetracycline and, particularly, its hexamethaphosphate, appeared to be slightly superior to others, possibly because of its increased stability; (7) effects of tetracyclines on the infectivity of insect vectors were not as distinct as in plants."

Effect of Tetracyclines on Mycoplasma Diseases of Fruit Trees and Woody Plants

Tetracyclines caused remission of peach X disease in peach and cherry (Nyland and Sachs, 1974; Sands and Walton, 1975; Rosenberger and Jones, 1977). In the case of eastern X disease of peach, injections of tetracycline-HCl and oxytetracycline-HCl were administered in small, but concentrated, doses in diseased and apparently healthy branches. It prevented the appearance of symptoms in apparently healthy branches. The injections were administered in autuum because the fruits harvested after 10 months would not have traces of antibiotics remaining and the vector *Scaphytopius acutus* would not be present in the field from November through May, thereby considerably reducing the chances of reinfection. Up to 95% protection was obtained in August in the branches treated the previous October (Sands and Walton, 1975). Rosenberger and Jones (1977) also used oxytetracycline-HCl to control X disease in peach. Injections of 1.25, 2.5, and 3.5 gm of the antibiotic per tree in September induced remission of symptoms for one year. Spring, summer, or fall injections of 0.5 or 0.9 gm were less effective. In the month of October and November, injections of 1.25 and 2.5 gm of the antibiotic were phytotoxic. Tetracycline also suppressed spike disease of sandal (Raychaudhuri, 1977). Of the four tetracyclines tried, oxytetracycline (Terramycin), chlortetracycline (Aureomycin) and dimethylchlortetracycline (Ledermycin) were more effective than tetracycline (Achromycin).

McCoy (1972, 1973) obtained remission of symptoms and reduction in the severity of symptoms by injecting tetracycline-
HCl and oxytetracycline-HCl directly into the vascular system
of coconut trunk infected with lethal yellowing. In some of
the treated trees fruits were also set.

Nyland and Moller (1973) used oxytetracycline-HCl to control pear decline. The antibiotic was administered into the
trunks of the trees two or three times in the fall of the
year. This restored severely diseased trees to normal or near
normal condition.

Tucker *et al.* (1974) reported symptom suppression of tree
decline-affected citrus trees by drenching tetracycline at
300 ppm for over 20 months on five trees. The symptoms on
four trees were suppressed. The condition of three trees had
noticeably improved in that dieback ceased and new growth was
more vigorous and normal. Nariani *et al.* (1973) found suppression of greening disease of citrus by spraying 500 ppm of
tetracycline, Achromycin and Ledermycin. Injection of BP was
more effective in suppressing the disease.

A number of antibiotics other than tetracycline have also
been tested against plant and insect mycoplasmas including
chloramphenicol, oleandomycin, kanamycin, tylosin, carbomycin,
polymyxin, bacitracin, neomycin, sulfanilamide, penicillin,
vancomycin, cyclocirin, erythromycin, and streptomycin. Of
these, only chloramphenicol was found to suppress symptoms.
Tetracycline and tylosin together could render the insect
vector of aster yellows noninfective or reduce its infectivity.

Windsor and Black (1973) compared the effect of either
penicillin or tetracycline-HCl on clover club leaf disease.
Remission of symptoms occurred first in plants treated with
tetracycline-HCl. Recovery in plants treated with penicillin
was slower, but eventually these plants appeared greener and
taller than the ones treated with tetracycline-HCl. At equal
concetrations, tetracycline-HCl proved to be phytotoxic, while
penicillin was safe.

Permanent remission of citrus greening and grassy shoot of
sugarcane could be obtained by using BP-101, a chemical produced by Hindustan Antibiotics Research Centre, Poona, India
(Thirumalachar, 1973). The nonphytotoxic chemotherapeutic
agents developed by Thirumalachar (1978) brought about remission of diseases caused by mycoplasma and rickettsia, such as,
elm phloem necrosis, sandal spike, X disease of peach, citrus
greening. The activity was more pronounced and "cidal" than
tetracyclines, which only bring about temporary remission and
have to be used periodically. Nystatin was found effective in
remission of the Dutch elm disease in the field (Costouis *et
al.*, 1978).

Bowyer and Calavan (1974) studied the antibiotic sensitivity *in vitro* of mycoplasma associated with citrus stubborn disease. They used 19 antibiotics, 2 systemic insecticides, and 3 systemic fungicides to determine minimum inhibitory concentration (MIC) and minimum biocidal concentration (MBC) for mycoplasma by the dilution broth method. The macrolide antibiotics, erythromycin and tylosin were most active, each having MIC and MBC values of 0.2 and 0.8 μg/ml, respectively. Six members of the tetracycline group and carbomycin were inhibitory at low concentrations 0.01-0.4 μg/ml but MBC levels were higher than for erythromycin or tylosin. MIC values for oleandomycin, lincomycin, and filipin were much higher. Chlorotetracycline, chloramphenicol, and kanamycin were relatively ineffective. Streptomycin and antibiotic BP had the highest level of MIC and MBC values tested. Of the insecticides and fungicides, only thiobendazole in 5% (v/v) and dimethyl sulfoxide had an MBC value (10 μg/ml) similar to that of the effective antibiotics. It was concluded that erythromycin, tylosin, and several antibiotics of the tetracycline group are potentially suitable for the control of citrus stubborn disease.

III. CHEMICALS OTHER THAN ANTIBIOTICS

A number of sulfa drugs namely sulfadiazine, sulfisoxazole, sulfisomidine, and sulfanilamide were tried against plant and insect mycoplasmas. Recovery of aster yellows agent has been found with sulfadiazine, sulfisoxazole, and sulfisomidine (Davis and Whitcomb, 1970; Whitcomb and Davis, 1970). The recovery was better by the combined application of sulfadiazine and tetracycline-HCl than by either alone. The efficiency of the vectors of the aster yellows increased or remain unchanged when sulfa drugs were applied to infected plants. The efficiency of the vectors was considerably lower when fed on healthy plants treated with sulfadiazine and/or tetracycline-HCl. Suppression of aster yellows by sulfa drugs has also been reported by Klein *et al.* (1973).

In our study against grassy shoot disease, sulfaphenazole suppressed the symptoms for 150 days, ampicillin for 90 days and chloramphenicol for 120 days on root application. When doses of oxytetracycline-HCl and sulfadiazine were administered together, they suppressed the disease for longer period than either could suppress alone. A similar trend was found when oxytetracycline-HCl and sulfaguanidine, tetracycline-HCl and sulfaphenazole, and chloramphenicol and sulfadiazine were administered together. Doses ranging from 100 to 500 ppm of these chemicals were administered through root inoculation.

The effect of organophosphorus compounds such as *O, O*-die-
thyl-*O*-(2-pyrazinyl)phosphorothioate, *O-O*-diethyl-*S*-phosphoro-
thioate, and *O,O*-dimethyl-*S*-phosphorothioate in controlling
tobacco yellow dwarf has been reported (Paddick and French,
1964).

Raychaudhuri *et al.* (1972) treated spike disease of sandal
in field by benlate [(methyl-1-butylcarbamoyl)-2-benzimidazole
carbamate]. Foliar application of benlate had no noticeable
effect on the disease, but recovery of spiked shoot after
about 5 months was noticed when thick paste of the chemical
was applied to the trunk after removing a portion of bark.

Direct effect of systemic oxime carbamate insectiside 2,3-
dihydro-2,2-dimethyl-1-7-benzophonemyl methyl carbamate and
S-methyl *N*-thioacetimidate on the mycoplasma in tobacco fields
has been reported (Paddick *et al.*, 1971). Certain systemic
organophosphorus and oxime carbamate insecticides reduced the
incidence of yellow dwarf on tobacco, not only because of vector control but also by disease remission in tobacco plants
(Sinclair, 1973).

IV. METHODS OF APPLICATION OF CHEMOTHERAPEUTANTS IN TREES

1. A bottle containing antibiotic solution was hung in
inverted position on the trunk. At a lower level, a deep hole
at slightly downward angle was drilled in the main trunk of
the tree. Copper tubing was hammered deep into this hole and
its other end was connected to the hanging bottle by a plastic tube. The antibiotic solution was taken into the tree by
gravity flow. The intake was about a quart of solution in 24
hours.

2. In a modification of the earlier method, the solution
was injected into the tree through the hanging bottle in the
inverted position at about 150 psi pressure. By this method
16 ounces of solution could be administered into the tree
within 30 minutes.

3. An alternate method is the Manget injection (J. J.
Manget Co., California). A small disposable plastic vial
containing concentrated antibiotic solution is used. The antibiotic solution is introduced into the tree through an aluminum tube which is inserted into tree. This method was very
effectively used for coconut palm tree (McCoy, 1974a).

4. In another method, concentrated antibiotic solution was
rapidly injected into the tree through a Minute tree injector (Minute Tree Injector Inc., Miami, Florida), the needle of
which was inserted into the tree through a predrilled hole.
The solution was forced into the tree by means of a pump
attached to the needle.

5.   Another method was to drill holes of 5.5 mm diameter and 38 mm depth at slightly downward slant with sterilized drill. Concentrated antibiotic solution was immediately added to these holes. The solution was absorbed by the tree in 30 seconds to 2 minutes.

6.   Systemic fungicide was used in the form of a paste to seal the premade girdle on the tree trunk by removing the bark. This was effective in the case of sandal spike disease.

7.   Tetracycline applied by drenching citrus trees infected with citrus decline for over 20 months, was successful in suppressing the symptoms.

8.   Methods of root injection and root collar injection have also been reported by Nair (1979) for administration of other groups of chemotherapeutants into trees.

For further details of these methods, readers may consult papers by McCoy (1972, 1974a, b), Raychaudhuri et al. (1972, 1977), Nyland and Moller (1973), Nyland and Sachs (1974), Sands and Walton (1975), Rosenberger and Jones (1977), Nariani et al. (1973).

V.   DISCUSSION

An appraisal of the above provides proof that the generalization that mycoplasmas causing yellows diseases in plants and the cultured species of *Mycoplasma* and *Acholeplasma* can be distinguished from bacteria by their sensitivity to tetracyclines and their resistance to sulfa drugs and penicillin. Susceptibility of *M.neurolyticum* to penicillin (Wright, 1967) and *M.pneumonia* to gold sodium thionate (Marmion and Goodburn, 1961) has been shown. There are reports of the effect of sulfa drugs on mycoplasmal diseases of plants. However, the *in vivo* tests alone do not prove that these drugs had a direct effect on mycoplasma. To visualize the direct effect on mycoplasma, screening of drugs *in vitro* is essential.

The exact mechanism of action of antibiotics on mycoplasma is not known, but a few hypotheses have been advanced. Weber and Kinsky (1965) and Feingold (1965) reported the role of sterol in the activity of certain polyene antibiotics on *Mycoplasma gallisepticum* and *M. laidlawii*. *Mycoplasma gallisepticum* has an absolute requirement of sterol for growth (Lampen et al., 1963), while *M. laidlawii* will grow almost equally well in the presence and absence of exogenous sterol, e.g., cholesterol. When grown in the presence of cholesterol, *M. laidlawii* will incorporate sterol into the cell membrane. It was found that filipin and amphotericin B inhibited the growth and caused lysis of *M.laidlawii* cells, which had been cultured in the presence of cholesterol, where as they had no effect on cells grown in the absence of sterol. This is a strong evidence that presence of cholesterol is a prerequisite for polyene sensitivity.

Novobiocin has a direct effect on magnesium binding (Brock, 1962a, b). Strong inhibition of L forms (Brock and Brock, 1959), protoplasts (Shockman and Lampen, 1962), and pleuropneumenia-like organisms (PPLO) further confirms the role of magnesium. Magnesium ions are essential for the stability of cell membrane (Weibull, 1956; Hershko et al., 1961). The divalent cations aggregate lipid-protein subunits in case of cell membranes of *Mycoplasma* spp. (Razin et al., 1965).

Nothing is known about the mechanism of action of tetracycline, erythromycin, and chloramphenicol on mycoplasma. In bacteria these antibiotics affect bacterial protein synthesis at the ribosomal level. Aminoglycosides, such as, streptomycin, also bind to prokaryotic ribosomes. Some aminoglycosides such as kanamycin and neomycin may bind one type of ribosome, but may not bind the other type.

The effect of at least one systemic fungicide (benlate) a systemic organophosphorous and oxime carbanate insecticide, and a few sulfa drugs and their combinations with tetracyclines and chloramphenicol on curing plant mycoplasma diseases gave very encouraging results. Even tetracyclines applied prior to vector preacquisition and preinfection feeding could prevent recurrence of disease (Davis et al., 1968). Some of the recently developed chemicals by Thirumalachar (1978) and Costouis et al. (1978) had cidal effect on plant mycoplasma. The time of application of tetracycline in a year is very important and has given substantial benefit in the case of peach X disease in peach and cherry (Sands and Walton, 1975; Rosenberger and Jones, 1977). In view of these results, it is desirable to try a very wide range of chemicals and their combinations against plant mycoplasma diseases by various methods of application and at different periods of the year.

To obtain the best effect of a chemotherapeutant it should be highly water soluble at a low pH. It should efficiently enter into the vascular strand and, in the case of trees, in the xylem vessels of the current year's annual ring for a very wide distribution into the plant.

Natural degeneration of mycoplasma has been observed in yellows and grassy shoot diseases of aster and sugarcane (Hirumi and Maramorosch, 1972; Rishi et al. 1973). The factors responsible within the plant to induce mortality of mycoplasma should be investigated since this would be very helpful in the effective control of mycoplasma diseases.

It is appropriate to mention here, that a number of mycoplasma diseases have been cured by heat (Kunkel, 1941, 1945, 1952; Nariani et al. 1973; Singh, 1977). A more extensive study should be made on heat therapy. Wherever heat therapy or chemotherapeutant alone does not control the disease, a possible combination of these two may be successfully exploited.

REFERENCES

Asuyama, H. and Iida, T. T. (1973). *Ann. N. Y. Acad. Sci.* *225*, 509.

Bowyer, J. W. and Calavan, E. C. (1974). *Phytopathology 64*, 346.

Brock, T. D. (1962a). *Science 136*, 316.

Brock, T. D. (1962b). *J. Bacteriol. 84*, 679.

Brock, T. D. and Brock, M. L. (1959). *Arch. Biochem. Biophys. 85*, 176.

Costouis, A. C., Becker, E. I., Garret, W. T., and Julien, I. M. (1978). *Proc. 3rd Int. Congr. Plant Pathol. Munchen, (Abstr.).*

Davis, R. E. and Whitcomb, R. F. (1970). *Infection Immunity 2*, 201.

Davis, R. E., Whitcomb, R. F., and Steere, R. L. (1968). *Science 161*, 793.

Feingold, D. S. (1965). *Biochem. Biophys. Res. Commun. 19*, 261.

Hershko, A., Amos, S. and Mager, J. (1961). *Biochim. Biophys. Acta 5*, 46.

Hirumi, H. and Maramorosch, K. (1972). *Phytopathol. Z. 75*, 9.

Ishiie, T., Doi, Y., Yora, K., and Asuyama, H. (1967). *Ann. Phytopathol. Soc. Jap. 33*, 267.

Klein, M., Frederick, R., and Maramorosch, K. (1973). *Ann. N. Y. Acad. Sci. 225*, 522.

Kunkel, L. O. (1941). *Amer. J. Bot. 28*, 761.

Kunkel, L. O. (1945). *Phytopathology 35*, 805.

Kunkel, L. O. (1952). *Phytopathology 42*, 27.

Lampen, J. O., Gill, J. W., Arnow, P. M. and Magana Plazq, A. (1963). *J. Bacterial. 86*, 945.

McCoy, R. E. (1972). *Plant Dis. Rep. 56*, 1019.

McCoy, R. E. (1973). *Proc. Fla. St. Hort. Soc. 86*, 503.

McCoy, R. E. (1974a). *Proc. Fla. St. Hort. Soc. 87*, 537.

McCoy, R. E. (1974b). *Fla. Agr. Expt. Sta. Circ. S-228*, p. 1.

Marmion, B. P. and Goodburn, G. M. (1961). *Nature (London) 189*, 247.

Nair, V. M. G. (1979). *Proc. 3rd IUFRO Conf. W. P. Mycoplasma Diseases, Rutgers (Abstr.)*, p. 13.

Nariani, T. K., Ghosh, S. K., Kumar, D., Raychaudhuri, S. P., and Viswanath, S. M. (1973). *Symp. Mycoplasmal Diseases, Chandigarh (Abstr.)* p. 24.

Nyland, G. and Moller, W. J. (1973). *Plant Dis. Rep. 57*, 634.

Nyland, G. and Sachs, R. M. (1974). *Colloq. Inst. Nat. Santa Rech. Med. 33,* 235.

Paddick, R. G. and French, F. L. (1964). *Proc. Aust. Tobacco Res. Conf.,* p. 304.

Raychaudhuri, S. P. (1977). *Eur. J. Forest Pathol. 7,* 1-5.

Raychaudhuri,S. P., Chenulu, V. V., Ghosh, S. K., Varma, A., Rao, P. S., Srimati, R. A., and Nag, K. C. (1972). *Curr. Sci. 41,* 72.

Razin, S., Morowitz, H. J. and Terry, T. M. (1965). *Proc. Nat. Acad. Sci. U. S. 54,* 219.

Rishi, N., Okuda, S., Arai, K., Doi, Y., Yora, K. and Bhargava, K. S. (1973). *Ann. Phytopathol. Soc. Jap. 39,* 429.

Rosenberger, D. A. and Jones, A. L. (1977). *Phytopathol. 67,* 277.

Sands, D. and Walton, G. S. (1975). *Plant Dis. Rep. 59,* 573.

Shockman, G. D. and Lampen, J. D. (1962). *J. Bacteriol. 84,* 508.

Sinclair, J. B. (1973). *Symp. Mycoplasmal Dis. Chandigarh,* p. 22.

Singh, K. (1977). *Sugar News 9,* 81.

Thirumalachar, M. J. (1973). *Symp. Mycoplasmal Dis. Chandigarh, (Abstr.),* p. 6.

Thirumalachar, M. J. (1978). *Proc. 3rd Int. Congr. Plant Pathol. Munchen, (Abstr.)*

Tucker, D. P. H., Bistline, F. W. and Gonsalvez, D. (1974). *Plant Dis. Rep. 58,* 895.

Weber, M. M. and Kinsky, S. C. (1965). *J. Bacteriol. 89,* 306.

Weibull, C. (1956). "Bacterial Anatomy," p. 111. Cambridge Univ. Press, Cambridge.

Whitcomb, R. F. and Davis, R. E. (1970). *Infection Immunity 2,* 209.

Windsor, I. M. and Black, L. M. (1973). *Phytopathology 63,* 44.

Wright, D. N. (1967). *J. Bacteriol. 93,* 185.

CONTROL OF TREE DISEASES BY CHEMOTHERAPY

V. M. G. Nair

Plant and Forest Pathology
College of Science and Environmental Change
University of Wisconsin-Green Bay
Green Bay, Wisconsin

The role of forests as an integral part of a nation's re-
newable natural resources is being felt increasingly these
days. Forests not only support various primary wood-using in-
dustries but are also very essential as a habitat for the
world's wildlife. Forests also play an important role in re-
creation, in the maintenance of high water quality, and protection of agricultural land. Still, forest diseases due to
fungi, mycoplasma, viruses, bacteria, and insects take a substantial toll of wood production. These tree diseases are considered more important than many of the annual crops because
when a tree dies 50 to 200 or more years of the land is lost
or wasted within a single year or less, depending upon the
pathogen. Although exclusion of various pathogens, such as
white pine blister rust *(Cronartium ribicola)*, chestnut blight
(Endothea parasitica), Dutch elm disease *(Ceratocystis ulmi)*,
citrus canker *(Xanthomonas citri)* through international quarantine regulations have been attempted, many have been imported to new countries and continents.
Early investigations in chemotherapeutic control of plant
diseases were directed toward the control of insect pests and
overcoming the nutrient deficiencies in various plants (Ray,
1901; Mokrzecki, 1903). The discovery of the antibiotic nature of *Penicillium* by Fleming in 1940 followed by the isolation of a powerful bacteriotoxic substance "penicillin" and
the realization of its outstanding value in chemotherapeutic
control of bacterial infections of humans led to the revival
of interest in the chemotherapeutic control of plant diseases
by antibiotics. Early records of chemotherapeutic treatments
of plant diseases with varying degrees of control included:
8-quinolinol sulfate soil treatment against *Rhizoctonia* (Stod-

MYCOPLASMA DISEASES
OF TREES AND SHRUBS

325

dard, 1952); sodium *p*-dimethylaminoazobenzenesulfonate treatment against *Phytophthora cinamomi* (Zentmyer and Gilpatrick,
1960); externally seed-borne *Helminthosporium* species on cere-
al grains with antibiotics such as mycothricin (Rangaswami,
1956), antimycin A, helixin B (Leben and Arny, 1952; Leben *et
al.*, 1953, 1954); internally seed-borne *Ascochyta* pisi on peas
with chloropicrin (Kennedy, 1961); vascular pathogen *Ceratocystis ulmi* with 8-quinolinol (Fron, 1936).

The role of various fungal toxins in the development of
various plant diseases attracted special attention during the
1950's (Dimond and Waggoner, 1953). Since then compounds capable of neturalizing these toxins by altering the metabolism
of the host led to the use of plant growth regulators, such as
2,4-D(Ibrahim, 1951; Davis and Dimond, 1953, 1956) as chemotherapeutants. Further studies have shown that some very active growth regulators such as S-carboxymethyl compounds also
exhibit chemotherapeutic value during their metabolism within
the host. Although many powerful antifungal antibiotics prove
effective against some pathogens *in vitro,* they may prove ineffective against the same pathogens during pathogenesis due
to the antagonizing effect of plant metabolites produced during the course of host-parasite interactions.

The major research interests in chemotherapy of forest
trees, in general, and systemic fungicides, in particular,
have been directed in the past toward fungus disease problems
in forest nurseries or Christmas tree plantations and, to a
limited extent, toward entire forest plantations. However,
recent efforts in the control of (a) root and foliage diseases,
(b) stem cankers, and, especially, (c) vascular wilts ("Dutch
elm" and "oak wilt") have attracted renewed research interest
in the understanding of the various interrelated complex phenomena which are playing a major role in the efficiency of
various systemic compounds, including: (1) toxicity of chemotherapeutants and their derivatives; (2) solubility of chemotherapeutants; (3) methods of application including improved
injection systems; (4) host response to injection wounds;
(5) understanding of various steps in the translocation process of the chemotherapeutants; (6) role of root grafts in
chemotherapy; (7) understanding of host physiology during
host-parasite interactions while or after treatment; (8) understanding of environmental influences during the time of
application; (9) the effect of seasonal variations on host
physiology and pathogenesis.

All these important aspects should be taken into consideration in developing an effective systemic compound against
various diseases of plants in general, and trees with woody
tissues, in particular.

I. TOXICITY OF CHEMOTHERAPEUTANTS AND THEIR DERIVATIVES

The realization of the delicate balance in our natural
ecosystem, especially in recent years, is responsible for the
creation of various environmental protection units. During
the early stages of development of various agricultural chemicals, mammalian toxicity was given prime importance. However,
currently we are not only concerned about our safety, but also with the safety of the fauna and flora of the environment
including microorganisms, insects, wildlife, fish, and domestic animals.

Serious consideration should be given to those compounds
for their possible penetration routes into test organisms and
those taken in by respiratory passages (Woodcock, 1971). However, most of the fungicides are considered relatively safe as
far as mammalian toxicity is concerned. These include antibiotics, (streptomycin, cycloheximide, griseofulvin, kasugamy-
cin), carboxylic acid anilides, [pyran and furan analogs (2,3-
dihydro-6-methyl-5-phenylcarbamoxyl-4-pyran; 2,5-dimethyl-3-
phenylcarbamoylfuran)], heterocyclic compounds, (thiabendazole,
fuberidazole, benomyl, the breakdown product of benomyl methyl-
benzimidazol-2-yl carbamate, azepines, Triforine, Dodemorph),
ammonium compounds, [G.N.64 (diemethyl, diethyl derivatives
of benzylammonium chloride)], and organophosphorus compounds
(triamiphos, phosphoric esters of pyrazolopyrimidines).

Many of the compounds already developed and those in various stages of development may prove to be safe and every
effective as systemic fungicides. However, manufactureres are
always concerned about the safety of the enzymatic, nonenzy-
matic, and possible conjugate derivatives of these compounds.

II. SOLUBILITY OF SYSTEMIC FUNGICIDES AND CHEMOTHERAPEUTANTS

Good solubility of compounds is very essential for the
good distribution of any systemic chemical. Research workers
have used 91% glacial acetic acid, 42.5% lactic acid, and HCl-
acetone for solubilizing benomyl in the case of Dtuch elm disease and oak wilt control (Smalley *et al.*, 1973; Nair and
Kuntz, 1973). Unfortunately, some of these solvents are phytotoxic. At present, we have only a very limited number of
water-soluble compounds now in various developmental stages
and they are being tested for their efficiency in the control
of tree diseases. These include (a) water-soluble benomyl
derivative carbendazim-HCl formulation "Lignasan;" (b) car-
bendazim-H_3PO_4 "BLP"; (c) thiabendazole hypophosphite; (d)
M & B 21914 [1-methoxycarbonyl-3-(dimethylaminoacetamidophen-
yl)thiourea-HC1]; (e) G.N.64(dimethyl, diethyl derivatives of
benzylammonium chloride).

III.   METHODS OF APPLICATION

A.   *TRUNK INJECTION SYSTEMS*

In trunk injection systems the systemic chemotherapeutant should be introduced into the actively conducting vesssel for the proper distribution of the compound. (Figs. 1-4).

   1. Mauget Injector System

In this system 5-mm diameter aluminum tubes were introduced into the current year's xylem vessels and at the free end of each tube a 65-ml plastic capsule filled with the chemical was attached. By this method the fungicide was introduced directly into the conducting vessels. This system partially failed in the treatment of elms and oaks against Dutch elm disease and oak wilt, respectively, due to the difficulty in placing the end of the tube precisely in the xylem tissues of active conduction. In tropical trees, where there are no definite annual rings, this system might prove still efficient.

   2. Pressurized Trunk Injection System

   a.   In this system the systemic fungicide was introduced through metal tubes which are connected with plastic tubes to the pressurized tank filled with the solubilized fungicide. At the time of treatment a 1 3/8-inch diameter hole is cut through the bark into the cambial area and at the center of the exposed surface a 1/2-inch hole is cut covering two to three annual rings deep into the xylem. [The metal tubes were inserted into these holes at various injection sites of the treated trees and the liquid was injected at 96 psi (McWain and Gregory, 1971).] Undoubtedly the pressurized system reduced the time of injection when compared to gravity flow.
   b.   The Elm Research Institute, Harrisville, New Hampshire, marketed a simple pressure injection system consisting of metal pipe nipples in the form of the letter "T" — the horizontal arm being introduced through the bark into the actively conducting xylem ring by drilling a hole through the bark into the xylem. The "T's" were then placed 6-9 inches apart around the trunk and the horizontal ends of each were connected with the next "T" by plastic tubing in order to form a complete circle of injection system. This is then connected to a simple hand-operated, pressurized backpack sprayer tank. Liquid was injected between 20-30 psi. This system is inexpensive and one of the best.

Fig. 1. Trunk injection with Mauget Capsules.
Fig. 2. Trunk injection through plastic "T" by gravity flow.
Fig. 3. Pressurized root collar - buttress root injection.
Fig. 4. Exposed root injection.

c. Recently Nair and Kuntz have modified and improved
this system by replacing all the metal "T's" with plastic
"T's" which are widely used as "maple taps" for drawing maple
sap for the preparation of maple syrup. This system is the
cheapest of all and leak proof. The Elm Research Institute
also has developed a very similar system with plastic "T's"
which are also leak proof and efficient.

B. ROOT COLLAR AND BUTTRESS ROOT INJECTION SYSTEM

Based on their bioassay studies, Nair and Kuntz have come
to the conclusion that root collar injection distributes the
chemicals more widely than injection of the trunk at 4 feet
above ground level. In this experiment over 3000 large elms
and oaks were injected with various chemotherapeutants (using
the leak-proof plastic maple tap T's) for the control of Dutch
elm disease and oak wilt. In this system anastamoses of the
conducting tissues at the root collar region might be playing
an important role in the wider distribution of systemic chemotherapeutants.

C. ROOT INJECTION SYSTEM

In this system (Kondo, 1972) many freshly cut 1 to 1½ inch
roots around the tree (elm) were connected to the McWain and
Gregory injection system. However, exposure of roots is still
a problem in many species. Moreover, root grafts between trees
will definitely interefere in this system.

D. TRUNK (BARK) BRANCH AND FOLIAR SPRAYS

Bioassay studies should prove the efficiency of these various methods of application. Of the 763 mature elms sprayed
with mist-blown sprays at 4.8 g/liter in 1971 and 1972, only
four developed new Dutch elm disease cases while in the case
of 838 unsprayed controls, 24 elms became infected (Smalley
et al., 1973). Smalley concluded that practical protection
on a mass basis was achieved, although bioassays of branches
from selected sprayed elms in these trials indicated erratic
distribution of active material in the xylem.

E. SOIL APPLICATION

Soil applications of chemotherapeutants are not practical
or economical in the control of any forest tree diseases except under nursery conditions.

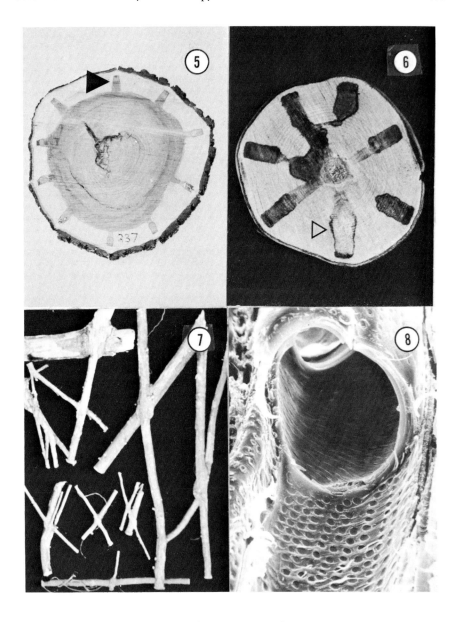

Fig. 5. Compartmentalization of discolored areas associated with deep injection wounds (elm).

Fig. 6. Compartmentalization of decay associated with injection wounds (maple).

Fig. 7. Intraspecific root grafts (oak).

Fig. 8. Open xylem vessel showing pits on its wall (sandalwood).

IV.   HOST RESPONSE TO INJECTION WOUNDS

Recently, development of discolored and decayed wood
associated with injection wounds have attracted the attention
of various research workers. These include (1) *type of wounds,*
such as, shallow wounds into sapwood and deep wounds into
hearwood; (2) *position of wounds,* namely trunk wounds above
soil line and root wounds at the ridges and valleys of fluted
roots (Shigo and Campana, 1977).

Whenever a tree is wounded for the introduction of chemotherapeutants, the tree will manifest some response in the
form of built-in systems that confine, wall off or compartmentalize injured and infected tissues (Shigo *et al.,* 1977) (see
Figs. 5-8). Some trees have a stronger wound response than
other and can compartmentalize invaded tissues more effectively than other trees of the same species (Shigo, 1979). Thus
a compartmented tree compartmentalizes the injured and infected tissues and this is the major reason why the trees are not
invaded rapidly by various organisms after they have been
wounded.

Presence of discolored wood associated with every type of
injection wound described above was noticed on 80 large American Elms subjected to chemotherapeutic treatments (Shigo and
Campana, 1977). In addition, Shigo and Campana also noticed
that the injection wounds made in repeated annual therapeutic
treatments caused severe internal injuries. Decayed wood was
also associated with some wounds. Most frequently *Collybia*
sp. and *Coprinus* sp. of the hymenomycetes were isolated from
the decayed wood.

Examination of limited elm trees (3) injected with chemotherapeutants, through 5 to 8 cm deep injection holes in 1974
and 1975, revealed extensive compartmentalization of discolored wood associated with these injection holes made on the prominent ridges of roots. Extensive discoloration without
compartmentalization was noticed on the flat side of the roots.
Based on this observation Shigo and Campana suggested that the
drilling of holes should be avoided in the valleys of fluted
roots (Shigo and Campana, 1977).

V.   TRANSLOCATION OF SYSTEMIC COMPOUNDS

Effective translocation of systemic chemotherapeutants or
their derivatives is essential both in protective and curative
treatment of various hosts against pathogens. Proof is lacking in most cases to confirm whether the actual injected compound is responsible for the disease control or its fungicidal
metabolite or metabolites.

Important attention should be given to these various steps in the translocation process:
1. *Apoplastic movement* — movement in the nonliving tissues such as open xylem vessels and tracheids;
2. *Symplastic movement* — movement in the living parts of the cells, the protoplasm, and the functional sieve tubes;
3. Movement in the free space, i.e., diffusion into the cell wall. Systemic chemotherapeutants applied as a foliar spray or bark paint-spray could penetrate the tissues. Ectodesmata in the cuticle of the leaf could act as a direct pathway from the surface to the protoplast of the epidermal cells (Franke, 1967).

A. *APOPLASTIC MOVEMENT*

In apoplastic movement transpiration pull-stream plays an important role. The main bulk of the movement is a mass flow from roots to leaves in a transpiration stream via xylem vessels (Crowdy, 1972). However, there are reports that although basic and neutral substances flow easier in the vessles, acidic and basic substances are adsorbed to the negative charges of xylem vessels. This has been reported for many amino acids (Hill-Cuttingham and Lloyd-Jones, 1968), basic antibiotics (Crowdy and Pramer, 1955), quartery ammonium compounds (Edington and Dimond, 1964), although, addition of cations in the form of calcium could reverse the action. In apoplastic movement of chemotherapeutants, the chemicals might accumulate in the margins of the leaves causing foliage injury. However, in the case of "vascular diseases" caused by vascular pathogens, such as, *Ceratocystis fagacearum*, *C. ulmi*, *Verticillium alboatrum*, *Fusarium* sp., and *Cephalosporium* sp. introduction of systemic chemotherapeutants into the site of multiplication and distribution of the pathogen (xylem vessels), is very essential for the control of this group of diseases. In addition, the compound must cross the plasmolemma and enter the protoplasm (Crowdy, 1972). Clemons and Sisler (1969) have reported that benomyl and its fungicidally active metabolic product MBC (methyl ester of 2-benzimidazole carbamic) are transported in apoplastic system. In this respect special precautions should be taken not to introduce the systemics too deeply into the nonconducting xylem vessels of the older rings because they might act as reservoirs holding the compound preventing distribution.

B. *SYMPLASTIC MOVEMENT*

Since symplastic movement occurs within the living parts
of the cell, the chemotherapeutant or its metabolites must coexist with the living protoplasm and plasmodesmata will play
an important role. Living xylem parenchyma cells, uni- and
multiseriate ray cells, and sieve tubes composed of rows of
living cells form a continuous living network of protoplasm
through which symplastic movement takes place (Figs. 9-12).

Special attention should be given during the development
of systemic chemotherapeutants especially for the control of
any obligate parasites including diseases caused by viruses.
Pathogenic organisms which do not possess independent metabolic systems must coexist with the living protoplasm of the
host and, as such, systemic chemotherapeutants or their metabolites, which are transported by symplastic movements, should
have a profound influence in the control of these pathogens at
various degrees.

Chemotherapeutic studies in the control or inhibition of
virus diseases have shown that many growth regulators (indole-
butyric acid, indoleacetic acid, 2,4-dichlorophenoxyacetic
acid, gibberellic acid, phenylpropionic acid, DL-2-aminobuty-
ric acid), antibiotics (Chloromycetin, tetramycin, blastici-
din, cytovirin, lidermycin), and other substances (acridine
orange, crystal violet, aflatoxins, tannins) have a wide range
of influence on various viruses (Raychaudhuri, 1977).

Our knowledge is sparce with regard to the movements of
chemotherapeutants and their metabolites in the phloem tissues
from the site of application. Further investigations in this
field will reveal a better understanding of phloem movement
and hopefully should aid in the development of systemic chemotherapeutants especially designed for symplastic movement as
bark applicators.

VI. SIGNIFICANCE OF ROOT GRAFTS IN THE SPREAD OF TREE DISEASES
     AND THEIR ROLE IN THE DISEASE CONTROL WITH SYSTEMIC FUN_
     GICIDES OR CHEMOTHERAPEUTANTS

In the case of a forest stand of one species or a mature
stand of trees of the same species planted under urban conditions, a single tree should be regarded only as a member of
the community of trees because it usually establishes many
intraspecific root grafts. Root grafts between different
species are rare (Nair and Kuntz, 1973). On the other hand,
Santalum album can establish root grafts with more than 100
species of unrelated hosts. These root grafts can act as
"pipe"lines" between the individual trees transporting nutri-

Fig. 9. Simple pit between ray cells (oak) along with en-doplasmic reticulum and plasmodesmata.

Fig. 10. Plasmodesmata in cell wall (oak).

Fig. 11. Slit like pits between xylem vessel and xylem parenchyma (sandalwood).

Fig. 12. Round pits on xylem vessel wall (sandalwood).

ents and can help them in their interdependency. However,
these root grafts will also transport systemic chemotherapeutants if injected into one tree and also vascular pathogens
under the conditions of natural tree-to-tree spread of vascular diseases.

VII. SYSTEMIC COMPOUND'S EFFECT ON HOST PHYSIOLOGY AND HOST-PARASITE INTERACTIONS

Systemic compounds used in the control of plant diseases
when applied could (1) kill the pathogen, (2) retard or arrest
its multiplication and growth or (3) alter the physiology of
the host and the host, in turn, may exhibit resistant host reactions.

A. *PRODUCTION OF FUNGITOXIC MOLECULES FROM COMPOUNDS*

Systemic fungicides, such as benomyl, when injected into
plants, are converted by hydrolysis to fungitoxic molecules,
methyl 2-benzimidazole carbamate (Clemons and Sisler, 1969),
which in turn could stop the pathogenesis.

B. *PRODUCTION OF PHYTOALEXINS*

Fungitoxic phytoalexins are produced by plants (1) when
they are treated with certain chemicals; (2) in the presence
of an invading pathogen; (3) when they undergo injury. Cruickshank (1966) suggested the development of systemic compounds
which induce the production of phytoalexins for the control
of plant diseases.

C. *ALTERATIONS OF HOST ANATOMY-XYLEM VESSELS*

Xylem vessels plan an important part in the pathogenesis
of vascular diseases (oak wilt, dutch elm, mimosa wilt, per-
simmon wilt, fusarium wilt, verticillium wilt) as a site of
multiplication and distribution of the pathogen, presumably
through transpiration streams, especially during the early
stages of disease development.
Systemic compounds which could reduce the "vessel group
size" (=average vessel diameter and diameter of contiguous
vessels in one annual ring)(McNabb *et al.*, 1970) in elms could
enhance the resistance of trees to *C. ulmi*. Aminotrichloro-
phenylacetic acid, when applied to elms, reduced the vessel

group size and reduced the severity of Dutch elm disease (Edgington, 1963).

Nair and Kuntz (1976) suppressed the production of vessels
in highly susceptible Northern pin oaks (to oak wilt) which
became resistant to the pathogen when treated with growth re-
gulator, sodium 2,3,6-trichlorophenyl acetate (TCPA). Vessel
production was eliminated not only in the year of treatment
(Venn et al., 1968), but also in the two succeeding years after treatment (Nair and Kuntz, 1976). Vessels formed prior
to treatment were filled with tyloses after treatment. Wood
formed after treatment consisted mainly of xylem parenchyma,
fibers, and fiber tracheids. The objective here being the
elimination of the site of multiplication and distribution of
the vascular pathogen, especially in its early stages of disease development by chemotherapeutic treatment (Figs. 13-16).

D. PRODUCTION OF TYLOSES, GELS AND GUMS

Growth regulating substances such as auxins could induce
tyloses formation in vessels of oaks attacked by the Oak wilt
fungus (Beckman et al., 1969a,b)(Figs. 17-20). In Dutch elm
disease studies, trees treated with 2,3,6-trichlorophenylace-
tic acid as a chemotherapeutant showed occlusion of large
vessels with tyloses. Smalley (1962) attributed the resistance to the prevention of spore movement due to the presence
of tyloses. Also, 2-chloroethyltrimethylammonium chloride
(Cycocel)-treated tomato plants produced tyloses and reduced
the severity of the vascular wilt caused by Verticillium alboatrum (Sinha and Wood, 1964). In addition, vascular gels and
gums produced by the host during pathogenesis could effectively
block the distribution of vascular pathogens inside the tree
(Sachs et al., 1967).

E. SUPPRESSION OF TYLOSES

Optimum dosage of cytokinin 6-benzylaminopurine (BAP) arrested or suppressed tyloses formation and wilt development
in innoculated (with Ceratocystis fagacearum) oaks although
the fungus was widely distributed in the xylem vessels (Nair
et al., 1969). Water conduction continued in these trees in-
spite of the presence of the fungus in vessels.

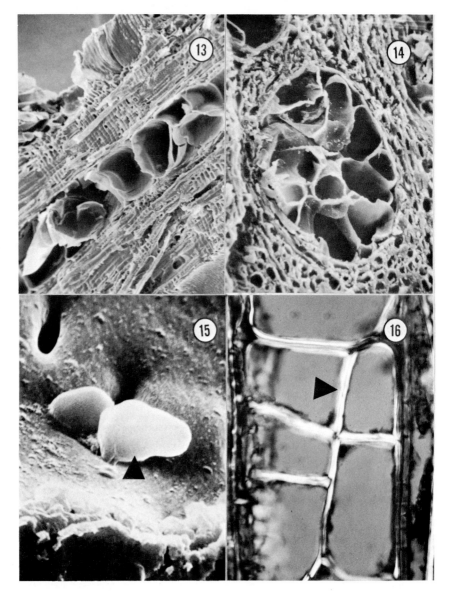

Fig. 13. Xylem vessel filled with tyloses in oak wilt (L·S).

Fig. 14. Xylem vessel filled with tyloses in oak wilt (C·S).

Fig. 15. Early development of tyloses from xylem parenchyma into xylem vessel in oak wilt.

Fig. 16. Blocked xylem vessel showing tyloses wall (L·S) in oak wilt.

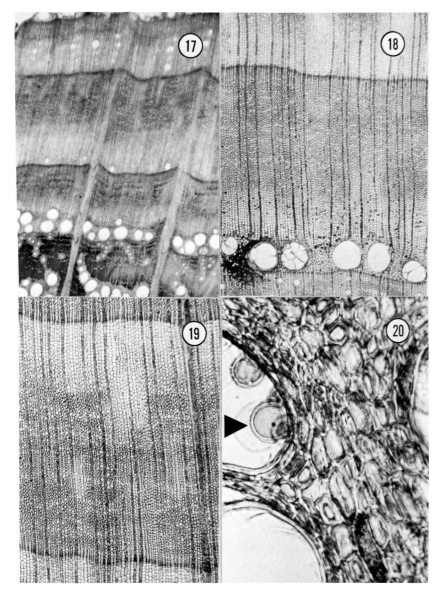

Fig. 17. Partial suppression of xylem vessel formation in oak treated with TCPA.

Fig. 18. Complete supression of xylem vessel formation for two years in oak treated with TCPA.

Fig. 19. Complete supression of xylem vessel formation for three years in oak treated with TCPA.

Fig. 20. Supression of tyloses formation in xylem vessels of inoculated oak (oak wilt) treated with cytokinin (BAP).

*F. DETOXIFICATION OF FUNGAL TOXINS AND INACTIVATION OF PECTO-
LYTIC ENZYMES*

Treatment of rice plants with ferulic acid caused detoxification of the toxin pyricularin (Tamari *et al.,* 1966) in the
case of rice blast control.

Since the hydrolytic enzymes play a very important role
during the pathogenesis of various diseases, inactivation or
changing the sequence of enzyme production of the pathogens by
the application of compounds could lead to disease control.

VIII.   ENVIRONMENTAL INFLUENCES DURING THE TIME OF APPLICATION

Various environmental factors such as high and low humidity, temperature, wind velocity, and cloud cover have a profound influence in the uptake and distribution of chemotherapeutants in trees. Recently, based on 900 large American elms
variously injected (pressure or gravity) primarily with Arbo-
tec, Nair came to the conclusion that the percentage of hy-
dration of these trees immediately before treatment had a significant influence on the uptake and distribution. In those
trees exposed to higher temperatures (80°-86°F) and low humidity (34%) along with high wind conditions (25 mph) during the
days immediately prior to treatments, the uptake and distribution was rapid compared to those exposed to lower temperatures (40°-42°F), high humidity (100%), and low wind conditions (5 to 8 mph)(Nair, 1979).

IX.   EFFECT OF SEASONAL VARIATIONS ON HOST PHYSIOLOGY

Seasonal variations play a major role in the seasonal susceptibility of trees to various pathogens. Nair and Kuntz have
shown that *Quercus macrocarpa* and *Q. ellipsoidalis* were highly
susceptible to artifical inoculation with *Ceratocystis fagacearum* during spring and early summer months, i.e., April to
June (Nair and Kuntz, 1963a,c). Further studies also showed
that the wound susceptibility of these trees to natural infection was very high during the same period (Nair and Kuntz,
1963b). In both these instances the susceptibility of oaks to
Ceratocystis fagacearum coincided with the formation of large
spring-wood vessels followed by small summer-wood vessels.

Based on recent bioassay studies, Nair came to the conclusion that the uptake and distribution of systemic chemotherapeutants was better in elm trees treated after full leaf expansion than during the bud break or leaf expansion periods
(Nair, 1979). Histological studies revealed the presence of

Forest tree disease	Causal organisms	Compound	References
ROOT ROTS			
Root Rot – red pine and white pine	Cylindrocladium scoparium	Benomyl	Berbee and Hadi (1977)
Root Rot – Norway spruce	Cylindrocladium scoparium	Benomyl	Berbee and Hadi (1977)
LEAF DISEASES			
Brown spot – scotch pine	Scirrhia acicola	Benomyl	Willis (1972)
Needle cast – sctoch pine	Lophiodermium pinastri	Thiophanate methyl	Witcher et al. (1975)
		Thiabendazole	Witcher et al. (1975)
Brown spot – longleaf pine	Scirrhia acicola	Cycloheximide	Snow et al. (1964)
Needle blight – Austrian pine and Monterey pine	Cothiostroma pini	Benomyl	Gilmour and Vanner (1971, 1972)
Needle cast – blue spruce	Rhizosphaera kalkhoffi	Benomyl	Skilling and Waddell (1975)
Needle rust – grand fir	Pucciniastrum goep-pertianium	Oxycarboxine	Davison (1969)
Phomopsis blight – eastern red cedar	Phomopsis juniperovora	Thiophanate methyl	Otta (1974)
Leaf spot – balsam poplar	Septoria musiva	Benomyl	Drorun and Kusch (1974)
Marssonina blight – big-tooth aspen	Marssonina populi	Benomyl	Dockinger (1975)
Leaf rust – cottonwood	Melampsora medusae	Benomyl	Filer (1975)
Anthracnose – sycamore	Gnomonia platani	Benomyl	Neely (1975)
		Triarimol	Hock and Berry (1975)

Disease – host	Pathogen	Chemical	Reference
Black walnut – anthracnose	*Gnomonia leptostyla*	Triarimol	Berry and Hock (1975)
STEM DISEASES			
Fusiform rust – slash pine	*Cronartium fusiforme*	Oxycarboxine	Hare (1973)
Blister rust – white pine	*Cronartium ribicola*	Cycloheximide	Phelps and Weber (1970)
		Benomyl	
		Triforine	
Chestnut blight – chestnut	*Endothia parasitica*	Benomyl	Anagnostakis and Jaynes (1971)
			Jaynes and Van Alfen (1974)
Hypoxylon canker – aspen	*Hypoxylon pruniatum*	Carbendazim-HCl	Brown and Silverborg (1970)
		Cycloheximide	
Cutting bed canker – poplar hybrids	*Marssonina*	Thiophanate methyl	Carlson (1973)
		Benomyl	
Canker stain – sycamore	*Septoria sp*	Benomyl	Carlson (1973)
	Ceratocystis fimbriata		Davis and Peterson (1975)
WILT DISEASES			
Dutch elm – American elm	*Ceratocystis ulmi*	Sodium 2,3,6-trichloro-phenyl acetate (TCPA)	Smalley (1962)
		Benomyl	Smalley et al. (1973)
		Carbendazim-HCl	Gregory and Jones (1975), Prasad (1974), Campana (1975)
		Thiabendazole	Epstein (1977)

Forest tree disease	Causal organisms	Compound	References
Dutch elm disease - mature English elm	Ceratocystis ulmi	Thiophanate methyl	Epstein (1977)
		Carbendazim (lactic acid)	Gibbs and Dickeson (1977)
Oak decline, "Cephalosporium wilt" - oak	Cephalosporium diospyri	Benomyl	van Arsdel and Bush (1972)
Verticillium wilt - Maple	Verticillium dahliae	Carbendazim-HCl	Sinclair (1975)
Mimosa wilt - Mimosa	Fusarium oxysporum f. sp. perniciosum	Benomyl	Phipps and Stipes (1975)
Oak wilt - oak	Ceratocystis fagacearum	2,3,5-Triiodo-benzoic acid	Geary and Kuntz (1962)
		Sodium 2,3,6-trichloro-phenyl acetate	Venn et al. (1968); Nair and Kuntz (1976)
		6-Benzylamino-purine	Nair et al. (1969)
		Thiabendazole, TCPA	Nair (1977)
Sandal spike disease - sandalwood	Mycoplasma (MLO)	Terramycin Benlate	Raychaudhuri and Srimathi (1977)
		Ledermycin Benlate	Raychaudhuri and Srimathi (1977)

FOREST NURSERY DISEASES

Leaf spot – sugar maple	Phyllosticta minima	Bravo, Dithane M-45, Benlate, Phaltan, E1 222	Worf et al. (1977) (1977)
Leaf spot – Green Ash	Cylindrospdium fraxini	Benlate, Bravo	Worf et al. (1977)
Leaf spot – gray dogwood	Septoria cornicola	Dithane M-45, Bravo, Benlate, Phaltan	Worf et al. (1977)
Powdery mildew – multi- flora rose	Sphaerotheca pannosa	E1 222, Bay- leton Triforine, Benlate (+adjuvant)	Worf et al. (1977)
Leaf spot – highbush cranberry	Cercospora varia	Bayleton, Benlate, Bravo, E1 222 Phaltan	Worf et al. (1977)
Rust – hawthorn	Gymnosporangium globosum	Dithane M-45, Bravo	Worf et al. (1977)
Downy mildew – wild grape	Plasmopora viticola	Dithane M-45, Bordeaux	Worf et al. (1977)
Leaf spot – nanny berry	Romularia viburoni	Benlate, Phaltom	Worf et al. (1977)
Powdery mildew – nanny berry	Microsphaera alni	Benlate, Phal- tom, Triforine, Bravo, Kocide 101	Worf et al. (1977)
Leaf spot – hazelnut	Cylindrosporium vermiformis	None tested	

large spring-wood vessels after full leaf expansion which apparently aided in the better uptake and distribution of chemotherapeutants.

In regard to tropical trees such as *Santalum album* it is important to note that the conducting patterns vary as these species produce more than one growth ring annually. This knowledge is essential before developing efficient systemic injection systems for the control of various diseases of tropical trees through systemic chemotherapeutants.

X. CONCLUSION

Control of tree diseases by chemotherapy is a growing field of study. The more we understand about the mechanisms involved in the host-parasite interactions during pathogenesis, the better equipped we will be to develop improved methods of control through chemotherapy. Chemotherapeutic control measures of various tree diseases are currently being explored. Chemotherapeutic control of particular disease becomes more complicated when alternate hosts of that particular pathogen enter into the disease control program. Various possibilities still exist for the future development of effective systemic compounds. Further investigations on chemotherapeutic control of tree diseases should include (a) improved methods of application of systemic compounds; (b) improved water solubility of compounds; (c) understanding of the host reaction to injection wounds; (d) the influence of environment during and after chemotherapeutic treatments; (e) understanding of the time and season of application favorable for the host in order to enhance the host resistance to the invading pathogen. Future chemotherapeutic control of tree diseases should be treated as an integral part of an integrated control program based on epidemiology of various tree diseases.

REFERENCES

Anagnostakis, S. L. and Jaynes, R. A. (1971). Inhibition of
    Endothia parasitica by benomyl in field-grown American
    chestnuts. *Plant Dis. Rep. 55,* 199.
Berbee, J. G. and Hadi, S. (1977). Control of Cylindrocladium
    root rot and stem canker of conifers with benomyl. *Phyto-
    pathology 67.*
Beckman, C. H., Elgersma, D. M., and McHardy, W. E. (1972).
    The localization of fusarial infections in the vascular
    tissue of single-dominant-gene resistant tomatoes. *Phy-
    topathology 62,* 1265.

Berry, F. H. and Hock, W. K. (1975). *In* "Walnut Anthracnose in Forest Nursery Disease in the United States" (G. S. Peterson and R. S. Smith, eds.). USDA Forest Service Agr. Handb. No. 470, p. 92.

Brown, D. H. and Silverborg, S. B. (1970). Effect of cycloheximide on Hypoxylon cankers of aspen. *Plant Dis. Rep. 54,* 462.

Campana, R. J. (1975). Evaluation of pressure-injected, solubilized benomyl as a therapy for American elms, naturally inoculated by native elm bark beetles with the Dutch elm disease fungus. N.C.R. 32 Report. University of Illinois.

Carlson, L. W. (1973). Poplar cutting bed canker (hybrid poplar)(Septoria sp. and Marssonnia sp.). *Fungicide Nematicide Tests 29,* 124.

Clemons, G. P. and Sisler, H. D. (1969). Formation of a fungitoxic derivative from benlate. *Phytopathology 59,* 705.

Crowdy, S. H. (1972). *In* "Systemic Fungicides" (R. W. Marsh, ed.), Chapter 5. Longman, London.

Crowdy, S. H. and Pramer, D. (1955). Movement of antibiotics in higher plants. *Chem. Ind.,* p. 160.

Cruickshank, I. A. M. (1966). Defense mechanisms in plants. *World Rev. Pest Contr. 5,* 161.

Davis, D. and Dimond, A. E. (1953). Inducing disease resistance with plant growth-regulators. *Phytopathology 43,* 137.

Davis, D. and Dimond, A. E. (1956). Site of disease resistance induced by plant growth regulators in tomato. *Phytopathology 46,* 551.

Davis, S. H., Jr. and Peterson, J. L. (1975). A tree would dressing to prevent spread of the Ceratocystis causing cnaker stain of the plane tree. *Plant Dis. Rep. 57,* 28.

Davison, A. D. (1969). Grand Fir needle rust *(Pucciniastrum goeppertianum). Fungicide Nematicide Tests 25,* 106.

Dimond, A. E. and Waggoner, P. E. (1953). On the nature and role of vivotoxins in plant disease. *Phytopathology 43,* 229.

Dockinger, L. S. (1975). "Marssonina Blight of Big Tooth Aspen in Forest Nursery Diseases in the United States" (G. W. Peterson and R. S. Smith, eds.). USDA Forest Service Agr. Handbook No. 470, p. 95.

Drorun, J. A. and Kusch, D. S. (1974). Balsam poplar leaf spots *(Septoria musiva, Marssonina populi). Fungicide Nematicide Tests 30,* 127.

Edgington, L. V. (1963). A chemical that retards development of Dutch elm disease. *Phytopathology 53,* 349.

Epstein, A. (1977). Soil application of systemic fungicides
    for suppression of Dutch elm disease infections. *Phyto-
    pathology 67.*
Filer, T. H., Jr. (1975). "Melampsora Rust of Cottonwood in
    Forest Nursery Diseases in the United States (G. W. Peter-
    son and R. S. Smith, eds.). USDA Forest Service Agr.
    Handbook, No. 470, p. 99.
Franke, W. (1967). Mechanisms of foliar penetration of solu-
    tions. *Annu. Rev. Plant Physiol. 18,* 281.
Geary, T. F. and Kuntz, J. E. (1962). Effect of growth regu-
    lators on oak wilt development. *Phytopathology 52,* 733.
Gibbs, J. N. and Dickson, J. (1975). Fungicide injection for
    the control of Dutch elm disease. *Forestry 48,* 165.
Gilmour, J.W. and Vanner, A. L. (1971). Radiata pine needle
    blight *(Dothistroma pini). Fungicide Nematicide Tests 27,*
    137.
Gilmour, J. W. and Vanner, A. L. (1972). Radiata pine needle
    blight *(Dothistroma pini). Fungicide Nematicide Tests 28,*
    128.
Gregory, G. F. and Jones, T. W. (1975). An improved apparatus
    for pressure-injecting fluid into trees. *U. S. Forest Ser-
    vice Res. Note NE-214.*
Hare, R. (1973). Soil applications of systemics for prevention
    and eradication of fusiform rust. *Plant Dis. Rep. 57,*
    776.
Hill-Cottingham, D. G. and Lloyd-Jones, C. P. (1968). Relative
    mobility of some nitrogenous compounds in the xylem of
    apple shoots. *Nature (London) 220,* 389.
Hock, W. K. and Berry, F. H. (1975). "Sycamore Anthracnose in
    Forest Nursery Diseases in the United States" (G.W. Peter-
    son and R. S. Smith, eds.). USDA Forest Service Agr. Hand-
    book NO. 470, p. 88.
Ibrahim, I. A. (1951). Effect of 2,4-D on stem rust develop-
    ment in oats. *Phytopathology 41,* 951.
Jaynes, R. A. and Van Alfen, N. K. (1974). Control of American
    chestnut blight by trunk injection with methyl-2-benzimi-
    dazole carbamate (MBC). *Phytopathology 64,* 1479.
Kennedy, J. (1961), Fumigation of agricultural products. XVII.
    Control of Ascochyta blight of peast by fumigation. *J. Sci.
    Food Agr. 12,* 96.
Kondo, E. W. (1972). A method for introducing water-soluble
    chemicals into mature elms. *Can. Forest. Serv., Sault
    Ste. Marie, Ont. Inf. Rep. O-X-171.* 11 pp.
Leben, C. and Arny, D. C. (1952). Seed treatment experiments
    with helixin B. *Phytopathology 42,* 469.
Leben, C., Arny, D. C., and Keitt, G. W. (1953). Small grain
    seed treatment with the antibiotic, helixin B. *Phytopatho-
    logy 43,* 391.

Leben, C., Arny, D. C., and Keitt, G. W. (1954). Effectiveness
of certain antibiotics for the control of seed-borne diseases of small grains. *Phytopathology 44,* 704.

McNabb, H. S., Heybroek, H. M., and MacDonald, W. L. (1970).
Anatomical factors in resistance to Dutch Eelm Disease.
Methods J. Plant Pathol. 76, 196.

McWain, P. and Gregory ,G. (1971). Solubilization of benomyl
for injections into trees for disease control. *N. E.
Forest. Exp. Sta. Upper Darby, Pa., USDA Forest Serv.
Res. Paper NE-234.* 6 pp.

Mokrzecki, S. (1903). Uber die innere Therapie der Pflanzen.
Z. Pakrankh. Papath. Paschutz. 13, 257/

Nair, V. M. G. (1977). Chemotherapy of tree diseases. First
working party meeting of IUFRO on mycoplasma diseases.
*IUFRO World Conf. Program Ministry Agr. Forest. Div. Gov-
ernment India.* Held at the University of Agricultural
Sciences, Bangalore, India, p. 17-18.

Nair, V. M. G. (1979). Chemotherapy of Dutch Elm Disease.
*Res. Progr. Rep. Univ. Wis. North Central Res. Committee
32.* (Diseases of Forest and Shade Trees).

Nair, V. M. G. and Kuntz, J. E. (1963a). Wound susceptibility
of bur oaks to artificial inoculation by *Ceratocystis
fagacearum. Univ. Wis. Forest. Res. Notes 93.*

Nair, V. M. G. and Kuntz, J. E. (1963b). Wound susceptibility
of but oaks to natural infection by *Ceratocystis fagacearum. Univ. Wis. Forest. Res. Notes 94.*

Nair, V. M. G. and Kuntz, J. E. (1963c). Seasonal susceptibility of bur oaks to artificial inoculation with the oak
wilt fungus, *Ceratocystis fagacearum. Univ. Wis. Forest.
Res. Notes 97.*

*Nair, V. M. G. and Kuntz, J. E. (1975). "Advances in Mycology
and Plant Pathology"* (S. P. Raychaudhuri, Anupam Varma,
K. S. Bhargava and B. H. Mehrota, eds.). pp. 231-240.
Sole distributors - M/S Harsh Kumar at Sagar Printers, 8
Jantar Mantar Road, New Delhi, India.

Nair, V. M. G. and Kuntz, J. E. (1976). Physiological control
of vascular wilt diseases of deciduous trees. "Invited
Paper" - Forest Pathology Section of the XVI World Contress
of IUFRO (IUFRO/FAO), Norway.

Nair, V. M. G. and Kuntz, J. E. (1976). Anatomical changes in
xylem tissues associated with oak wilt. "Invited Paper" -
Section on plant physiology at the XVI World Congress of
the Intl. Union of Forestry Res. Org. (IUFRO/FAO), Norway.

Nair, V. M. G., Wolter, K. E., and Kuntz, J. E. (1969). The
inhibition of tylosis and oak wilt development by the cy-
tokinin-6-benzylaminopurine. *(BAP) Phylopathology 59,* 1042.

Neely, D. (1975). Treatment of foliar diseases of woody ornamentals with soil injections of benomyl. *Plant Dis. Rep.
59,* 300.

Nicholls, T. H. (1973). Fungicidal control of Lophodermium pinastri on red pine nursery seedlings. *Plant Dis. Rep.* 57, 263.

Otta, J. D. (1974)., Benomyl and thiophanate methyl control Phomopsis bligh of Eastern red cedar in a nursery. *Plant Ids. Rep.* 58, 476.

Phelps, W. R. and Weber, R. (1970). An evaluation of chemotherapeutic treatment of blister rust cankers in Eastern White Pine. *Plant Dis. Rep.* 54, 1031.

Phipps, P. M. and Stipes, R. J. (1975). Control of Fusarium wilt of mimosa with benomyl and thiabendazole. *Phytopathology* 65, 594.

Prasad, R. (1974). Translocation of benomyl in elm *(Ulmus americana* L.*)*.VII. Application of the trunk injection method for suppression of the Dutch Elm Disease *(C. ulmi)* in landmark and historical trees. *Chem. Contr. Res. Inst., Ottawa, Can. Inf. Rep. CC-X-72.*

Rangaswami, G. (1956). A preliminary report on the use of mycothricin complex in plants. *Plant Dis. Rep. 40,* 483.

Ray, J. (1901). Les maladies cryptogamiques des vegetaux. *Rev. Gen. Bot. 13,* 145.

Raychaudhuri, S. P. (1977). "A Manual of Virus Diseases of Tropical Plants," p. 299. Macmillan Company of India, Ltd. New Delhi, India.

Raychaudhuri, S. P. and Srimathi, (1977). "Mycoplasma diseases of Trees."

Sachs, I.B., Nair, V. M. G., and Kuntz, J. (1967). Penetration and degradation of cell walls in oak sapwood by *Ceratocystis fagacearum. Phytopathology 57,* 827-828.

Shigo, A. L. (1979). "Tree Decay - An Expanded Concept." U. S. Dept. of Agriculture, Forest Service. *Agr. Information Bull. No. 419,* p. 72.

Shigo, A. L. and R. Campana (1977). Discolored and decayed wood associated with injection wounds in American Elm. *J. Arboricult. 3*(12), 230-235.

Shigo, A. L., Money, W. E., and Doods, D. I. (1977). Some internal effects of mauget tree injection. *J. Arboricult. 3*(11), 213-220.

Sinclair, W. A. (1975). Verticillium wilt of maples. *Res. Progr. Rep. Cornell Univ. Cooperative Rep. Regional Project* NE-25 (wilt diseases).

Sinha, A. K. and Wood, R. K. S. (1964). Control of Verticillium wilt of tomato plants with cycocel [(2-chloroethyl) ammonium chloride]. *Nature(London) 202,* 824.

Skilling, D. D. and Waddell, C. D. (1975). Control of Rhyzosphaera needle cast in blue spruce Christmas tree plantations. *Plant Dis. Rep. 59,* 841.

Smalley, E. B. (1962). Prevention of Dutch elm disease by
    treatments with 2,3,6-trichlorophenyl acetic acid. *Phy-
    topathology 52,* 1090.
Smalley, E. B., Meyers, C. J., Johnson, R. N., Fluke, B. C.,
    and Vieau, R. (1973). Benomyl for practical control of
    Dutch elm disease. *Phytopathology 63,* 1239.
Snow, G. A., Czabator, F. J., and Sorrels, S. S. (1964). Cy-
    cloheximide derivatives for controlling brown spot on
    long leaf pine. *Plant Dis. Rep. 48,* 551.
Stoddard, E. M. (1952). Chemotherapeutic control of Rhizotonia
    on greenhouse stock. *Phytopathology 42,* 476.
Tamari, K., Ogasawara, N., Kaji, J., and Togashi, K. (1966).
    On the effect of a piricularin detoxifying substance, fe-
    rulic acid, on tissue resistance of rice plants to blast
    fungal infection. *Ann. Phytopathol. Soc. Jap. 32,* 186.
van Arsdel, E. P. and Busch, D. L. (1972). Uptake of dimethyl-
    formamide solutions of benomyl in sycamore, live oak and
    port oak. *Phytopathology 62,* 807.
Venn, K. I., Nair, V. M. G., and Kuntz, J. E. (1968). Effects
    of TCPA on oak sapwood formation and the incidence and
    development of oak wilt. *Phytopathology 58,* 1071.
Willis, W. G. (1972). Scotch pine needle disease control.
    *Res. Progr. Rep. Kan. Cooperative Rep. Regional Project
    NCR-43.* (Diseases of Landscape Plants).
Witcher, W., Arnett, J. D., Baxter, L. W. and Cocke, M. L.
    (1975). Control of needle cast of Scotch pine in South
    Carolina. *Plant Dis. Rep. 59,* 881.
Woodcock, D. (1971). Chemotherapy of plant disease - progress
    and problems. *Chem. Brit. 7,* 415.
Worf, G. L., Kuntz, J. E., and Camp, R. F. (1977). "Chemical
    Control of Foliage Diseases of Deciduous Trees and Shrubs
    in Forest Nurseries," Bull. 19, Sept. Dept. of Plant
    Pathology and Forestry and the Wisconsin Department of
    Natural Resources.
Zentmyer, G. A. and Gilpatrick, J. D. (1960). Soil fungicides
    for prevention and therapy of Phytophthora root rot of
    avocado. *Phytopathology 50,* 660.

Index

A

Acacia, MLO in diseases of, 238
Acertagallia spp., *Spiroplasma citri* culture
  from, 121
Acholeplasma
  biochemical characteristics of, 22
  coconut lethal yellowing and, 194–195
  plants infected with, 2
  taxonomy of, 3
  tetracycline sensitivity of, 321
Acholeplasma axanthum, in rotting coconut
  tissue, 267
Acholeplasma laidlawii, biochemical
  characteristics of, 22
Acholeplasmataceae, taxonomy of, 3
Acorn fruit disease, of oranges, 98, 104
Acridine orange, effect on viruses, 334
Aflatoxins, effect on viruses, 334
Agrimycin, in control of, mulberry dwarf
  disease, 157
Alfalfa dwarf, serology of, 55
Almond leaf scorch disease, 53
  fastidious bacteria in, 37–38, 290
    detection, 38
  serology of, 67
  symptoms, detection, and transmission
    of, 41
Ambutyric acid, as chemotherapeutant, 334
Ammonium compounds, safety of, 327
Amphotericin B, mycoplasma sensitivity to, 23
Anaeroplasma, taxonomy of, 3
Antibiotics
  in control of plant mycoplasma
    diseases, 22–23, 122, 315–319
  toxicity of, 327
Antigens, of spiroplasmas, 24–25

Antimycin A, in plant disease control, 326
Apoplastic movement, of chemothera-
  peutants, 333
Apple chat fruit disease
  MLO associated with, 289
  symptoms of, 41–42
Apple proliferation disease, 35
  MLO etiology of, 135, 281, 289
  quarantine for, 285
  symptoms and transmission of, 42–43
Apple rubbery wood disease
  MLO associated with, 289, 290
  symptoms of, 43
Apricot peach rosette, 68
  MLO with, 289
Apricot chlorotic leaf roll disease,
  MLO with, 289
Areca yellow leaf disease, 288
  coconut lethal yellowing and, 45
  symptoms and transmission of, 43–45
Arginine
  as *Mycoplasma* requirement, 4
  in *Spiroplasma* growth medium, 8
Ascochyta pisi, chemotherapy of, 326
Ash dieback syndrome, ash witches'-broom
  and, 46
Ash witches'-broom disease
  MLO associated with, 289
  symptoms and transmission of, 45–47
Aspen marssonia blight, chemotherapy of,
  341
Aster yellows disease
  antibiotic control of, 315
  colonial appearance of, 15
  natural MLO degeneration in, 322
  ultrastructure of, 16, 20

Chestnut blight
  chemotherapy of, 342
  quarantine for, 325
Chestnut yellows disease
  MLO associated with, 288
  symptoms and transmission of, 51, 52
China aster, *Spiroplasma citri* infections
  of, 117
Chinese cabbage, *Spiroplasma citri* infections
  of, 117
Chloramphenicol, mycoplasma disease
  symptom suppression by, 318, 319, 322
2-Chloroethyltrimethylammonium chloride,
  as chemotherapeutant, 337
Chloromycetin, effect on viruses, 334
Chlortetracylcine
  in chemotherapy of mycoplasma
    diseases, 156, 282, 315-319
  structure of, 316
Chokecherries, X-disease of, 77
Cholesterol
  mycoplasma–antibiotic effects of, 321
  as mycoplasma requirement, 4
Circulifer tenellus
  as citrus stubborn vector, 99, 115, 294
    experimental transmission, 120, 121
  as tree disease vector, 291
Citrus canker, quarantine for, 325
Citrus greening disease, 35, 239
CGD
  agent associated with, 281
  chemo/heat therapy of, 236, 237, 318
  diagnosis of, 235
  differential hosts of, 310
  electron microscopy of, 235-236
  epiphytological studies of, 234
  history of, 299-300
  in India, 234-238, 299-313
  MLO associated with, 288
  strain and strain interactions in,
    299-313
  symptoms of, 234
Citrus little leaf disease, 97, 288
Citrus stubborn disease, 35, 97-134
  control or prevention of, 122
    by antibiotics, 122, 319
    by clean stock, 123
    by noncitrus host, 124
    by quarantine, 123-124
    by roguing, 125-126
    by vectors, 125
  diagnosis of, 110-114
  economic importance of, 99-100

environmental effects on, 116
epidemiology of, 115-121
  experimental, 118-121
  natural, 115-118
geographical distribution of, 99-100
historical aspects of, 97-99
in noncitrus plants, 117
Spiroplasmi citri as pathogen of,
  5, 36, 99, 107-115, 260, 288
symptoms of, 100-107
  on fruit, 102-105
  on leaves, 102, 103
  on trees, 101-102
temperature effects on, 106-107
transmission of
  by graft, 118-120
  by leafhopper, 116, 120-121
trap plants for, 118, 124-125
variability of, 106-107
varietal susceptibility of, 105-106
variety and age of citrus hosts for,
  116-117
Citrus tree decline, chemotherapy of, 318
Citrus tristeza virus, quarantine for, 285
Citrus young tree decline (YTD), serology
  of, 55
Clover, *Spiroplasma citri* infections of, 117
Clover club leaf disease, chemotherapy of, 318
Coconut lethal yellowing disease, 35, 185-229
  Areca yellow leaf and, 45
  causal agent (MLO) of, 192-195
  chemotherapy of, 318
  control of, 198-200
  etiology of, 213-214
  history of, 187-190, 212-213
  hosts of, 289, 290, 291
  MLO agent for, 288, 289
  palm species susceptible to, 189
  relationship to other coconut diseases,
    200-202
  resistance to, 198-200
  spread of, 195-196, 214
  symptomatology of, 186, 190-192, 213-214
  synonyms for, 187
  transmission of, 39, 211-213, 214-223
  vector studies on, 197, 214
Coconut palm
  importance of, 185, 211-212
  lethal diseases of, 186
  mycoplasma site of, 266, 267
Coconut stemnecrosis, 186
  coconut lethal yellowing and, 201
Coelidea indica, see Jassus indicus